Y0-BRS-986

Pueblo Indian AGRICULTURE

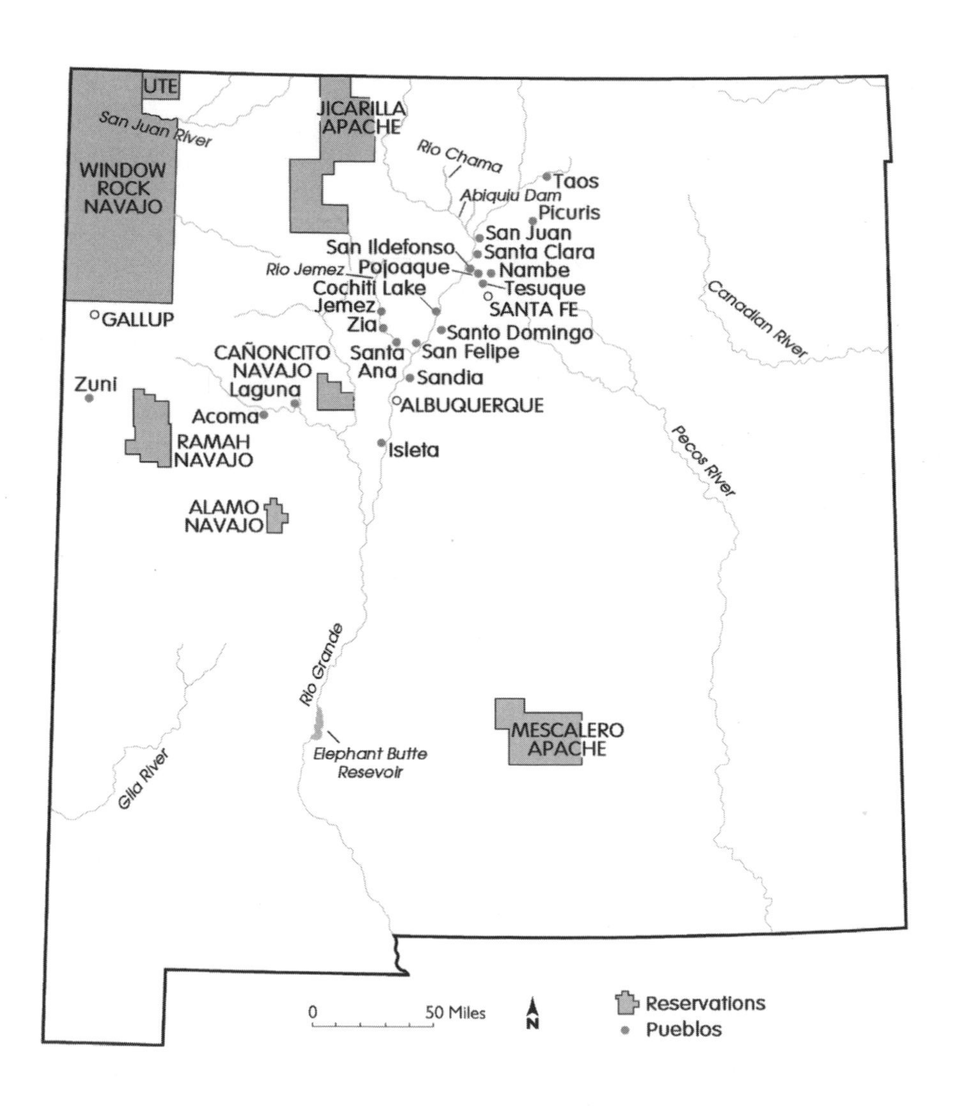
UTE
San Juan River
WINDOW ROCK NAVAJO
JICARILLA APACHE
Rio Chama
Abiquiu Dam
Taos
Picuris
San Juan
San Ildefonso
Santa Clara
Rio Jemez
Pojoaque
Nambe
Cochiti Lake
Tesuque
SANTA FE
Jemez
Zia
Santo Domingo
GALLUP
CAÑONCITO NAVAJO
Santa Ana
San Felipe
Sandia
Laguna
Zuni
Acoma
ALBUQUERQUE
RAMAH NAVAJO
Isleta
Canadian River
Pecos River
ALAMO NAVAJO
Rio Grande
Elephant Butte Resevoir
MESCALERO APACHE
Gila River
0
50 Miles
N
Reservations
Pueblos

To Denis,
Always a good friend on some great walks.
All the best,
Jim

Pueblo Indian Agriculture

JAMES A. VLASICH

University of New Mexico Press
Albuquerque

Printed in the United States of America

09 08 07 06 05 1 2 3 4 5

Library of Congress Cataloging-in-Publication Data

Vlasich, James A.
Pueblo Indian agriculture / James A. Vlasich.
p. cm.
Includes bibliographical references and index.
ISBN 0-8263-3504-7 (cloth : alk. paper)
1. Pueblo Indians—Agriculture. I. Title.
E99.P9V59 2005
978.9004'974—dc22

2004030092

Book design and composition by Damien Shay
Body type is Minion 10.5/14
Display is Bremen and Exotic 350

For My Parents
Hank and Emma Vlasich
Who Always Supported Me
In Every Way Possible

read 02/19/07

CONTENTS

LIST OF ILLUSTRATIONS

PREFACE

The term "Pueblo Indians" is an all-inclusive label first designed by the Spaniards to describe various groups of native village dwellers in the far reaches of the northern frontier of New Spain that came to be known as New Mexico. Today they represent nineteen different reservations whose people generally prefer to be recognized by their individual village names—Acoma, Cochiti, Isleta, Jemez, Laguna, Nambe, Picuris, Pojoaque, Sandia, San Felipe, San Ildefonso, San Juan, Santa Ana, Santa Clara, Santo Domingo, Taos, Tesuque, Zia, and Zuni. In location they range from Taos (above Santa Fe) in the north to Isleta (below Albuquerque) in the south and extend as far west as Zuni near the Arizona border.

Despite their common name, the Pueblos are actually distinct in many ways. They have different religious clans and styles of government and the languages they speak are of three different linguistic stocks—Keresan, Tanoan, and Zunian. However, they all have similar recorded histories and all have been influenced by the rule of Spanish, Mexican, and U.S. governments. Realizing their common bonds, the Pueblos also banded together at various times in the past to combat problems that were similar in nature and threatening to their existence.

The particular aspect of Pueblo culture that made them unique as a group was that they embraced agriculture as the mainstay of their economy just as it had been for their prehistoric Anasazi ancestors. Agriculture has been a major concern for the Pueblo Indians from time immemorial. It is an activity inextricably interwoven with religion, government, and, in essence, their very existence. Of course other Native American groups grew crops, but in the arid Southwest, irrigation was a prerequisite to a successful agricultural program. It was no coincidence that most of the Pueblos were located on the Rio Grande and its tributaries, and while these streams provided a fairly reliable water flow, its transference to their fields had rather dramatic consequences, including total village cooperation, permanent local housing, and a government to control the maintenance and distribution of this lifesaving resource. There were no guarantees in this rather risky business so the Pueblos also developed a sophisticated religious ceremonialism to accompany the labor in the fields with the hope that this powerful combination would provide sustenance.

Early Spanish explorers who came into New Mexico in the sixteenth century noted the efficient irrigation canals and abundant crops of the Pueblos. The explorers' reports encouraged settlers who came with institutions, seeds, and tools to transplant their own agricultural program on the northern frontier. Their impact on Pueblo farming was limited to the introduction of new crops and some implements and the codification of legal systems governing irrigation farming. Still, the fact that this group of Indians represented an atypical lifestyle among natives on the northern frontier greatly influenced the Spanish policy governing the Pueblos. Following the famous revolt of 1680, brought on to a great extent by the suppression of Pueblo religious practices and demand for tribute in the form of labor and food, the Spaniards became more accommodating and united with the Pueblos in defense against nomadic invaders.

An important aspect of Spanish culture introduced in the New World was their legal system. The *Recopilación de Leyes de los Reynos de las Indias*, published in 1681, was designed to deal with various aspects of colonial society. Although the Crown hoped that these laws would prevent encroachments on native land and protect the Indian prior rights to water, such concepts were difficult to enforce on the distant northern frontier. Throughout the Spanish colonial period, the Pueblos were involved in frequent litigation concerning trespass on their farmlands and usurpation of their water rights. As the Hispanic population increased and the Pueblos declined, they continued to be involved in water and land conflicts even after the Mexican government took control over their area in 1821. An inefficient government in Mexico City and competition for arable land in New Mexico resulted in a double blow to native rights.

When Mexico opened the Santa Fe Trail to the United States, American observers wrote about the Indian farmers in New Mexico. Since most Americans viewed western Indians as nomadic hunters and gatherers, the Pueblos represented a unique group in the minds of many Anglos. They were often praised for their multistory adobe buildings, their industry, their self-sufficiency, and, most importantly, their agriculture. The American government saw something unique in the Pueblos and therefore sought to protect them, but because the number of Hispanic and Anglo settlers along the Rio Grande increased greatly following the American takeover in 1846, Pueblo rights were soon jeopardized. This situation worsened in 1876 when the Supreme Court ruled in the *Joseph* case that the Pueblos were not Indians according to congressional definition and thus not entitled to federal protection. That ruling immediately resulted in increased encroachment on Pueblo land that still affects them today. Around the turn of the century, government agents tried to bolster the Pueblo farm program by introducing modern techniques, agricultural education, and machinery. This action met with limited success because it often conflicted with Pueblo traditions. More importantly, in the *Sandoval* decision of 1913, the Supreme Court overturned its previous ruling and granted the Pueblos the same status as other Indians. However, that meant that non-Indians who had moved onto Pueblo land were trespassing and something had to be done about it.

The new Pueblo status and their resulting land claims were soon challenged by the introduction of the Bursum Bill, designed to quiet title the areas in question in favor of non-Indians. Suddenly, the little-recognized tribes of northern New Mexico became a focal point for a progressive movement, led by John Collier, to champion the cause against the land grab. This action would catapult the reformer into the Office of Commissioner of Indian Affairs and the head of a revolutionary transformation of federal Indian affairs in the 1930s. Before that significant change, Collier, along with the newly formed All Indian Pueblo Council, helped to initiate legislation that would set up a governing agency to deal with encroachment on Pueblo land. The Pueblo Lands Act, passed in 1924, created a Lands Board, which intended to deal with trespass on Indian land. Through its decisions, Pueblos were either to regain lost property or to be compensated for the loss of land, water rights, and improvements.

Various federal reports made during the 1920s helped to increase the general awareness of the need to improve Indian agriculture. As a result, the Pueblo agency started to improve Indian development projects and introduced large machinery and modern methods to those native farmers willing to accept them. Slowly these attempts at modernization began to take hold in certain villages and they were bolstered by agricultural education that took into account Pueblo beliefs.

The pressures of the Great Depression and World War II made the Pueblos more willing to accept modernization. As their population grew and money became tight, a culture heretofore resistant to change gradually adjusted. New Deal institutions were extended to the Pueblos and generally helped them to make better use of their land and water without threatening their culture. However, while the farm program was modernized, the number of people involved in agriculture eventually declined. Drained off by the war effort, these people gained new skills in the service or war-related industry or took advantage of the GI bill later on. The result was a farm program that no longer held its valued place in a culture that once knew little else. In its place were relocation programs designed to open the Pueblos to new job opportunities away from their traditional homelands.

The postwar world also introduced a new government policy known as "termination," which served as a death knell for Indian agriculture. Long the hallmark of the American assimilation program, the new dictates from a conservative Congress guaranteed the failure of farming among the country's native people. Even the help of extension agents and Great Society programs could not revitalize a sagging agricultural economy. The postwar population boom in the Southwest also put a strain on water resources and increased demands for more to meet the urban growth. Combined with the lack of assistance by state agencies, especially the Middle Rio Grande Conservancy District, Pueblo farmers soon found themselves in a fight for the water that they so long enjoyed and to which they supposedly had superior rights. A federal government that originally sought to protect a group admired for its industry and self-reliance eventually let down its guard. The result was a struggle for water rights that is still unresolved today. For the Pueblos, it also raised the question of the future of their most traditional occupation and the religious ceremonialism that accompanied it.

Agriculture defined the Pueblos throughout their history. It became a symbol for their culture in much the same way that buffalo identified the Plains Indians. To Europeans and later Americans who first observed their villages, agriculture and its irrigational adjunct represented industry, stability, prosperity, legality, and technology—all elements of a civilized society. To many non-Indian minds, these underpinnings were too often missing in native cultures. The fact that the Pueblos embodied them colored the perceptions of numerous outsiders and garnered their respect. This was not enough to stop trespass and water loss, however, but it also attracted outside support so necessary for the preservation of traditional Pueblo life during times of crisis.

Throughout this drama, the Pueblos moved from being a people in total control of their farm program to being a group that was continually challenged by the rule of outsiders as the dominant role shifted from Spanish to Mexican and finally to the U.S. government. In all of these shifting circumstances, they could have been overwhelmed by sheer numbers and military strength, but their tenacious character led them to rise to any challenge that threatened their traditional lifestyles. While they sometimes lacked the legal wherewithal to cope with the more sophisticated judicial systems imposed on them, they were always able to find the resources to preserve. This book demonstrates how they took matters into their own hands and, with the help of concerned non-Indians, managed to hold onto an agricultural program that stretches back to a time before European arrival.

ACKNOWLEDGMENTS

Thirty years ago I found myself in the middle of a career change. Gone were the days of being a computer programmer, and the search began for a more fulfilling occupation. I found my way to Durango, Colorado, where I enrolled at Fort Lewis College. In the conversion process I met Dr. Robert Delaney, originator and director of the Center for Southwest Studies, and began taking courses on that region of the country. Always a great lecturer, he fascinated me with his stories on this unique part of the country and the native people within. This was also my first introduction to Pueblo Indians, who made up a sizable portion of the student body. Many friendships were formed, and they still exist today. One of them includes Calbert Seciwa of Zuni Pueblo, director of the American Indian Institute at Arizona State University. He supplied much needed information that a non-Indian could never obtain.

Moving on to graduate school at the University of Utah, I studied under Dr. S. Lyman Tyler, who was the director of the American West Center. Realizing my interest in the Pueblos, he suggested the topic of agriculture. The seed was planted, so to speak, and he also suggested that I conduct my research at the Federal Records Center in Denver, Colorado. Their research staff did an excellent job of finding the materials that became the basis for my doctoral dissertation.

The project lay dormant for about a decade as I pursued other academic interests. Then a sabbatical leave allowed time for further research at the Bureau of Indian Affairs, the Pueblo Cultural Center, and the Southern Pueblos Agency in Albuquerque. Additional investigation took place at the Northern Pueblos Agency in Santa Fe. Diligent and cooperative employees at each of these facilities aided my quest for more information.

In addition to further research, numerous revisions were necessary. I was fortunate to be assisted by people who do the difficult task of editing rather well. One was Laura Bayer, who served as an editor for the American West Center and is currently an architectural planner and programmer. She did a yeoman's job of pointing out the problem areas in the first seven chapters and was personally responsible for two articles I published. The other was Durwood Ball, editor of the *New Mexico Historical Review*. He did a remarkable job of helping me put together information from the last three chapters to produce an article for his very prestigious journal. Durwood also added the kind of encouragement that one needs to pursue a manuscript of this nature. Finally, Michael Cohen, a colleague and English professor at Southern Utah University, read the entire manuscript and offered suggestions.

Perhaps no one is more personally responsible for the publication of this book than Dr. Veronica Tiller of Tiller Research, Incorporated, a member of the Jicarilla Apache tribe in New Mexico. Initially, she was a history professor at the University of Utah, and I had the privilege of studying under her during my time in graduate school. While taking a class from her, I wrote a paper that eventually became the seventh chapter of this book and the first article I ever published. Six years ago, when I continued my research on the Pueblos, she provided materials that make up most of the water rights information. She and her family also accommodated me in every way possible during my stay in Albuquerque; I shall never forget her kindness and consideration.

All non-typists should appreciate the effort of people who do this laborious task. Since my work on this project spanned a quarter of a century, there were many involved. My sister-in-law, Francis Hong Wylie, typed my dissertation, and her sister (and my wife), Grace Hong Vlasich, finished the last three chapters. Even my mom, Emma Parola Vlasich, got in on the act.

Finally, Sandi Levy, department secretary for the Social Sciences Department at Southern Utah University and a good friend, typed numerous corrections and made sure that the manuscript was computer correct. This was no small task. I should also thank Grace for helping with the arduous task of indexing and Jamie Orton, student secretary in our department, for typing it. To all of these people, I extend my deepest appreciation.

INTRODUCTION

THE IMPACT OF IRRIGATION ON SOCIETY

Agriculture has been a key element in the development of civilization throughout the world. Its application allowed various bands of aboriginal people to settle in permanent villages. While ancient tribes supplemented their farming activities with fishing, gathering, and hunting, their crops generally proved to be a more reliable source of food, particularly in arid regions where the ecological system could not support large quantities of animal and plant life. To farm in these dry regions, however, the villagers had to learn to irrigate their fields. Developing a system to provide water needed to produce crops became a necessity for those groups who settled in less productive locations.[1]

Irrigation may take a variety of forms. In the purest sense, the term refers to the act of watering or moistening an area for the purpose of plant growth.[2] This action can include floodwater farming, dry farming, or canal farming. The first method gained popularity among native Southwesterners

Figure 1:
Vidal Casaquito gathering corn, Jemez Pueblo, October 21, 1936. Courtesy of the Museum of New Mexico, neg. no. 42074.

because it involved the least amount of work. To irrigate by this method, a farmer must select an area that receives runoff water. Sometimes a plot of land at the foot of the mountains is chosen because it receives water during spring thaws. Areas that absorb the runoff of natural springs can also be farmed in this way. In addition, land on the broad valley floors of perennial rivers receives enough water and silt to allow for agriculture. This type of irrigation was needed and used only in the arid regions of the world. But floodwater farming cannot succeed without dependable rainfall. In an extended drought, crop failures were a disastrous consequence. For people who farmed in alluvial fans close to a water supply, a drought was less critical. As long as the drought persisted, however, the lands had to be watered by artificial means and that often involved extensive labor.

Dry farming is more precarious than other forms of irrigation and requires the talent of an expert agriculturist. This process results in the creation of a dust mulch around individual plants. The mulch would retain water brought by intermittent showers or carried to the fields by the farmers. During periods of drought, areas once farmed with floodwater irrigation became, in effect, dry-farming regions. This change in categorization was more than a technicality for ancient farmers who tried to eke out an existence in the arid Southwest. It sometimes meant the difference between living in their established homelands and having to find a more favorable environment.

The use of canals to supply life-giving water to nearby fields is the most dependable form of irrigation. It provided the foundation for many great civilizations including those of Egypt, China, Greece, Rome, Peru, Mexico, and Spain.[3] In each of these regions, the rudiments of a civilized society developed because the communities devised a method to feed an expanding population. With less pressure to obtain food through time-consuming hunting, fishing, gathering, or raiding, the people had more leisure time. As a result crafts developed and individuals began to specialize in specific pursuits. Newly created items such as jewelry might be used for barter to supplement the economy. Pottery could be used to store food or carry water. In addition to artistic pursuits, these societies had more time to devote to religious life and often developed ceremonies designed to accompany agricultural production. The dependability of irrigation ditches also allowed the creation of larger villages or cities, which led, in turn, to the formation of more highly developed governments.

ANCIENT AGRICULTURE IN NEW MEXICO

Agriculture in the New World diffused from Middle America northward into modern New Mexico. At first it had little impact, but the prehistoric residents of the Four Corners region adopted it centuries before the birth of Christ. For the Anasazi, who accounted for most of the ancestors of the modern Pueblos, crop growing became the basis of a subsistence economy. Hunting and gathering supplemented their food supply, but eventually neither played as important a role in their economy as cultivated crops. Gradually, their increasing population became more dependent on agriculture and thus more threatened by anything that disrupted a good harvest. Although anthropologists debate the exact date of the introduction of corn into the Four Corners region, it was at least being cultivated by 1500 B.C. For the next millennium and a half, the Anasazi were in a period of transformation from the hunt and gather to true horticultural activity.[4] This transition period allowed for experimentation in agricultural techniques. By 1000 B.C., squash was added to the diet, and beans were introduced by about 500 B.C. This gave the ancient people an almost complete diet, but it also coincided with the beginning of a period of unstable rainfall. These periods of drought resulted in larger campsites and storage areas. By the first century A.D., farmers were overtaking hunter-gatherers as the major providers. The agricultural tradition had begun its dominance.[5]

Anthropological excavators have discovered numerous metates and manos used to grind the corn as well as cists in which it was stored. Corn has remained an important crop in the Pueblo economy and is still revered among the Indians today. A member of the pueblo of San Ildefonso stated that the area is swept clean before the corn crop arrives "because corn is just the same as people and must have the plaza clean, so that the corn will be glad when we bring it in."[6]

Although implements and techniques of the Ancient Ones sometimes appeared crude, their production was not necessarily inefficient. The size of ancient villages discovered throughout the Southwest indicates that the Indians were able to support a substantial population. Over time the ancient residents also learned to sun-dry food in order to store it for future use. Storage areas in these sites evidence that the farmers were able to grow large enough surplus crops to support people in lean years. While pottery, first introduced in the area between A.D. 300 and 400, did not completely replace basketry for storage, it revolutionized how food was cooked. The Anasazi methods of irrigation were very sophisticated, including canals, reservoirs, and dams. Some of the canals in this region were so well built that their remnants have been incorporated into modern irrigation systems. Lacking any metal tools, the Anasazi used digging sticks to plant their crops. Corn was planted about a foot deep, where it could draw on subsurface water and would be protected from animals.[7]

The Anasazis' most advanced technology was the use of check dam gardens to conserve water and retain soil in small plots. To build these dams, they stacked rocks at strategic points on intermittent streams. Whenever the stream flow stopped, the organically rich streambeds remained for planting. The consequence was a large increase in land that was not merely arable but rich enough to provide a consistent yield.[8] While it is usually assumed that the Anasazi knew little about soil depletion, the check dam plots indicate that soil conservation was a part of their agricultural program. However unconscious their efforts, the Anasazi also grew corn and beans together, which aided soil nutrition because beans added nitrogen to the soil that corn exhausted.[9]

During the so-called Basketmaker period (A.D. 200–700), the Anasazi life changed more rapidly than any time before. Agriculture became more efficient and supported rapidly growing communities. Pottery reduced fuel needed for cooking and made food storage more secure. The adoption of the bow and arrow made hunting more efficient than in the days of the *atlatl* or sling-type device, thus allowing more time for work in the fields. With the construction of larger underground structures to accommodate population growth, the earliest kivas, or ceremonial centers, developed and religious sophistication enhanced to bolster crop growth. Finally, trade increased to outlying agricultural communities thus expanding the Anasazi world.[10]

Growth slowed during the Pueblo I period (A.D. 700–900) as precipitation declined and people left lower elevations for more moisture in the higher areas. Their communities also grew sharply in size in reaction to the need for protection during possible times of warfare over scarce food supplies. Anasazi adaptation also played a role in the new environment as the shorter growing season in the higher elevations necessitated the development of new types of corn. The number of farming hamlets expanded dramatically from A.D. 850 to 1000 and competition over farmland reached a peak. Farm and population expansion was not totally without design as it insured that at least some land bore crops, but it also set up conditions for economic collapse.[11]

Weather conditions turned in favor of the Anasazi at the end of the first millennium as the midsummer rains became more predictable—a condition that would last for the next 130 years. This situation led to a cultural apex in the Chaco region of northwestern New Mexico. The result was not only the construction of what would become great house ruins but also the foundation of widespread farming areas in the Four Corners region that remained connected by trade, cultural, and religious similarities. The Anasazi fought off droughts that commenced in A.D. 1090 and occurred again 40 years later as the heart of their cultured elites moved to the Mesa Verde area of southwestern Colorado. The great Chaco society had simply overreached itself in its economic development and once-prosperous farmers simply left for ancestral homes or new areas in higher elevations. This situation sometimes led to violent confrontation, such as in the Taos area where refugees clashed with long-time residents who had broken away over three centuries before. By the mid-thirteenth century, new pueblos were being constructed in the area of Los Alamos, and after A.D. 1275 declining rainfall turned into serious drought.[12]

At the height of Anasazi development, the Great Pueblo or Pueblo III period (A.D. 1100–1300), there was a dramatic shift in the population. Beginning in the late thirteenth century, the Anasazi started to abandon areas along the San Juan River drainage. This movement continued for more than two centuries. At first, many of the new immigrants into the Rio Grande region spread out into peripheral areas, but in the 1400s the resources of these mountainous areas could not support the expanding population. Gradually, the modern pueblos were formed as people consolidated their high-elevation villages into individual farming communities. For example, from 1350 to 1550 the inhabitants of Nambe left all but one of their villages in the high foothills and occupied the farmland at a lower elevation.[13] By 1400, approximately thirty major villages existed along the Rio Grande, Rio San Jose, Jemez River, Rio Puerco, Rio Chama, Zuni River, and Rio Pescado along with other significant creeks in the region.[14]

The explanation for this displacement is purely conjectural, but most of the proposals suggest a connection with Anasazi agricultural pursuits. The most common theory holds that a drought that occurred throughout the Southwest in the late 1200s forced the Anasazi to abandon their villages. Some observers theorize that Athabascan-speaking peoples or other raiders were preying upon the well-stocked Anasazi food supplies.

Both of these ideas, while certainly significant, have been discounted as the sole cause because no proof has ever been found.[15] One of the more sophisticated possible solutions given for this problem involves arroyo cutting. According to this theory, steep channels were cut into streambeds during times of deficient rainfall. Through the years, the deep arroyos that were cut made the surrounding area inaccessible for floodwater farming and also limited the construction of irrigation canals. This theory, however, is not applicable to all regions and therefore does not provide a complete answer to the problem.[16]

Other theories dwell on intra-Anasazi conflict, overuse of farmlands, climatic changes, and dietary deficiencies. Numerous observers feel that the Ancient Ones moved for a combination of these reasons. Whatever the cause of this displacement, many of the Indians moved to an area that had a very reliable water source—the Rio Grande. The river valley was also at a lower elevation and therefore had a longer growing season than most of the Anasazi villages. The new location, a dependable water supply, and a different climate had a dramatic effect on cultural development as prehistoric Anasazi transformed into the historic Pueblos.

The new agricultural societies along northern New Mexico's river valleys stood in stark contrast to their prehistoric ancestors near the San Juan drainage. The Anasazi, unlike their Hohokam neighbors in southern Arizona, never created a pure hydraulic technology. They were more floodwater farmers than pure irrigationalists.[17] Upon their geographic relocation, the new Pueblo society basically broke down into eastern and western branches. Both utilized some combination of dry farming and irrigation, but those along or near the Rio Grande emphasized canals while their western neighbors relied more on the irregularities of rainfall.[18]

Changes in the late 1300s up to the mid-1500s were rather dramatic. With a more dependable water source and the protection offered by aggregation into larger villages, riverine farmers increased varieties of standard crops (including gourds) and also raised cotton, tobacco, and turkeys (mostly for feathers). The canal system expanded as the population increased along the Rio Grande and its tributaries. Correspondingly, forms of government to control the use of these ditches had to become more sophisticated. With the advancement of their farming program came more craftwork and further religious ceremonies from planting season to harvesttime. In accordion-like fashion, families scattered from the main pueblos in the spring to perform the rituals of planting and harvesting only to return in the wintertime for festivities and socializing.[19]

By 1540, the Pueblos and their Anasazi ancestors had experienced approximately three thousand years of agricultural tradition. The very fact that they survived and expanded from their rather humble beginnings in the arid and unpredictable Four Corners region is a tribute to their inner strength, work ethic, innovation, resourcefulness, tenacity, and adaptability. So successful were they that by this time an estimated one hundred sizeable villages north of modern Socorro thrived.[20] A comfortable and rather secure routine set in, and the Pueblos could enjoy the fruits of their labors. However, they had no idea that their daily behavior would soon be threatened by an invader. Spanish forces would soon enter New Mexico, and Indian life in this region would never again be the same. The Spaniards carried with them their own agrarian tradition and tried to transplant their crops, implements, scientific techniques, and laws among the natives of the Americas. Sometimes these innovations took hold, but generally speaking only the most expedient blended in with native institutions.

CHAPTER ONE

New Spain Comes to New Mexico (1540–1700)

THE SPANISH ENTRADA

The Spanish empire in the New World spread rapidly after initial contact and came to embrace various pockets of Indian civilization. The rumored riches of these aboriginal dominions lured the treasure-seeking Spaniards in every direction. Information concerning New Mexico came from Álvar Núñez Cabeza de Vaca and three companions who wandered into Culiacán in 1536. His reports to Viceroy Antonio de Mendoza mentioned people who lived in permanent villages. These natives resided in a fertile valley and raised maize, squash, and beans. Cabeza de Vaca also claimed that the Indians traded some of their own items with neighbors to the north who lived in large houses. Finally, he reported many signs of precious metals in the mountains surrounding the area.[1]

It is difficult to assess the accuracy of Cabeza de Vaca's statements, but apparently he contacted agricultural Indians. We cannot pinpoint the exact route of Cabeza de Vaca's journey. However, we do know that the climatic conditions of the Southwest are relatively uniform or more specifically, generally arid.[2] This would mean that he and his companions probably contacted Indians who were employing irrigation. We know that village-type farmers lived along the lower reaches of the Rio Grande, and the possibility of the wanderers encountering these people is a good one.

In 1539, Fray Marcos de Niza led an expedition north from Culiacán in order to reconnoiter the situation in the far reaches of the frontier. Although he never really visited any of the pueblos in New Mexico, the exaggerated accounts of his travels helped to encourage further exploration of the region. In the next year, Francisco Vásquez de Coronado led a group of 336 friars and soldiers into the American Southwest. Accompanying them were some of the soldiers' wives and several hundred Indian servants.[3] The King of Spain instructed Coronado to be just in his treatment of the native people. Not only was his mission one of discovery and pacification but also one of Christianization. The implication here is that Spanish authorities anticipated that the Coronado party would encounter a settled, village-dwelling group of Indians. Descriptions from previous travelers indicated that they were not nomadic. Spanish authorities specifically advised Coronado to protect and defend the Indians and their land.[4]

The first pueblo visited by Coronado's party was Zuni, which was then made up of six different villages. Captain Juan Jaramillo, a member of this group, reported that the Indians stored significant amounts of maize, beans, and squash. He also remarked on the sandy condition of the soil and the lack of grass in the area.[5] Pedro de Castañeda indicated that the Indians' corn was planted in holes. Although the stalks were not very tall, they contained three or four ears, each of which had approximately eight hundred grains.[6] Coronado sent Captain Hernando de Alvarado on ahead to Acoma. Here, Alvarado reported, there was a large area set aside for growing crops. The captain also noticed cisterns that were used to hold water. The Pueblos' food included maize, bread, and piñon nuts.[7]

The advance guard then moved into the Rio Grande valley. Here Alvarado observed the wide and fertile valley of New Mexico's largest river. He remarked on the cornfields of the nearby province of Tiguex and commented about the abundant supply of maize, beans, melons, and turkeys. Concerning the last item, the Indians used turkey feathers for decoration rather than the turkeys themselves as a source of food. Alvarado also said that they wore clothes made of cotton, which they grew near the river. Noting the peaceful nature of the people, the captain said that they were more devoted to agriculture than war.[8] After the Tigua Indians described the numerous villages that existed along the river, Don Hernando journeyed upstream. Finally he arrived at the pueblo of Taos. Because of the altitude of this region, the Indians grew no cotton and were clothed instead in buffalo robes. A similar situation existed at Pecos. Of the various pueblos that the group encountered, the Spanish chroniclers were most impressed with Zia, Taos, and Pecos. At these and other pueblos along their journey, they noticed that the Indians were growing the usual crops to be found in this region.[9]

Castañeda remarked about the fertility of the soil in the region. He claimed that the Indians only cultivated once a year during the planting season and that the crops they grew for one year were enough to last them for seven. Indeed, the Indians often found corn in the field during planting season that could not be picked the previous year because of an overabundance of crops. In total, Don Pedro counted sixty-six pueblos that the Spanish encountered with an approximate total population of twenty thousand men.[10]

Although it was Coronado's original intention that his army would stay in Cibola, Alvarado's report changed his mind. The Zuni villages were considered to be too small and poor to support the large Spanish force, and Tiguex was chosen for winter quarters. In order to accommodate the intruders, the Indians were forced out of one of their twelve pueblos. From here, the Spaniards could pressure the Indians to supply them with food, and sometimes their methods were exceedingly harsh. Supposedly, Coronado and his men purchased these goods from the Indians, but regardless, their demands for food must have forced the Pueblos to deplete a large portion of their supplies. Open resistance to the invaders soon followed, but the Spaniards were able to put down the rebellion. Coronado's men thus destroyed some of the pueblos and ruined many of their fields in the process.[11]

The Coronado party stayed among the Pueblo Indians for two years. In the many places they visited, the Spaniards were given food by the locals.

Since they did not receive any supplies from Mexico, it was necessary for them to rely totally on the abundant crops of the Pueblos. Their presence also caused a demographic shift at Tiguex and a considerable strain on the food supplies due to the destruction of native fields. It is remarkable that the Indians were able to feed a large body of men for such a lengthy stay. This was a tribute to their irrigation systems and agricultural methods, which, although they may seem crude by modern standards, were extremely productive.

FURTHER EXPLORATION IN THE SIXTEENTH CENTURY

Due to the discouraging reports of the Coronado expedition, the authorities in New Spain ignored the northern frontier for almost four decades. Religious zeal spurred the next Spanish expedition. In 1581, Fray Agustín Rodríguez and Captain Francisco Chamuscado led nine soldiers, three priests, and nineteen Indian servants into New Mexico to convert what they considered to be the heathen natives. They entered the region from the south and stopped at the now defunct Piro pueblo of San Felipe near the site of Fort Craig. Here they found supplies of corn and cotton in addition to fields of corn, beans, and calabashes.

The small caravan journeyed into Tigua country, and they remarked about the typical Indian food. Hernán Gallegos, the chronicler for the trip, indicated that the natives ate buffalo meat, cultivated large cotton fields, and raised a sizable number of turkeys. He also explained that the men did the work in the fields all day while the women stayed home to prepare the food. In addition, Gallegos provided the first description of a *mitote* or ceremonial rain dance. They visited other modern-day Pueblos including Zia, Cochiti, Pecos, Santo Domingo, Acoma, and Zuni.[12]

Two of the friars of the aforementioned expedition remained in the area, and in 1583 a relief expedition sent to rescue the missionaries entered New Mexico. Captain Antonio de Espejo was in charge of this expedition, and he led his men along a similar route as the Rodríguez caravan had taken the previous year. Both expeditions received large amounts of food from the Indians as they passed through their territory. When Espejo journeyed through the Piro nations, he stopped and admired their irrigation facilities. Since this is the first extensive description of Pueblo agricultural practices, Espejo's remarks are included here.

> They have fields of maize, beans, gourds, and piciete [*piciere*—an herb] in large quantities, which they cultivate like the Mexicans. Some of the fields are under irrigation, possessing very good diverting ditches, while others are dependent upon the weather. Each one has in his field a canopy with four stakes and covered top, where he takes his siesta, for ordinarily they are in their fields from morning until night, after the Castilian custom. In this province are many pine forests which bear pinenuts [*piñones*] like those of Castile, and many salines on both sides of the river. On each bank there are sandy flats more than a league wide, of soil naturally well adapted to the raising of corn.[13]

The Espejo expedition continued up the Rio Grande until they reached the Jemez valley. In their journey they passed the Tigua nation, including San Clemente, which was located where modern-day Isleta stands. The group also encountered ruins near Sandia and the Keres pueblos of Cochiti, Santo Domingo, San Felipe, Zia, and Santa Ana.[14] In the pueblos where they stopped, the Indians gave the travelers more than ample food supplies. Eventually the group made their way to Acoma. When they left Sky City, they traveled four leagues up the Rio San Jose where they found irrigated cornfields and dams that were constructed as well as any in Spain. They also remarked about the cultivated lands in Zuni. The date was April 7, and the Zunis were beginning to plant their fields. At Hawikuh, one of the Zuni villages, the Spaniards noted water holes and canals that were used for irrigation.[15] On their return trip from Hopi, the Espejo expedition encountered hostile natives near modern-day Acomita. The Spaniards destroyed some of the Indian cornfields. One of these areas was a field at the foot of a sierra that received water from spring thaws. They also set fire to the pueblo of Puaray, which was the predecessor of Sandia, and executed sixteen Indians there.[16] The Indians rejected the Spanish request for food, and Espejo decided to return through Pecos, where they had to force their way into the pueblo in order to get supplies.

The expedition of Gaspar Castaño de Sosa in 1590–1591 demonstrated the zeal for colonization created by the promising reports of Espejo. He led a group of approximately 170 men, women, and children into New Mexico. Unfortunately, Don Gaspar did not license his intent, and eventually he was arrested and returned to Mexico. However, he did find additional evidence of Pueblo Indian farming expertise. Since his group entered the region via the Pecos River, the first pueblo they encountered was one named after that stream. Everyone marveled at the supply of maize that the Indians kept in storage. Some people estimated the amount to be thirty thousand *fanegas* (approximately forty-eight thousand bushels). They also held an abundant supply of beans in storage. Some of the corn was said to be two or three years old. In addition to these crops, the Indians stored herbs, chiles, squashes, and their farming implements.[17] Castaño continued toward the Rio Grande and encountered various Pueblo villages. In Tesuque, Nambe, and Pojoaque, he marveled at the canals that were used for irrigation. At San Ildefonso, he commented on the vast valley that was under ditch. He also visited San Juan, Cochiti, Santo Domingo, and San Felipe. In spite of the punitive actions taken by Espejo ten years earlier, the Castaño caravan obtained ample provisions from the Indians.[18]

The age of exploration in New Mexico came to a close with the Castaño expedition. Other exploratory ventures occurred later on, but they would originate from established Spanish towns in the northern frontier. These early chronicles reveal a great deal about Pueblo agriculture. First, we have conclusive proof of the extensive number of irrigation canals that were employed throughout the region. We also have evidence of areas that were farmed with floodwaters. The Indian crops were typically corn, beans, and squash, but they supplemented their diet with gathered piñon nuts and wild herbs and hunted buffalo meat along with meat from other game animals. The Spaniards were impressed with Pueblo farming expertise and praised their irrigation construction. Equally impressive was the food held in storage, for the various invading parties relied on it during their travels. Ditch construction and maintenance required communal efforts by the Indians as witnessed by the many villages that existed in their region.

The impact of Spanish exploration on Pueblo agricultural systems was considerably less than that of the colonization period. Tragically, a number of Indians were killed, a few settlements were abandoned, and food supplies diminished, but time could ease the pain of these wounds. As of yet, Spanish law and institutions had not been applied, no encroachment existed, Indian water rights remained unchallenged, and their foods were unaltered. All of this would change with full-scale Spanish occupation of the region. The Pueblos had yet to realize that previous investigations were only the harbinger of a permanent Spanish presence that would leave their world forever changed.

THE COLONIZATION OF NEW MEXICO

As Spanish authorities instructed their explorers in matters of Indian relations, they also devised legislation to control colonization. These laws, which will be dealt with in detail later on, reveal the deep concern on the part of Spanish rulers to lessen the impact on aboriginal communities. The first consideration was to gain legal authority to conduct a colonization venture. The Laws of Settlement of 1573 also stated that Spanish colonization should not be injurious to the Indians, and this specifically meant the homes and other improvements that they had already established. The laws also prohibited wars of conquest or involvement in intertribal conflicts.[19]

The sensational reports by various Spanish explorers concerning New Mexico helped to intensify competition among those who hoped to lead the first colonization party. As early as 1592, Don Juan de Oñate demonstrated his eagerness to gain this authority and he received his contract three years later. An undertaking of this nature involved more than a military expedition. As a settlement company, the Oñate colony carried many items not previously found in exploratory ventures. The group brought its own flour, maize, wheat, jerked beef, and animals. They also had iron tools such as plowshares, picks, hoes, axes, and sickles. In addition Oñate received permission to collect tribute from the Indians for himself, the Crown, and other men of stature.[20]

Before Oñate left, the viceroy gave instructions to him in October of 1595. The viceroy emphasized the need to pacify and convert the native inhabitants and to follow the royal orders of 1573 concerning colonization. He warned Oñate that the Indians should not be forced to perform tasks for the Spaniards, but rather that they should be encouraged to live in Spanish homes in order to learn the techniques of Iberian husbandry at a slower pace.[21] The vanguard of the Oñate caravan finally settled at San Juan Bautista on August 11, 1598, and work on acequias, or irrigation canals, began immediately. Oñate felt that the Indians would copy the Spanish actions and enjoy the benefits of their agricultural advancements. Thus, Indian lifestyles and economy would be enhanced by the new livestock, grain, vegetables, and fruits of the Spaniards. When the leaders of various pueblos met with Oñate and capitulated to his request that they become vassals to the king, he explained that they must now follow the laws of their conquerors.[22]

As might be expected in the early stages of colonization, all did not go well. In some places, like Acoma, the Indians received harsh treatment from the Spaniards. Also, the Spanish colony suffered from a lack of provisions. In 1601, Captain Don Luis de Velasco wrote to the viceroy to inform him about conditions in New Mexico. By then, the colony had established new headquarters at San Gabriel, a former Indian pueblo on the Rio Chama. This was an appropriate place because of the proximity to water and forests. In spite of the good location, the new colony struggled for survival. Lacking the wherewithal to raise crops in this arid region, the settlers solicited relief from the Indians in the form of maize, and this action drew many complaints from the native farmers. Velasco tried to justify the failure of Spanish farms at this time. He complained that the land was not fertile and the water was insufficient.[23]

Don Francisco de Valverde conducted an investigation in July of the same year in order to get an accurate account of conditions in the area. Many Spaniards testified, including one Marcelo de Espinosa. The witness said that the Indians were devoted to agriculture and grew and stored maize, beans, calabashes, and melons. He remarked about their irrigation canals and the areas that they dry-farmed and how the Indians planted crops in May and harvested in August, the period of greatest rainfall. Espinosa claimed that while their cornstalks were small, the ears were big. He also said that the Pueblos traded their goods with nomadic tribes in the plains for buffalo skins, meat, fat, and tallow.[24] The inference here is that while Spaniards were suffering economic deprivations, the Pueblos were able to maintain themselves rather well.

Another witness, Gines de Herrera Horta, claimed that the new variety of vegetables brought up from New Spain were doing well in New Mexico. However, he added that due to a lack of fish and vegetables, the Spaniards were given a special dispensation to eat meat during Lent. Horta said that Oñate ordered his soldiers to go out every month to collect corn from the Pueblos, but upon their arrival, the Indians fled to the mountains. He also claimed that the Indians stored their corn for three or four years in order to combat drought conditions. Due to the lack of cotton grown in the area, the Spaniards resorted to stealing blankets from the Indians in order to deal with the cold winter climate. While the witness saw Spaniards plowing their own fields, he also admitted that some of them employed Indians to cultivate their land.[25]

Consideration was given to total abandonment of the region. Franciscan friars held a meeting to discuss this serious matter. Fray Francisco de San Miguel, vice commissary of the region, said that the Indians had abandoned many of their pueblos for fear of ill-treatment because they were being tortured or killed in order to induce them to supply food for the Spaniards. He claimed that the Pueblos were also short on supplies and resorted to eating roots and whatever they could find in their fields. He stated that their fields were in the worst condition ever and that many of them died from hunger. Fray Francisco de Zamora added to this, saying that the Spaniards had pilfered Indian storage bins until the locals went without food.[26]

From this experience, it can be seen that disagreements between church and state existed early in New Mexico. Fray Juan de Escalona wrote to the viceroy concerning the affairs in New Mexico. Unlike the glowing reports of the sixteenth century, his account accentuated the plight of both colonists and Indians. He attacked Onate as an incompetent who lacked the ability to establish any kind of a settlement. Escalona claimed that after Oñate's arrival, the governor commenced to take corn that the Indians had stored during previous years. Escalona considered this a grievous offense to commit against the Pueblos because to them "corn is God, for they have nothing else with which to support themselves." In addition, the Spaniards forced the Indians to give tribute in the form of blankets before they received the word of God. He felt this action caused great resentment and often forced the Pueblos to abandon their villages and move to the mountains.

Escalona believed that the governor also made a mistake by not planting a community plot for the settlers. The friars encouraged Oñate but to no avail. The ground was finally planted by Indians so that they would not have to give up their own food supply. This action proved fruitless, and the Spaniards entirely depleted the Pueblo corn supply, which resulted in starvation for many of the native people. Two major conflicts stemmed from the colonists pilfering native food supplies, and as a consequence more than eight hundred Pueblo Indians died and three villages were burned. These actions alienated both Indian and Spaniard, and many settlers leaned toward total abandonment. Earlier witnesses also complained of the lengthy eight-month winters in New Mexico, and Escalona added to this with his claim that a frost had recently damaged fields in the area after months of summer drought.[27]

Contrary to those reports from the religious sector, civil authorities also filed claims. Officials selected Captain Gerónimo Márquez to report to the viceroy, and for this purpose he summoned witnesses to testify concerning conditions in New Mexico. One of the men selected, Captain Alonso Gómez Montesinos, claimed that the Spaniards would harvest fifteen hundred fanegas that year that, when added to the supplies brought in by the Indians, would enable the settlements to survive. Each Spaniard received one-third fanega of corn each week, and Montesinos felt that since the colonial population had decreased by about one-half, there was an ample supply of grain and cattle to support them. He also noticed that gardens of the settlers produced bountiful amounts of fruits and vegetables. The captain testified that the amount of grain planted and harvested increased in each of the three years of the colony's existence. Others claimed that this increase was also true for Castilian vegetables. Alferez Martín Gómez of San Gabriel stated that if he were furnished fourteen Indians and plowshares, he could provide enough sustenance for Pueblos and Spaniards alike.[28]

Events during the early years of colonization point out some significant aspects of the initial impact of Spanish settlement. Oñate's attitude concerning Indian agriculture was paternalistic as he felt that the Pueblos could only benefit from Iberian technology. He also expected that they would follow his dictates and this included the requirement of tribute. Concerning the new crops brought into the region, it is apparent that, while they were doing well in the New Mexican environment, they were not yet adopted by the Pueblos.

Whereas Oñate believed that Indian agriculture could benefit from Spanish advancements, quite the opposite seemed to be true. Pueblos' grain and cotton mantas were first demanded and finally stripped from them in order that the colony could survive. It should be noted that some witnesses claimed that Indian fields were not doing well at this time either.

Adverse climatic conditions could have been a cause of low Pueblo crop production. Many witnesses in 1601 complained about the eight-month-long winters in the region, but in the next year, Juan de León testified that the winters could last six months and maybe longer.[29] Perhaps the first three years of settlement were hampered by unusually harsh weather.

It is a tribute to Pueblo agriculture circa 1600 that it was so productive. In addition to giving grain and cloth as a form of tribute, the Indians were required to perform agricultural chores for the settlers. Taking care of these tasks obviously detracted from the amount of time that they could use on their own fields. Conflict over the Spanish demand for tribute had vastly depleted their numbers, and this would also have an adverse effect on agricultural production. Finally, due to tortures imposed by the Spaniards over the tribute issue, some of the Indians simply abandoned their pueblos, and this action meant that still fewer native farmers were available to maintain the colonial economy. The situation was not at crisis stage yet, but the forces set in motion during the early years of settlement would eventually unleash the fury of the Pueblos—a crisis made worse by a conflict within the Spanish community.

CHURCH, STATE, AND THE PUEBLOS

Starvation and possible abandonment were the immediate results of the early crisis in the Spanish-Pueblo relationship. More significant in terms of future Indian affairs in the seventeenth century, the situation in 1600 points out the beginnings of a struggle between church and state for Pueblo labor. Certain friars felt that the Indians should be relieved of the burden of tribute so that the priest could utilize this time toward their conversion. Civil authorities believed that this action would result in total domination of Indian labor by the Franciscans and that the unstable economic conditions in New Mexico would worsen. In either case, the Pueblos lost valuable time and effort that could have been spent tending their own crops.

Many of the colonists and friars deserted New Mexico in 1601, and the region was ignored for the next eight years. Oñate resigned in the face of possible removal in 1606. He was constantly in search of new riches to be mined rather than preparing land to be farmed. His unconcern with the colonization process resulted in almost complete failure for the settlement. Only the desire to convert the so-called heathens kept Spanish interest in New Mexico alive at this time. Oñate's replacement, Pedro de Peralta, hoped to end the economic stagnation so he brought horses, cattle, and necessary implements for farming. He also established the capital at Santa Fe in order to separate the Spaniards from the Indians.[30]

Peralta continued to enforce the collection of tribute among the Pueblos. In addition, he required their labor on various construction projects in the new capital. Supplies of maize were collected from pueblos that had not already been assigned to the early Spanish settlers. This action almost led to revolt in Taos in the summer of 1613, and the attempted resolution of this conflict furthered the split between civil and religious authorities.[31]

Juan de Eulate became governor in 1618, and he furthered the church-state schism. He denied the missionaries military escort into frontier pueblos even though Indians there were required to pay tribute to these same soldiers. Although he did not lend support to the church's campaign to end Indian idolatry, his defense of the Pueblos lacked any altruistic motives. Rather, he desired to win their favor with the ultimate motive of exploiting their labor.

Complaints by civil and ecclesiastical leaders led to the promulgation of the viceregal decrees of 1621. These orders included a statement that the friars were not to hinder the collection of tribute. Friars would be given military escort to the frontier, but the laws limited the kind of labor that could be performed by the Pueblos. By these restrictions, the friars could only employ Indians as farmers, the number of native laborers was strictly limited, and payment for their work became a necessity. The instructions of 1621 marked a turning point in the conflict between church and state. Clearly, the viceroy demonstrated his support for the civil authorities, and secular rights came to dominate affairs in New Mexico.[32]

The laws of 1621 also touched on livestock in the area. The Spanish introduction of sheep, horses, and cattle initially caused problems with the native farmers. These range animals required large areas to graze, and lacking fencing materials, they tended to drift toward the better lands along the river. Inevitably, the untended stock began to encroach on Pueblo farmlands, eating their crops and damaging irrigation canals. The viceroy decreed that Spaniards could not pasture their herds within a three-league radius of the Indian villages except under peculiar circumstances. During the first century of Spanish occupation, the Pueblos remained basically farmers, but stock raising became an adjunct to their agricultural pursuits. However, the introduction of livestock caused problems for the Pueblo farmers. The colonists wanted to affix their European lifestyles on the locals, and this included teaching them the principles of breeding and herding. The Indians learned these methods by serving as herdsmen for the missionaries as a form of tribute.[33] Naturally, this activity took away from the time that Pueblo farmers could spend in their fields.

Like other matters at this time, stock tending as a form of tribute got caught up in the battle between church and state. In 1639, the town council or *cabildo* of Santa Fe reported to the viceroy that the ecclesiastics caused the seculars great harm by not allowing them to raise cattle on their farms, which were located two or three leagues away from the Indian villages. The civil authorities were particularly upset because they claimed that the friars kept extremely large herds within the pueblos. In addition, the missionaries required Indians to guard their fields and gardens from livestock intrusion.[34] In that same year, Governor Francisco de la Mora y Ceballos wrote to the viceroy concerning the problems of stock grazing in New Mexico. He complained that the religious heads encouraged the Indians to fence off the friars' fields to protect them from wandering herds. Mora felt that the friars' concern stemmed from greed, for in this manner, they could assure that the Indians paid them tribute in the form of food. He claimed that sometimes the Pueblos planted their fields in reserved areas with the knowledge that Spanish herds were not allowed in the near vicinity. In this manner the Indians prevented the establishment of *estancias*, land grants used to herd animals.[35] Earlier, Fray Esteban de Perea had raised counter charges and claimed that the governor allowed estancias to be established on the Pueblo farms. At the same time, Mora allegedly took stock from soldiers in order to increase his own herds. To compensate the Spaniards, he allowed them to take over Indian farms, and this deprived the Indians of a major source of food. One particular soldier drove his herds onto the cotton fields of three neighboring pueblos and built his corrals and barns right on their farm.[36]

Apparently the instructions of 1621 neither resolved the tension between civil and religious authorities nor ended the problems of Pueblo farmers. Whether they paid tribute to friars or the government, the Indians were constantly under the yoke of Spanish taxation. The continuing struggle between these two groups caused deep resentment among the Indians. Where once they had been independent and concerned only with the lives of their own people, they now faced exploitation in two directions. Gradually, this dual imposition pushed the Pueblos beyond accommodation and into the stage of direct rebellion.

SPANISH INSTITUTIONS IN NEW MEXICO

After the first few decades of Spanish occupation, the institutions they brought to New Mexico became entrenched in their new environment. Many of these aspects of Iberian culture had a direct bearing on agriculture in the new colony. It was inevitable that some of the colonial innovations would be adopted by the Pueblos. This does not mean that the Pueblos became assimilated into the mainstream of Spanish culture. On the contrary, they resisted dramatic changes in their lifestyles and integrated only those changes that did not cause major alterations in their ancient cultural traditions.

The *encomienda* system was one of the first of the Spanish institutions to have a direct impact on the Pueblos. The system had its roots in medieval Spain and was carried to the New World by early Spanish explorers. Essentially, the encomienda was a grant of land and the labor of the people on it to those who had participated in the conquest of a particular region. Its purpose was to reward those who helped to expand the empire, to encourage the settlement of new regions, and to set up a defense system of the recently conquered area. The encomienda carried with it the right to utilize Indian labor, and this was known as the *repartimiento*. Thus *encomenderos* were entitled to products produced by Indians and that could mean either service or materials. Legally, the grant gave its holders the right of Indian labor in lieu of tribute. Those who held these grants were not allowed to live within the boundary of the pueblos for fear of exploitation, but some people in New Mexico ignored this law. The original concept behind the encomienda system was to protect the Indian and give him spiritual instruction. Often, there were abuses to the system, and it became a method to exploit Indian labor as the encomenderos came to regard the native farmers as their own serfs.

The encomienda system in New Mexico began with the initial colonization under Oñate, but the pueblo of San Juan was apparently exempt because of aid given to Spanish settlers. Because of demographic differences and geographic isolation, the system in the northern frontier was unlike the one in the central valley of Mexico. Lacking any element of a political power base, it became a system of exploitation. Eventually, it also became a major source of rebellion and would not be completely destroyed until the Pueblo Revolt of 1680.[37]

The Spanish government knew of the possibility of abuses inherent in this system and passed laws to put limitations on it. The Law of Succession of 1536 held that the encomienda would only be valid for the lifetime of the grantee and his children. The New Laws of 1542 attempted to eliminate the program entirely. Arguments against the abolition of the encomienda stressed the need for the Indians to follow Spanish agricultural practices in order to obtain a bountiful harvest. An abundance of crops would then result in a peaceful ward who would be more willing to accept Christianity.[38] Because of the strong opposition to this legislation by the settlers, however, it was repealed.

Sometimes the Spaniards were given outright grants of land. The original conquistadores received *peonajes* (for the infantry) and *caballerías* (for mounted soldiers). Local governments set aside *propios*, which were cultivated areas that were rented in order to finance the local government. Villages also reserved *dehesas*, which were used as common grazing areas, and *montes*, or common woodlands. Since villages in New Mexico were more pastoral than agricultural, these grants tended to be considerably larger than in other areas of New Spain. Their vast size and ambiguous boundaries, however, caused encroachment problems for Pueblo farmers. Spanish law, as will be seen, was designed to protect Indian land and water, but concomitant with this notion was the overall economic goal of exploiting resources for the profit of the empire. Much to the chagrin of the Pueblos, their most valuable possessions were slipping away, and this loss was one of many factors that ultimately led to rebellion.[39]

Encroachment also plagued the Spanish settlers, and in order to stop these vexations, the colonists sought clear title to their land. A major step in this direction was the process of issuing *composiciones*. This process became essential after most of the good land in New Mexico disappeared rapidly due to the vast grants that were issued. By these *composiciones generales*, the authorities examined all deeds and set up a standard procedure to legalize land claims that were spurious or doubtful. Another practice, called *congregación*, consolidated villages that suffered large losses of population. Finally, through the practice of *denuncia*, a citizen could file a claim to lands known to be held without title. Through these institutions, Spaniards were able to strengthen their holds on New Mexican land claims in the seventeenth century.[40]

Although the colonists were forbidden to trespass on Pueblo property, they found the temptation irresistible. In order to curb injustices against the Indians, the Spanish government created the offices of *protector* and *defensor*. In the early years of colonization in New Spain, leading clergymen held this job. Later the viceroy appointed a *protector general* who in turn assigned *protectores partidarios* at the local level. While the protector acted as an Indian agent, the defensor appeared in court in all Indian legal affairs.[41]

Since the Pueblos were settled farmers with advanced irrigation techniques, the Spanish government felt that they deserved certain benefits of civilization, including property rights. This attitude contrasted with the Spanish approach to the more nomadic tribes that surrounded the Pueblos. Royal ordinances recognized the rights of Pueblo Indians to their land. These rights were confirmed by the awards from Domingo Jironza Petriz Cruzate of 1689. Unfortunately for the Pueblos, the Cruzate grants were investigated in the late nineteenth century by American officials and found to be spurious. However most people in New Mexico believed them to be real and, along with the small Spanish population in seventeenth century New Mexico, helped to protect the Pueblos from encroachment. However this situation changed after the reconquest of the 1690s.

From the very beginning of Spanish colonization, Pueblo lands were supposedly protected by grants from the king of Spain that measured four square leagues or 17,712 acres, which typically extended from the cross in the middle of the village cemetery. The Cruzate awards only helped to confirm this notion in the minds of everyone concerned—including the Pueblos. However some of the Indian pueblos were so close (like Pojoaque and Nambe) that they were unable to establish their own legal boundaries. More importantly, the very purpose of these grants was not to define Pueblo territory but rather to protect it from encroachment by outsiders. Therefore Pueblo land claims, lacking any real legal title, emerged from the vagueness of early traditions and came to rely on actual use rather than formal measurements. Precise definitions of property only emerged when lands and waters were challenged by encroachers.[42]

Tribute in the form of corn and other crops naturally caused a great deal of resentment in New Mexico. While the initial colonizers felt satisfied with Indian crops, eventually the Spaniards hoped to extract food from the native population that was more familiar and palatable. Also, the addition of Old World products would help to diversify and bolster the sagging economy of the area. Missionaries introduced the new crops to the Pueblos and they gradually adopted them. The introduction of new food items commenced with the Oñate colony. A captain in that group, Gaspar de Villagrá, reported that items such as lettuce, cabbage, peas, chickpeas, cumin seed, carrots, turnips, garlic, onions, artichokes, radishes, and cucumbers were immediately accepted by the Indians. Gradually, the Spaniards introduced wheat, alfalfa, oats, barley, peppers, watermelons, muskmelons, grapes, peaches, apricots, apples, pears, plums, cherries, and different varieties of beans. Tomatoes and chiles were two New World crops that the colonists brought up from Mexico.[43]

Pre-Spanish foods such as corn, beans, and squash predominated over the newly introduced crops. For example, fruit trees received limited acceptance among most of the Indian villages. Since the western Pueblos were far removed from the center of Spanish activity, they did not embrace the new items, but Rio Grande Pueblos adopted many of them. While a number of the Spanish crops gradually became Indian food, wheat and barley never replaced corn as the staple crop. Acceptance of the new items did not change the Pueblo standard of living since they continued limiting their crop production to community needs.

In order to supplement the Indian labor in the fields and on irrigation ditches, the colonists supplied some of the Pueblos with some advanced agricultural accouterments. Indian tools consisted of the digging stick and sometimes a plow that was made from a crotch or knee of a tree. While this rudimentary equipment involved more labor, many Indians were reluctant to part with their traditional tools. Missionaries usually introduced items like hoes and plows made of iron, but a lack of metal in the region impaired their impact on the Pueblo society. Spaniards also introduced new methods of food storage in order to preserve harvested crops through the harsh New Mexico winters.[44]

Both Spaniards and Pueblos knew the vexations of growing crops in an arid climate as both had employed irrigation methods for centuries. As previously mentioned, early Spanish explorers marveled at the native irrigational expertise. In order to cope with the vast problems of an irrigation society, the Indians worked communally on the construction, maintenance, and management of their ditches. Some of the modern Pueblos still own and operate their communal ditches while Spanish people acquired rights to others. When the Oñate colonists first settled in the region, they immediately began construction on the *acequia madre* or mother ditch, the main feeder to secondary canals called *sangrías*. The amount of frontage on the acequia madre determined the quantity of labor required by individual farmers. The main ditch provided water for areas ranging from three hundred to three thousand acres and normally measured fifteen feet in width and two to six feet in depth. Diversion of the water from the main stream by a wall of stones usually occurred two to four miles above the farms so that gravity could carry water to the most far-removed sangrías. Dams called *presas* impounded the water of small streams so that it could be used at a later date.[45] The Spaniards modeled their irrigation construction in New Mexico after the systems they previously constructed in Spain and Mexico. However, the works in New Mexico were primitive due to a lack of engineering expertise, surveying apparatus, and effective tools. For example, Pueblo laborers constructed sangrías along with floodgates to control the flow of water in small streams, but the crudeness of their construction reduced the effectiveness of these devices.

To control the utilization of irrigation ditches, the invaders initiated the office of *mayordomo* or ditch boss, and most of the Pueblos adopted it. The mayordomo's responsibilities included the overseeing of construction, repair, and management of acequias and the administration of water distribution. Men of the community elected the mayordomo and determined his salary. Because he governed an organization that was essential to the life of the community, the mayordomo held a position of power and esteem. Part of the responsibility of the ditch boss included settling disputes that arose over infractions of the laws governing irrigation. For instance, in times of short water supply, he issued permits that limited the time landowners could water their property. Because of these limitations, some families were forced to labor through the night in order to maintain their very existence.[46]

Generally speaking, the Pueblos owned their land communally rather than individually. Villages divided their fields so that different clans or family groups had large enough areas to feed their members adequately. The entire community helped to form societies that handled the construction and maintenance of the ditches. Due to erosion caused by swift currents or overflowing streams, annual repairs on the acequias were necessary. Farmers had to provide workers for this maintenance and failure to do this could result in a fine. In addition, the Pueblos employed community workers to tend the fields of the resident priest and the *cacique* or chief religious leader of the Indians. While ditch cleaning required communal labor, individuals worked on their respective plots. Indian villages owned their land communally, but an individual farmed areas assigned to him and his family. Any parcel of land left unattended reverted back to the community to be reassigned.[47]

Some Pueblo lands adjoined continuously flowing streams while others bordered intermittent ones. Both required specific hydraulic techniques. The first required the control of running water into canals and laterals while the latter relied on husbanding floodwaters for maximum utilization. Floodwater farming is often ignored in historical writing, but it was a fairly common practice throughout the growing season as weather conditions allowed. As the flow slowed during drier times, temporary dams constructed with local materials could redirect the flow from the main stream and into the canals. Areas relying solely on floodwater had to be carefully selected on a flood plain.[48]

Unlike their Spanish counterparts, private ownership of land was unrecognized by the Pueblos. Tribal leaders distributed land to various clans responsible for farming and the maintenance of their irrigation systems. Kinship groups held the tracts as long as they continued to be productive. User rights would pass to others if the fields lay fallow. The size of these plots varied from about a half acre to possibly two or three acres depending on local conditions. Due to the unlevel conditions of the Pueblo surroundings, the resultant farmlands were irregular in shape and thus inefficient for maximum utilization of space.[49]

During the seventeenth century, Spanish innovations had some impact on aboriginal Pueblo lifestyles. New crops diversified the native diet, and new implements helped to ease the burden of labor when they were available. Land grants caused few encroachment problems as long as the Spanish population was small, but the encomienda system resulted in tremendous resentment among the Indians. In general, the irrigational and agricultural innovations from Spain only enhanced the Indian farm systems that already existed. However, Spanish institutions initiated at this time would continue to grow in the decades to follow, and so would their land holdings. With arable ground and water such precious commodities in this arid region, it would only be a matter of time before conflict arose over their use.

PUEBLO REBELLION

The church-state conflict that commenced with colonization widened by mid-century. The instructions of 1621 put limitations on the use of Indian labor, but missionaries only paid lip service to these laws. Although the laws made it clear that Indian labor could be employed to raise crops for the missions, civil authorities complained that the extensive fields of the clergy could not be justified. On the other hand, the friars said that their vast storage bins helped to feed the Indians during years of famine. Indeed, a shortage of food occurred in 1623 at Jemez.[50] However, before the Spanish intrusion, the Pueblos survived seasons of low production with their own supplies. Because of this fact, Governor Bernardo López de Mendizábal suggested that the Indian labor could be better spent in their own fields.[51]

The major issue between the clergy and the Pueblos remained the question of native religious practices. In addition to their conversion to Christianity, the friars sought to end the native ceremonies, which they considered idolatry. So zealous were they in this pursuit that the priests often punished the Indians for their religious crimes by whipping them. The governor reacted negatively to this treatment. In the late 1650s, he declared that Pueblo attendance at mass was not mandatory and allowed them to revive their traditional cults.[52]

Pueblo Indian ceremonialism dealt largely with their agricultural practices. They believed that these rites would guarantee a bountiful harvest. Religious leaders held ceremonies throughout the farming season but not necessarily for all crops. In Picuris, for example, they only conducted rites for traditional native crops and excluded those of Spanish origin.[53] The cause of Christianity might have been aided by an excessive crop yield, but climatic conditions worked against the friars in the last two decades before the Pueblo Revolt of 1680. In 1659, famine struck the region, and Indians had to sustain themselves on grasses and herbs. In addition, the Spanish military departed from their settlements on punitive missions against nomadic Indians who invaded the region, and the unattended crops of the region were left open to wandering livestock.[54]

A variety of problems plagued the Pueblos in the next few years. Another period of famine occurred from 1665 to 1668, and the lack of food proved devastating for Indians and Spaniards alike. Many Indians perished, including 450 from one pueblo alone. The Spaniards ate the toasted hides of cattle in order to keep from starving.[55] Lack of food continued to be a problem in 1670, and in the next year, a great pestilence gripped the area, killing both animals and residents. In 1672, hostile Apaches attacked the region and stole or killed many sheep and cattle.[56]

Famine, disease, attacks, tribute, and the suppression of native religious practices climaxed in one of the most successful revolts in Native American history. Previously, some of the villages demonstrated their disgust with the Spanish invaders, but in 1680 a movement resulted from a united effort by the Pueblos. The Indians suffered heavily under the invaders' rule, and their desire to return to ancient traditions culminated in the overthrow of the Spanish government in New Mexico. In addition to murdering friars and citizens, the Pueblos also destroyed their fields.[57] Santa Fe had to be abandoned along with other Spanish towns and ranches. Refugees gathered at Isleta in an attempt to protect lives and hold out against the thrust of the Pueblo attack. However, the Spaniards had few munitions and little food to sustain themselves. On the other hand, the Indians captured over 150 guns and enough food to supply their forces for months.[58] The scanty supply of sustenance at Isleta and the realization that no food existed in the area forced the refugees to flee south out of New Mexico.[59]

Not all of the Indians shared a hatred for the Spanish colonists. One of the Christian Indians appeared before Governor Antonio de Otermín to testify about the rebellion. He claimed that he was hoeing in a field that belonged to his master when he received warning about the revolution. When questioned about the nature of the rebellion, he said that the Pueblos suffered many hardships. Also, he emphasized that the Indians never had the time to devote to their own fields because their efforts had to be spent on Spanish projects. A few days after this admission, Otermín interviewed another Indian refugee named Don Pedro Namboas, a former resident of the pueblo of Alameda. He claimed that the suppression of native religious practices caused deep resentment among the Indians. According to Namboas, this attitude against the Spaniards had been harbored since the first colonists arrived in New Mexico.

Another Indian added significant detail to the story behind the revolt. He claimed that a native of the pueblo of San Juan named Popé instigated the rebellion. He went to the various Pueblos and told them to destroy the churches, rosaries, crosses, and other symbols of Catholicism. He also told the natives to stop speaking Spanish and to destroy all of the seeds that the colonists brought into the region. He ordered the Indians to plant nothing but native crops, but some of the Pueblos decided to retain Spanish foods. Another Indian claimed that Popé commanded the Indians to take over Spanish fields so that they could enlarge their areas to grow crops. Popé felt that the Pueblos could devote more time to these areas because now they did not have to perform duties for the Spaniards.[60]

Otermín attempted to reconquer the area, but his forces lacked military might since most of their food and munitions now belonged to the Indians. During their return to the area, the Spanish forces noticed that most of the natives abandoned their pueblos after taking what food they could carry. The Spaniards sacked some of these areas and burned many of the Pueblo homes. The Indians either moved up to the mesas above their farmlands or completely left the area in fear of Spanish retaliation. Otermín's forces halted at Isleta, where they were accepted by the locals. The Indians gave them fifteen bushels of corn and warned them that a famine plagued the area due to a drought and that Indian food supplies were low. They also claimed that the Pueblos to the north planned to attack their village in order to obtain their food.[61]

When the Indians to the north heard that the Spaniards had returned to the area, they fled to the mountains. This group included San Felipe, Santo Domingo, Cochiti, Santa Ana, Zia, and Jemez.[62] The Spanish forces discovered food at Sandia and Alameda and at the latter pueblo, they found plowshares and axes. Evidently, the Indians left in great haste for these items were important in their farm program. As Otermín entered this area, he found that the Indians harvested corn that grew on former Spanish fields. However, quantities of corn and beans remained in storage, and the Spaniards took as much of these supplies as they could carry. All corn that could not be taken was scattered throughout the pueblo and destroyed in a fire. The governor personally saw to it that the granaries in the pueblo were burned.[63] Otermín continued this procedure as his party moved north, and in this process, he wiped out vast supplies of Indian grain. He retreated back to El Paso, taking almost four hundred Isleta Indians with him. While he damaged eight of the pueblos south of Santa Fe, he did not come in contact with the main body of the rebels.

The burden of Spanish institutions proved to be repugnant to the Pueblos. They saw virtually no benefits from the intrusion by the new settlers. Indeed, famine consistently plagued the Indians during the last two decades of Spanish occupation. Added to this were the tactics employed by the friars to suppress native religious ceremonies. Also, nomadic tribes increased their raids on Pueblo villages after Spanish occupation, thereby depleting the Indian food supply. It is highly probable that the Pueblos felt that these depredations and natural disasters resulted from the suppression of their religious customs. The result of their action helped to unify the Pueblos and eliminate the traces of Spanish existence. However, the impact of the revolution on their irrigation and agriculture must have been devastating. Crops were destroyed, fields laid to waste, and many of the villages abandoned. Throughout the period of Spanish removal, agriculture continued to be practiced, but undoubtedly the Indians worked with a watchful eye cast in the direction of a possible reconquest.

SPANISH RECONQUEST

The Pueblos held out against possible reentrance by the Spaniards for twelve years. During this period, El Paso became a focal point of Spanish control. Menaced by Indian wars and threatened by a possible French intrusion into New Mexico, the authorities spent large sums of money in the late 1680s to recapture the Pueblo territory. Finally in 1692, Don Diego de Vargas led an expedition that returned New Mexico to the Spanish fold.

During his four-month campaign, Vargas recaptured the area without firing a shot against the Pueblos. However, when the main party of settlers finally entered Santa Fe, they encountered stern resistance and Vargas stormed the city. Only the pueblos of Pecos, Santa Ana, Zia, and San Felipe remained loyal to their promises of a year earlier. Most of the others still lived on the mesa tops and resisted a Spanish takeover. The vanguard of the Spanish forces who entered the area in 1692 noticed many of the abandoned pueblos. However, they found large fields of maize, melons, and squash at Cochiti. The natives of San Felipe also lived above their former settlement. When the Spaniards entered their old pueblo, Vargas made sure that nothing was done to their fields or property.[64]

Because of initial Indian resistance to Spanish rule, Vargas began to attack the recalcitrants. When they entered Pecos, the troops found plentiful supplies of vegetables and maize. Evidently, the Indians still worked in their fields right up until the time that Vargas and his men entered the region. Wanting to stress the peacefulness of his mission, Don Diego invited the Pecos natives to return to their pueblo and promised not to destroy their fields. Vargas then journeyed east and contacted the natives at San Ildefonso. Here the people submitted to his rule, but Vargas stayed overnight because he wanted to meet with the Indians who worked late in their fields. The natives of Picuris and Taos still lived in their villages, and in this latter pueblo, Vargas found much resistance. When Vargas entered Taos, he found that the Indians had deserted, but in a short while, he convinced them to return. At Zuni, Vargas found the Indians more friendly as he was given watermelons, tortillas, and mutton. They were still in the process of gathering their corn crops.[65]

The Spaniards desired to return not only their people to the region but also their customs. Vargas asked that Spanish seeds be brought into the area in order that the Pueblos could plant them. He also asked the Spaniards who came into the region to bring supplies so as not to cause trouble with the Indians. He tried to resettle Spanish people in good farming areas, and he encouraged the Pueblos to reestablish themselves in their former villages. If the Indians did not return, he wanted the Spaniards to settle in the old pueblos such as Jemez and Sandia. He also wanted to restore Isleta to those who had abandoned that village for El Paso. He assured them that the lands here were still capable of being irrigated.[66]

Another revolt broke out in 1696, but Vargas did not chase the rebels. Instead, he let starvation take its toll. He even burned cornfields at Acoma and Taos in an attempt to end the second rebellion. By the end of that year, the major Pueblo revolts were over. Spanish leaders could now concentrate on the reestablishment of their previous lifestyles in New Mexico. However, the revolt had taught the Spaniards a lesson, and their approach to colonization changed. No longer would they encourage the suppression of native religious practices or the oppressive taxation that had weighed heavily against their success. The battlefields of Pueblo-Spanish conflicts now shifted to the courts. Here the Indians hoped to preserve rights to their land and water and bring some semblance of order to a lifestyle that had been upended over the past century.

CHAPTER TWO

Spanish Law in New Mexico (1700–1765)

THE ORIGIN OF SPANISH LAW

With Spain's reoccupation of New Mexico, the Pueblo Indians once again became wards of a foreign government. However, the restrictive atmosphere of the seventeenth century gave way to a more lenient policy in the latter period of Spanish rule. Native ceremonialism returned in the eighteenth century as Spain's religious leaders tried to create a less oppressive milieu. A new cultural blend, based on proximity and fostered by intermarriage and exchange, resulted in mixed communities in the eighteenth century. With the exception of Zuni, Taos, Picuris, Isleta, and Acoma (all on the extremities of the central Spanish core), pueblos had changed locations in the past, but in post-1700 New Mexico the situation stabilized. Realizing that the lack of military protection in New Mexico threatened the existence of the renewed colony, the authorities of New Spain encouraged the Pueblos to become an auxiliary fighting force.[1] These new attitudes demonstrated a spirit of cooperation between the races that had not existed in the first century of occupation. Undoubtedly, the Pueblo Revolt helped to bring about this change in Spanish attitude by threatening their very existence in the area.

In order to avoid a repeat of the revolution, Spanish authorities took necessary measures to protect the economic well-being of the Pueblos. Since farming provided the basis of the native subsistence economy, the preservation of Pueblo rights to operate this system was crucial to their very existence. As the Hispanic population increased during the eighteenth century, the competition for land and water created numerous disputes. Therefore, the most important laws for the preservation of the Pueblo economy and the continuation of Spanish rule were those concerning agriculture and irrigation.

The implementation of a Spanish legal system had a significant impact on New Mexico, both before and after the revolt. Fortunately for the Pueblos, Spanish law regarding irrigation and agriculture had evolved in an area in Europe whose climate and terrain largely resembled that of New Mexico. The physical conditions of Spain offered more diversity, but the dry regions of the Iberian Peninsula greatly outnumbered the humid ones. In addition, most of the New World immigrants came from this arid section of Spain.

Spanish law, like other elements of Spain's society, emerged from various contacts with the Romans, Visigoths, and Moors, as well as local Spanish customs. Throughout the early history of Spain, regional customary laws ruled the various sectors of the country. Geography dictated this diversity. Countless valleys cut off from one another made communication difficult. The customs that developed thus had a strong local character and resulted in a wide variety of ruling traditions. In time these traditions grew into laws governing the use of arable land, the enforcement of grazing rights, and the distribution, apportionment, and sale of irrigation water. Although these customs originated in pre-Roman times, they persisted throughout the centuries of foreign rule whenever practical. Eventually these traditional customary laws were codified in the *fueros*.[2]

No code of laws has had such a wide influence throughout the world as that of ancient Rome—the first invaders of Iberia more than two centuries before the birth of Christ. Roman civil law pervaded the entire European continent and spread throughout the various colonial empires. Our knowledge of Roman civil law is based on the work of commissioners under the Roman Emperor Justinian. This codification was completed in A.D. 533. Its first principle stated that "by natural law these things are common to all: air, running water, the sea, and as a consequence, the shores of the sea."[3] By this definition, running water could not belong to any person, but the civil law did recognize the use of water as a right of property.

According to Roman law, the waters of unnavigable streams could be used by anyone who gained legal access to them for the purpose of supporting life, and this use included diversion for irrigation projects. Since these waters were subject to common use, the civil law stated that the first person to appropriate water from a natural stream for a useful purpose held title over other users.[4] If water flowed between properties that belonged to different owners, each could use only that portion needed to irrigate his area. If a landowner possessed both banks of a stream, he could divert all the water he wanted, but he had to return the unabsorbed water to the original stream. In essence, the only water that was his property was that which he could make use of as it flowed along. Dams could be constructed, but they could not cause upstream overflow or downstream deprivation.[5]

Generally speaking, Roman water laws were strictly observed; any violation resulted in punishment, including the loss of rights. The tenacity with which the Iberian natives held to customary laws, however, limited the immediate success of the Roman civil law in Spain. Most Iberian people continued to live in isolated microcosms, and their lack of outside contact made them distrustful of the outsiders. In addition, the Romans, who had a great deal of experience with alien nations, were sagacious enough to realize that it was better to let foreign people abide by their own private customs when they were not contrary to the general interest.[6]

Some of the Roman law prevailed, and it was only natural that Latin legislation would conflict with that of the Visigoths who invaded Spain in A.D. 409. These Germanic peoples had three alternatives: they could impose their own laws on the natives, adopt the Roman law, or allow both laws to coexist. The invaders decided to choose the latter course. As a result, regional conflicts occurred for a time, but eventually the Hispano-Roman inhabitants and the Visigoths developed a pattern of mutual coexistence. Although some of the Germanic code was never totally abrogated, the Visigoth administration did not make the profound changes that the Romans had.[7]

Like the Romans, the Moors adopted the irrigation practices of the Egyptians prior to their invasion of the Iberian Peninsula in A.D. 711. Not only did the Moslems employ reservoirs, wells, and aqueducts to irrigate their arid lands, but they also developed laws and customs to administer their watering systems. Individual irrigators could acquire water rights from small rivers, canals, and wells, but only those who worked on the construction of a ditch were entitled to the use of its waters. Those who held these claims for a long period eventually gained the legal right to them. Usually, Moslem law provided that those rights would remain unchanged unless the holders agreed to make alterations.

Moorish water law, like all Moslem law, emphasized community cooperation in times of need. Due to the communal nature of irrigation, the Moors required that co-owners were responsible for canal maintenance. Laws changed from one region to another to fit the needs of a particular community, but, in general, local irrigators were responsible for the regulation and distribution of water. In Valencia, the Tribunal of Waters, a group of irrigation officials, performed administrative and judicial functions. Their primary concern was the equitable apportionment of water during times of drought.[8] Moorish law established precedents for irrigation law in New Mexico by emphasizing prior appropriation, beneficial use, and community regulation.

As the Iberian natives gradually threw off the shackles of foreign rule, modern Spain began to emerge, and this included the development of their legal system. There was really no school of law before 1200. Roman law, though taught in some cities, had little importance until the thirteenth century.[9] The crusade against the Moors in Valencia formally closed in 1244; seven years later regulations for irrigation canals of that kingdom were published. These laws contained stipulations that governed canal taxes, maintenance and cleaning procedures, and fines for negligence of duties.[10] As the reconquest continued, Moorish sections of Spain still implemented local laws and customs regarding water regulation. The role of the pueblo or municipality also contributed to the community control of irrigation practices. In previous centuries, the pueblos had gained control of land and water for defensive purposes. What emerged was a system in which local towns held water as a common property for domestic and irrigation use. Eventually these local customs and regulations contributed to the water laws of Spain and its colonies in the New World.[11]

Spanish kings also sought to determine the course of the legal system. After the partial expulsion of the Moslems, the royal authorities wanted to consolidate their power over the isolated sectors of Spain. In 1256, King Alfonso IX ordered some of the prominent lawyers of the kingdom to compile and codify all Spanish laws. The *Siete Partidas* was completed in 1265, but it did not gain formal recognition until 1348, when the *Ordenamiento de Alcala* was promulgated. While the codification mainly reflected Roman law, it also contained local Spanish water regulations. The partidas had provisions for irrigation, canal maintenance, community rights, and common ownership of running water. Later, as the central government emerged, it based the water laws of Spain on this document. The *Siete Partidas* remained in effect even after Spain began her vast colonization program, thus paving the way for the future application of the modified civil law in New Spain.[12]

Although the *Siete Partidas* contained general stipulations for the control of irrigation, local customs still influenced the communities' irrigation systems. A study of Valencia during the fifteenth century reveals how the canals of that area were administered. The head irrigation officers in the kingdom were the *sobrecequiers* and the *cequiers*. While the former was a municipal officer, the latter usually governed only one canal. Since the administration of the daily activities of the irrigation community required an intimate knowledge of local topography, rules, and traditions, it was necessary that the cequiers be cultivators of that region. Each community also had at least two *veedores*, assistants to the cequier. These inspectors usually represented the users' rights and counseled their boss before he made a decision, insuring the observance of local customs. Finally, these communities had financial officers who collected taxes and fines.[13]

SPANISH LAW IN THE NEW WORLD

The *Siete Partidas* was updated by the *Leyes de Toro* of 1505, the *Nueva Recopilación* of 1567, and finally the *Novisima Recopilación* of 1805. The revised version became the basis for Spanish law throughout the colonial period. A separate code of laws, however, governed Spain's colonial possessions in the New World. First published in 1681, the monumental *Recopilación de Leyes de los Reynos de las Indias* incorporated past Spanish legal theory in an attempt to provide a compilation of regulations covering various aspects of colonial administration. The *Recopilación* resulted from a lengthy and arduous codification that spanned the administrations of five kings. This voluminous work had to be divided into nine books for publication. Each of these contains some laws concerning Indians, but book four, which deals with discovery, pacification, and settlement, and book six, which is concerned with Indian affairs, contain the bulk of ordinances governing native land and water in the New World.

The significance of the *Recopilación* should not be underestimated. Sometimes, it has been criticized as being merely an enumeration of colonial laws that were exceptions to legislation already in effect in Spain.[14] This judgment reflects the code's incompleteness. The *Recopilación* provided that "in all cases, transactions and suits which are not decided nor provided for by the laws contained in this compilation . . . the laws of our kingdom of Castile shall be observed."[15] While this statement demonstrates that Castilian law could supplement the legislation designed for the Indies, it also shows that the *Recopilación* contained the laws for those kingdoms, as the *Leyes de Toro* included the laws for the kingdom of Castile. The crucial point is that New Spain was not part of the kingdom of Castile; rather, it was a separate kingdom that belonged to the king of Spain by papal declaration. Therefore, the *Recopilación* was to the New World what the Leyes de Toro and the subsequent updates to it were to Spain. Spain's royal authority stressed that the king's officials in the New World and the Council of the Indies should follow the dictates of the *Recopilación*. Its laws were the supreme commandments in the New World. Any revisions to it would be authorized by the council and explained to the local officials in New Spain.[16]

In 1573, Phillip II issued the *Ordenanzas para los nuevos descubrimientos*, a detailed set of instructions that governed the pacification and settlement of unexplored regions in the New World. Eventually this codification became the basis for book four of the *Recopilación* and the major regulations that guided the Oñate colonization effort in New Mexico. According to the ordinances of 1573, all discovery or colonizing missions had to obtain regal authority before setting out on their journeys. After the exploration of a region, settlements might be made, but only in places that would not be detrimental to the Indians. Colonists were advised to select areas close to water and wood, with enough available land for farming and grazing. Equipment for agricultural purposes would be distributed to the settlers. The person in charge of the settlement had the responsibility to allocate land to the colonists for the purposes of tilling and raising animals. The Indians were to share in the Spanish lifestyle; they would be allowed the use of certain Spanish items, such as cattle and tools. Tribute was not to be levied at once if the colonial authorities felt that its implementation at an early date would upset the Indians.[17]

Book four of the *Recopilación* further elucidates the Spanish procedure for discovery and colonization. The discoverers were told to avoid friction of any kind with the native people. The clergy was encouraged to do everything possible to see that the Indians were treated fairly. The discoverers were not to damage or steal their property. The Spanish government also cautioned settlers to avoid conflicts with the Indians. Crops were to be planted and animals grazed in places where these activities would not prove injurious to the aborigines.[18] In order to encourage settlement in the remote areas of New Spain, the viceroys distributed lands, house lots, and waters, but only so as no harm would come to a third party. While the viceroy or governors had the authority to apportion land to the settlers, the town council or cabildo advised them in these matters. Concerning the natives, the authorities were to leave "the lands [*sic*] cultivated properties and pastures of the Indians for the Indians in such a way that the Indians may not lack what they need, and that they may have all the relief and repose possible for the support of their homes and families." In addition, any assignment of lands later found to be "prejudicial or offensive" to the Indians would be canceled and the lands returned to the natives.[19]

As previously mentioned, the introduction of livestock affected Indian farming in New Mexico. Not only were the Pueblos forced to increase corn production in order to feed the animals, but they were also required to spend time watching the herds. As a result of a lack of both vigilance and fencing, livestock often wandered into Indian fields and canals. For this reason, the law required the Spaniards to construct their grazing fields at a distance from Indian farmlands and to provide enough herdsmen to avoid damage to their fields. Settlers were also required to remove their livestock from irrigable lands so that wheat could be planted. More significant in the long run was the damage done to the environment by animals who consumed large amounts of water and grass. As the number of animals greatly surpassed the human population in New Mexico, overgrazing contributed to the process of desertification, which led to periodic water shortages.[20] Unfortunately, these dire long-range consequences could not be envisioned by Spanish lawmakers.

The *Recopilación* also contained specific regulations regarding the clearing of land titles by *composición* or sale. For example, land titles would not be confirmed if the claimant had acquired property from the Indians by means other than those set out by royal decree. Waters and irrigation systems, as well as properties, were to be protected in this manner. Any land that benefited from Indian irrigation was to be reserved for them, individually or communally, in all circumstances. In order to decrease the alienation of Indian property, Spanish law held that civil authority had to be obtained before the natives could sell their property. In fact, a formal public hearing and auction was to be conducted for this purpose.[21]

Since the Pueblos practiced agriculture and lived in locally governed villages, the Spaniards perceived them as different from the nomadic tribes that surrounded Spanish settlements. Because of this distinction, the Pueblos and other "civilized" tribes were considered to be prime candidates for Christianization, tribute, and a limited amount of self-government. Since the Pueblos already conformed to the Spanish expectations of sedentary, agricultural wards of the government, the civil authorities were instructed to respect their municipal institutions. The *Recopilación* stated that "the laws and good customs that the Indians had in antiquity for good government and general welfare, as well as the . . . customs . . . retained since they became Christians, shall be respected and enforced." The law, however, made two major stipulations. The Indian traditions could not conflict with religious doctrine or Spanish law, and the terms "good customs" and "good government" were left open to the interpretation of the legal authorities of New Spain.[22]

Although servile work was not a typical Spanish forte, they expected the Indians under their vassalage to engage in the hardships of agriculture. The native tillers of the soil, however, were encouraged to acquire oxen in order to ease the burden of their farming labors. Although the civil authorities rejected the use of force to compel them to work in the fields, they thought it important to keep the Indians occupied. Idleness, it was felt, would deter them from achieving a prosperous life and all of its heavenly rewards. So that other duties would not conflict with Indian farm programs, the Spanish government required that the Indians be given enough time to cultivate both their own lands and those of the local community.[23]

Since the Pueblos were already settled in established villages when the Spaniards arrived in New Mexico, officials had little initial concern about the location of Indian homesteads. After the Reconquest, however, some of the Indians, like those of Isleta, had to be relocated. The *Recopilación* provided for the establishment of sites for Indian pueblos, which were to be located in areas that had ample water, woodlands, and farmlands. For Pueblos already in place, Spain would confirm their land holdings and make sure that they had enough water and acequias to irrigate their fields.[24]

As the Spanish civil authorities were instructed to promote native farms and protect native land, they were also required to preserve Indian water rights. According to the *Recopilación*, judges were to be appointed to allocate water for Indian farms, fields, and livestock. The native inhabitants were to receive an ample supply of water, and the judges were to make sure that no harm came to them in this process. After this allotment was made, the viceroy was to explain the procedure in the case to the authorities in Spain. If the judicial rulings were appealed, the *audiencia* would make a decision on the case. Realizing the necessity of immediate action on water rights cases, the law held that the decisions of the audiencia would be acted on directly. Only after the high court's judgment was executed could the parties involved make an appeal and have their cases reviewed.[25]

The law of the Indies, like Roman and Spanish law, contended that water was held in common by the inhabitants of a region. According to the *Recopilación*, water, forests, and pasturelands—including area held under grants—were to be enjoyed by all the residents of a region. The viceroy and audiencia decided what was good for the land and the farming practices of the people.[26] It should be noted that the good of the community as a whole was supposedly more important than individual rights. The law also specified procedures for Indian watering policies that strongly encouraged the customs of the native people. Indians who had controlled irrigation in pre-Columbian times were to remain in control. Anyone who took matters into his own hands would be reprimanded by being deprived of water until all other farmers had met their own needs. Spaniards who had been allotted lands were also to respect these methods of water distribution.[27] In this piece of legislation, the Spanish government demonstrated a concern for local provisions in much the same way that it had in Spain following the reconquest from the Moors.

Spanish law regarding land and water in the New World reflected the attitudes that had developed in Europe. It was fortunate for the Pueblos that the colonists shared with them centuries of experience in the practice of irrigation. Riparian water rights were not a major factor for either the Indians or the Spaniards in New Mexico. Since the Iberian laws resulted partially from local customs, this attitude was easily incorporated in New World legislation. Realizing the need for communal efforts in large-scale irrigation systems, the Spaniards stressed the idea that water was a property to be held in common. In this manner, water could be employed for the beneficial use of everyone.

The paternalistic attitude of the Spaniards in matters of Indian agriculture and irrigation is undeniable. The civil authorities, however, had practical, as well as philosophical, motivations for this policy. Indian crops could provide tax income, barter, food supply in times of scarcity, and sustenance for the local religious population. Settled wards also made better citizens, who could be mobilized to thwart the attacks of barbarous neighbors.

SPANISH LAW IN NEW SPAIN

It is one thing to draw up laws that attempt to regulate society, but the implementation of this legal system in a distant colonial possession is an entirely different matter. Decisions concerning irrigation often required immediate action. By their nature, conflicts over water rights during times when the crops needed irrigation could be detrimental to the very existence of the society. Naturally, the laws tried to cover all situations so that local authorities did not have to rely constantly on royal advice. The difficulty of communications during the Spanish period did not allow civil leaders in New Mexico the immediacy of contact that other settlements farther south enjoyed. Therefore, it was important that land and water rights be cleared up so that the everyday functions of a distant irrigation society could proceed without lengthy delays.

The legal phase of Spanish assimilation was only partially successful because of the distinct historical and geographical character of New Spain.

Separated by unique local traditions and the vast distance between locales on opposite sides of the Atlantic Ocean, Iberian legislation was altered. The failure to transplant Castilian legislation completely to the New World meant that a unique legal system would emerge in the colonies. In a conciliatory move, the Crown tried to incorporate the judicial customs of the Indians into the emerging jurisprudence of the New World. In addition, the native institutions related to irrigation and agriculture generally did not contradict the fundamental precepts of Spanish law. And, since initially the Indians possessed the best land and prior water rights, compatible regulations were necessary in order to prevent the inevitable conflicts between the natives and settlers.

The land question was an immediate problem for the original European colonizers in any region. Since, by European law, the physical attributes of the New World belonged to the king, the settlers could only gain possession of them by royal grants (*mercedes*) or purchase. Frequently, as in New Mexico, the Spaniards simply encroached on the desirable land of the Indians. Settlers often disregarded Pueblo property rights because of the difficulty of enforcing the law in the distant northern frontier. Still, settlers were better off if they could establish a legal foundation to their property. As previously mentioned, the process of composición was employed to clear up unstable or spurious land claims. According to the law, a person had to live on the land for ten years before the composición could be granted. In addition, composición was not to be granted to those who gained Indian lands illegally or possessed a faulty title to them. Indian communities were always to be given preference in matters related to the clearing of land titles.[28]

Another device that Spaniards employed to gain title to land was the practice of *denuncia*, which became part of the colonial law by a *cédula* of 1735. It was an aspect of the Bourbon reforms of the eighteenth century drawn up by the new monarchy in Spain to invigorate a sagging colonial economy. Initiated as a part of the reform measures, this process allowed citizens of New Spain to claim unoccupied royal lands or other property that was being held illegally. In the process of denuncia, the claimant could then go ahead with composición so that he could gain legal title to the disputed property.

A more successful piece of legislation under the Bourbons was another cédula of 1754, which had its foundation in the *Recopilación*. The viceroy and judges were personally responsible for land policies in New Spain. The king sought to protect the rights of Indians in order to create a class of small landholders. Realizing that the native people lacked the judicial wherewithal to prove their land claims, the king asked that his representatives in the New World proceed with a little moderation in their dealings with native inhabitants. Non-Indian owners who claimed and cultivated land before the beginning of the century (the start of the Bourbon rule) were to receive confirmation in titles to those areas. Those who held legal title to lands acquired after that date received immediate confirmation of their claims; others who had continuous possession of property since 1700, but no title, could normalize their claims through composición. In 1786 when a new political structure, the intendancy system, was introduced in New Spain, the intendants were instructed to follow this cédula and the *Recopilación* in their dealings with land claims. The cédula of 1754 continued to affect policy into the Mexican period.

No matter how equitable this law may have seemed to settlers, Indian areas were dwindling rapidly and more protection was required. Early in the colonial period, Spanish law gave Indians the right to dispose of their property, but only with a Spanish official present. Since this policy did little to protect native land from greedy buyers, a new ordinance was adopted in 1781. Accordingly, Indians could not conduct the sale of their property; rather, Spanish authorities were required to license all land transactions. In order to obtain a license, the purchaser had to prove that the sale was necessary, and if he did not obtain a license, he could never gain title. Although this practice continued into the American period of the Southwest's history, Spanish bureaucrats never enforced the law with any consistency. While the Bourbon land reforms seemed to have had little effect on affairs in New Mexico, it appears that composición and denuncia were accepted procedures for clearing up land titles in this region. However, it was generally the laws and customs of previous decades that governed the New Mexico land claims in the eighteenth century. Evidently local administrators depended on their own systems to govern the amorphous villages that grew up after the Reconquest.[29]

The appropriation and use of water for irrigation purposes was another major concern of the aborigines of New Spain. The only water right they recognized was that of using water in their irrigation canals. In this system, riparian rights were never a consideration.[30] The *Recopilación* encouraged this idea by providing that the waters of New Spain were common to all inhabitants and that local authorities should regulate the distribution of irrigation water to promote the general welfare. In addition to caring for their own canals, water users had to help with the maintenance of communal ditches according to their frontage. Finally, they were not supposed to construct new ditches upstream to the detriment of people who had prior water rights.[31]

Land and water were granted to both individuals and communities, which included Spanish municipalities and Indian villages. The usual farming mercedes (caballerías) stipulated that the land should be cultivated within a year, that the property could not lie fallow for four straight years, and that the grant could not be sold until the owner had held it for four years. Although livestock were not permitted on the farmlands during the cultivating process, they could be pastured on these fields after harvest. Finally, these grants were not supposed to cause any harm to Indian land and water rights. At first there was no real distinction made between land and water in a merced; the use of water was not specifically mentioned. By the time of the first colonization movement in New Mexico, however, water provisions within land grants had become a general rule. This policy continued through mid-seventeenth century. In some of these mercedes, the grantee received an assignment of water sufficient to irrigate not only the land already under ditch, but all irrigable acres.

Although most grants were farming and ranching mercedes, some dealt specifically with water. Characteristically, these water grants had to do with giving a certain party the rights to a stream or river as opposed to solving conflicts between various groups. Most of the early mercedes did not involve new grants, but rather confirmed existing Indian water rights based on prior use before Spanish colonization. In this manner, the Crown could support Indian agriculture and discourage Spanish encroachment. One expert in these matters feels that the low number of water grants indicates that adjudication of these matters was handled in a spontaneous manner. When enough landholders existed to cause conflicts, the procedures became more formalized.[32]

In northern New Spain, land classification determined the degree of water rights that went with a grant, and there were many different types of uses. The *Recopilación* recognized distinctions between grants based on land usage. Thus, grants were made for cropland, pasture, irrigation land, and cattle land. Some of these contained water rights while others did not. For example, grazing land did not include water rights and no use of water was implied. The use of water on farmlands is more complex. Some of these grants included water rights in order to cultivate the land, but a large number of them had no provisions for water. Farmlands governed by Spanish law fell into three classifications—those used for dryland farming, areas requiring a rainy season for growth, and land that could be irrigated. Although scholars are still in disagreement concerning implied water rights on all farmlands, it appears that the concept of unwritten rights applied only to irrigation land.[33]

Litigation over water rights in New Spain did not always involve formal procedures based on the legal doctrines of the *Recopilación*. Rather, the judgments in these proceedings often evolved from concerns with quick solutions between the parties involved. In this manner, the authorities hoped to avoid entangling lawsuits and the local unrest that would follow. During the eighteenth century, water litigation generally followed a standard procedure. Mercedes would sometimes establish existing rights, but, lacking this formal statement, officials sought to discover an authoritative recorded statement on the distribution of water in the region. In the absence of any official policy, authorities tried to establish rights through investigation. Witnesses were called and inspections conducted so that local officials could observe the availability of water and the prior rights of long-time users. After these inquiries were made, a *repartimiento de aguas* or distribution-of-water agreement was drawn up. Rights to the stream were thus established, and the waters were divided among local users.

Water rights in New Spain were generally based on title, prior usage, community or individual needs, intent, the amount of accessible water, and the protection of Indian villages. Ideally, the repartimiento was considered the ultimate method of solving water questions. The adjudication proceedings that led to this system of distribution took into account the use of water by Spanish farmers and Indian villages. So custom was an important factor in determining water usage. In this manner, local authorities hoped to promote the general welfare of the entire agricultural community. Unfortunately, Indians were hurt in these proceedings when they could not produce a title.

It would seem that the Spanish concern with prior rights would be extremely advantageous for Indian tribes who had employed irrigation methods since pre-Columbian times. However, the concept was interpreted to mean prior and uncontested beneficial use of water over a long time. Many Spaniards took advantage of this definition to establish their own rights to water. Normally, Indian communities were given enough water to sustain their members; the rest was doled out to other users based on need. Therefore prior usage, while important to the establishment of Indian water rights, did not preclude the possibility of Spanish rights where a surplus of water existed. It also did not mean an exclusive right in times of drought because Spanish officials attempted to prevent an injury to a third party.[34]

Since the public interest was a major concern of Spanish authorities, the needs of communities were a primary concern in the adjudication of water rights. Naturally, population along a given stream or river changed in time. Usually this meant a decrease in Indian population accompanied by an increase in Spanish population. As this demographic shift occurred, the needs of communities along the water source varied accordingly. The predominant needs of a given locality might necessitate a change in the distribution of water. Therefore, if sharp population shifts occurred, existing arrangements could be altered.[35] In spite of this apparent elasticity in Spanish water policy, the rights of an Indian pueblo to use all of the water necessary for its existence remained of paramount interest to Spanish authorities.

By the time the Bourbons replaced the Hapsburg dynasty, the Spanish empire in the Americas was on the decline. To prevent further decay the new rulers initiated measures designed to streamline the New World's administration and protect its outer reaches from intervention. As part of the Bourbon reforms of the eighteenth century, New Spain sought to strengthen its control over the northern frontier. To protect New Spain militarily against the advancement of other European countries, authorities established the Provincias Internas in 1776. Originally, four provinces fell within its jurisdiction: the Californias, Nueva Viscaya, Sonora, and New Mexico.

In 1789, the commandant's office drew up a settlement plan, known as the Plan de Pitic, for future towns in the Provincias Internas. In part, the plan was designed to regulate land and water policies within the region. Many of the ordinances depended heavily on the stipulations found in the *Recopilación* and ordinances of 1573. According to the Plan de Pitic, the governor was to supervise the founding of new communities. His duties included the distribution of water and land to the colonists, so that both would be held in common by Spaniards and Indians alike for their benefit and use. In addition, article seven stated that both groups should "enjoy equally the woods, pastures and waters and other features outside of those assigned to the new settlement in common with natives of adjoining and neighboring pueblos."[36]

The plan provided for future expansion by setting up *ejidos* or common plots that could later be subdivided. After the acequias madres and sangrías were laid out, irrigable plots were distributed to the farmers. The cabildo doled out land and water to families based on need and resources to farm the land. Rather than allowing a random selection of irrigable plots, the plan encouraged contiguous plots so that the farmers could practice intensive agriculture. Title to the land could only be gained after a person occupied and improved his acreage for a four-year period. Articles nineteen through twenty-three provided rules for water distribution and irrigation control. The leading official of the new community would oversee the distribution of water, and settlers were warned not use more than their share. An irrigation officer would control future distribution and accompanying system maintenance.[37] The Plan de Pitic was supposed to be a model for future Spanish towns on the northern frontier. Although there is no evidence that it was ever applied specifically in New Mexico, many of its principles, based on established law, were already being employed there. It is important both as a reaffirmation of the ideas that governed Spanish colonial land and water policy, and as the eventual basis for the California Pueblo Rights Doctrine.

SPANISH LAND AND WATER POLICIES IN NEW MEXICO (1700–1765)

While the Plan de Pitic represented the ideal situation, the real world of the northern frontier provided great challenges to legal application. Spanish law came to New Mexico early in the colonial period, but its enforcement lacked consistency. For example, the law strictly forbade encroachment on Pueblo lands, but many encomenderos ignored this stipulation. The small size of the settlement in New Mexico and its remote location allowed the authorities there to ignore the precise implementation of the law. Because of destruction of documentary evidence in the revolt of 1680, we do not know if the Pueblos were given formal grants by the Spanish government. It has often been held that the Pueblos received specific grants of four square leagues, but more evidence suggests that they were given the right to all of the lands that they occupied and used. After the reconquest of New Mexico, the protection of legal rights, like other aspects of Spanish-Pueblo relations, reflected the changed Spanish understanding. Throughout the eighteenth century, the Indians sought and received some protection from the courts on matters concerning land and water.[38]

After the Reconquest, Spanish farmers in New Mexico typically received small farms called "ranchos" in place of the larger hacienda grants that had predominated in the seventeenth century. The increase in the number of settlers produced heavy competition for the limited arable land available. Since the new colonists were also reluctant to seek Indian aid in their farming pursuits because of the havoc caused by forced native labor, the agricultural plots were small. The ranchos were frequently long and narrow and paralleled the river to give them maximum access to the limited water.[39]

The transition to a new relationship with the Pueblos following the Reconquest caused many problems for returning Spanish settlers. Before the colonists resettled New Mexico, they had suffered severe deprivation in the El Paso area. They complained about the lack of rainfall and arable land. They had too few oxen to expand their farmlands, and even if they had had more of these animals, they would have lacked pastures for them. Starving families had to sell household items to obtain food from the Indians. Fearing that they would never be able to regain the possessions they had left behind in New Mexico, these Spaniards asked for permission to colonize in the state of Chihuahua. Vargas, however, denied this petition, basing his opinion on a royal decree sent to his predecessor. This proclamation had encouraged the Spaniards to resettle their abandoned possessions and increase their claims if necessary. However, no harm was to come to the Indians in this process.[40] In 1705, Governor Francisco Cuervo y Valdés reiterated this negative attitude toward encroachment so that land grabbing would not encourage another revolt.[41]

In spite of repeated prohibition of encroachments on Indian land, Spaniards began to settle on Pueblo land immediately following the Reconquest. After Diego de Vargas reclaimed New Mexico for Spain, he began to parcel out land that had been abandoned during the revolt period. When he found that Tano Indians had relocated La Villa Nueva de la Santa Cruz de la Cañada after the rebellion, he evicted these natives and reserved the town for settlers. The new arrivals were fortunate to be able to use the houses, farms, and irrigation ditches that had been built by the former residents.[42] In this particularly blatant form of encroachment, all Indian rights were ignored.

Soon after the Reconquest, the Indians of San Ildefonso sought protection through the Office of the Protector, Captain Alonso Rael de Aguilar. According to the native people, Captain Ignacio de Roybal had acquired land on the opposite side of the Rio Grande from the pueblo. The Indians complained that this violated royal laws and caused them great harm. They stated that the area in consideration had belonged to them since before the Spanish arrival and that they planted squash and melons there. The protector petitioned acting governor Juan Páez Hurtado in September, 1704, asking that Captain Roybal display his grant for examination. Diego de Vargas had granted this land to Roybal in the previous March. At the time, Don Ignacio requested acreage for pasturing his animals but he did not immediately take possession of it. He claimed that his grant could not interfere with the pueblo and that he never witnessed the Indians farming on it.

The acting governor dispatched a group of investigators to inspect these opposing claims. They discovered an apparent irrigation ditch and some cultivated land. The Indians also pointed out a monument constructed by the Spaniards before the revolt as a method of marking off native property. After the governor heard this report, he decided that the San Ildefonso land boundaries should be firmly established and that Roybal should be given lands adjacent to the Indian claim. If the parties involved were not satisfied with the results, they could petition the governor or send their appeal to the viceroy.[43]

This was a case in which the community needs of the Indians received the benefit of the law, even though they admitted that they had not farmed the area for ten years. Their claim to the land, they felt, should be undisputed because, previously, leaders of the pueblo cultivated and encouraged the Indians to plant in this area. They had constructed a house and a tower for protection, and remnants of their ditches and farms still remained. The governor protected the San Ildefonso rights in spite of the

Spanish presence on land that had once been used by the Indians. In the previous century, it is doubtful that any of the Pueblos would have been able to hold on to this disputed property.

Throughout the post-revolt period, settlers sought to use Indian land. In 1743, Vincente Durán de Armijo petitioned the governor, Don Gaspar Domingo de Mendoza, concerning some farmland near the pueblo of Nambe. Armijo, an original settler with the Vargas company, complained that because of a drought, he had been unable to raise enough crops for subsistence that year. He asked the governor to grant him some surplus land near the Indian pueblo, claiming that his use of this tract would not interfere with the waters or pastures of the Indians or the Crown. Acting justice Juan García de Mora proceeded to Nambe in order to read the Armijo petition. The Indians opposed this grant, claiming that it would cause them great harm. Instead, they proposed that they be allowed to assign him a different piece of land adjacent to their pueblo. The new parcel was bounded by the Rio Nambe and also by an acequia. Mora took Armijo across the new section and Don Vincente agreed to the new terms. Governor Mendoza was pleased that the Indians were not harmed and consented to the agreement.[44] In this case, legal standards were observed, and Spanish encroachment was not allowed to be detrimental to Indian affairs. At the same time, however, the Spanish farmer legally gained title to some productive property in an area potentially dangerous to the Indian pueblo.

In one case the Indians requested payment in food from alleged trespassers. Rafael Sánchez complained to the *jefe político* that the alcalde of Laguna asked for one-third of the crops Sánchez had grown on lands near the pueblo, which were recently purchased from a Navajo Indian named Francisco Baca. The document of purchase supposedly had been witnessed by the former alcalde, Don Juan Chaves. Sánchez added that another tract of land had been ceded to the Laguna Indians in order to compensate for their loss, but the workers rejected this parcel because they claimed rights to the land sold by Baca.[45] While there was no further documentation on this case, Baca was evidently trying to sell Laguna land. The natives of the pueblo attempted to hold on to property they considered their own and added a fine for the alleged encroachers.

A report on New Mexican affairs in 1760 reveals the conditions of agriculture in the region. Fray Juan Sanz de Lezaun described the fertility of the soil and the productivity of corn, wheat, and vegetables. He also stated that the governor and the *alcaldes mayores* used Indians as weekly servants to till their lands and clean their ditches. The natives were not provided with food and received no compensation, in spite of a 1704 viceregal that Indians should be paid for their work.[46]

Lezaun's report also touches on the resettlement of the pueblo of Sandia, which had been burned in the rebellion of 1680. Many of the natives had moved to Hopi. After the Reconquest, Spaniards settled on their lands. In 1742, two Spanish priests returned these refugees to New Mexico. In April 1748, Fray Juan Miguel Menchero asked Governor Codallas y Rabál that certain lands taken away from various criminals be given to him for the reestablishment of the pueblo of Sandia. Lacking the authority to proceed, the governor suggested that the priest contact the viceroy.[47] Fray Menchero, who had already worked among the natives of New Mexico for six years, had previous instructions from the viceroy to construct pueblos with sufficient lands and water for the use of Indian converts. Therefore, he knew the royal provisions concerning property and water. In his petition, Menchero asked that the boundaries of the pueblo be determined and that Spanish land claims within the pueblo be given to the Indians if they so desired. In return the Spaniards would be given land elsewhere as compensation.

Governor Codallas y Rabál commissioned Don Bernardo de Bustamante to proceed to Sandia to conduct an examination of the site for the future pueblo. In addition, he was to "distribute the lands, waters, pastures and watering places sufficient for a regular Indian pueblo as required by the royal orders concerning the matter." West of the pueblo and across the Rio Grande, three Spaniards held land that would have been overlapped by a four-square-league measurement. These men were allowed to keep their property as long as the natives could graze stock on these pastures. Finally, a judge was appointed in order to protect the Indians from unlawful adjudications.[48]

The Indians were not always unwilling to sell their land to Spanish settlers. In 1732, Luis Romero of Picuris Pueblo petitioned the governor about some land that he desired to sell to Pedro Montes Vigil. Governor Cruzat y Góngora referred this note to the tribe for their approval. The selling of this property was not adverse to the general welfare of the Indians, and the deal was consummated.[49]

Confusion over land claims often recurred in the eighteenth century. In 1765, Bartolomé Fernández, the chief alcalde of the Keres pueblos, sent a note to the governor concerning apparent trespass on the summer pastures of the Indians by the Romero family. Governor Vélez Cachupín sent Fernández to visit the area and expel any violators. Miguel Romero said that he would obey the governor's order, but he claimed that he had purchased an interest in the land five and one-half years before, and nobody complained until they found out about his purchase. The tract lay halfway between the gardens of Cochiti and San Ildefonso. Don Miguel had a grant from his father that dated back to 1739, but it did not describe the exact boundaries. In spite of this condition, Governor Vélez Cachupín stated that Romero should neither have a house on the property nor cultivate the area. The Romeros were allowed to run stock there as long as they did not cause harm to the neighboring Indians. Fernández apparently agreed with the governor because he found certain discrepancies between the aforementioned will and other land holdings.

Romero appealed his case to the authorities, but he was denied on several grounds. For one thing, his father had not consulted with adjacent owners when the claim was initially made. Also, the Romeros did not attempt to make a settlement on the property in the period required by law; in fact, they had waited twenty-seven years. Although Miguel pastured his stock on the land in question, he had done so without just title. The area was declared part of the royal domain and deemed a common pasture for the local community.[50]

The Indians of San Ildefonso faced land problems from all sides in the mid-eighteenth century. In 1763, they made complaints about encroachment by various Spaniards. In one case, the Indians had loaned a house lot to Mathias Madrid, who began to cultivate lands around the area. This land was in turn sold to Juana Luján and her son, Juan Gómez, who continued to farm near the Indian pueblo. Marcus Lucero of Ojo Caliente also built a house on property purchased from the Indians, and the natives complained that the stock that grazed on this land trespassed on their property. Even after Lucero received the payment for the land ordered by the governor, he refused to leave the area. Finally, Pedro Sánchez had been granted some of the Indian land west of the pueblo and had built a house on this area. His stock also caused damage to the Indians.

Basically, the inhabitants of San Ildefonso asked for protection of their land claims. The governor commissioned Don Carlos Fernández, the chief alcalde of Santa Cruz, to examine the complaints. He measured boundaries and collected relevant documents. Since the Indian claims to the Luján property were countered by strong Spanish claims to possession from years of occupancy, the civil authorities felt that justice could best be served by giving the Indians a grant of unoccupied property. The governor's decision resulted in a grant of land, which included the Sánchez ranch, to the Pueblos. He also deemed that the Lucero house could be destroyed or used by the Indians as they saw fit.[51] These decisions reveal that laws guaranteeing the protection of Pueblo land rights by the Spanish authorities existed, but the Indians had to pursue legal recourse against trespassers or Spanish encroachment would continue unabated.

In addition to land claims, Pueblo water rights were also threatened by increased colonization. In 1708, Captain Baltazar Romero was given possession of a grant of land that had not been previously occupied. This tract bordered the Rio Grande just outside of Albuquerque. In addition to grazing and agricultural lands, Romero received the right to water to make the property useful. Governor Marques de la Penuela ordered that he be given all the land and water that he requested and that his claim be protected from usurpation by another party.[52]

Often Spanish claims to water conflicted with Indian interests. In 1724, Juan and Antonio Tafoya requested some land near the pueblo of Santa Clara, and the governor granted the property to them. The Indians, present during the scheduled transfer, complained that Spanish cultivation of the tract would be detrimental to Santa Clara's water supply. Cristobal Tafoya, father of the petitioners, claimed that the land would be used, not for agriculture, but rather for grazing. Under this condition, the Indians agreed to let them have the land.[53] Ten years later, however, the controversy over this land had not been resolved. The Tafoya brothers wanted to cultivate their property with water from a spring that they claimed would not interfere with the native water rights. In addition, they claimed that some areas on their property could be farmed without irrigation. Governor Cruzat y Góngora commissioned Don Juan Páez Hurtado to investigate the problem. He conducted his survey in the presence of the governor of the pueblo, five Indian leaders, and the Tafoya brothers. Páez Hurtado found the spring in question, but he also discovered the Spaniards using irrigation. Because of this, the governor denied the Tafoyas' request, ruling that the use of irrigation there would make farming at Santa Clara precarious if not impossible.[54] In this case the civil authorities were careful to protect the Indian water rights in spite of the fact that a decade had elapsed since the original settlement.

In general, the specific application of the *Recopilación* to water cases on the northern frontier was neither practical nor enforceable. Governors in New Mexico retained the power to make decisions concerning water disputes, but they preferred to rely on the advice of local officials. In this procedure the emphasis was on the need to pursue farming activities over community expansion. According to local custom, governors also emphasized Indian priorities when Spanish settlers sought to take over areas considered detrimental to the Pueblos' previous use of water.[55] The major question concerned rights to water at times of drought, and during these times of strife, it was custom that prevailed over law. This situation meant that the rigid legal words were bent in order to make pragmatic judgments. Therefore, prior use did not necessitate exclusive rights. In a system of balance between need and availability, a method of water sharing considering the needs of everyone prevailed over absolute water rights. Like their predecessors in Spain, judges in New Mexico leaned toward a custom emphasizing the community over the individual.[56]

The use of irrigation and the legal authority to protect the rights of long-time residents were absolute necessities in the Pueblo Indian agricultural program. The Spanish government sought to protect native water rights to encourage Indian farming. It was far easier to govern settled, peaceful wards of the government who were constantly busy with the tasks of irrigating than it was to rule the malcontents. In New Mexico, however, the amount of good farmland was limited. As the Spanish population increased in the eighteenth century, conflict over land and water use became a consistent and nagging vexation for the civil authorities. In the next century, this problem became even more acute as the Mexican authorities began to ignore encroachment on Indian land. The adjudication of native water rights, also a crucial issue, eventually became a major source of complaint in the early American period.

CHAPTER THREE

Age of Transition (1760–1850)

THE NEW ORDER

When the Pueblos emerged from the period of revolution against the Spanish government, a new order had been established. With the conflict over religious and civil oppression apparently over, the village Indians of New Mexico looked forward to better times. For some of them, however, there would be no accommodation with European colonizers. Fearing retaliation by the vanquished Spanish refugees, numerous groups of Pueblos simply left their traditional homelands. The decline of the Pueblo population was actually a continuation of the demographic changes of the seventeenth century, but now their numbers were drastically reduced. Of course war deaths and the taking of prisoners during the conflict took their toll. Many of those who remained behind were reluctant to stay in their former villages, and thus their croplands were fallow. The result was starvation for many or at least a significant decrease in their food supply.

The population shifts among the various groups of Indians demonstrated each village's reaction to the dramatic impact of the rebellion. The Piros, loyal to the colonizer during the revolt, were removed to the south of El Paso along with some captives from Isleta. The natives of Sandia left at the time of Vargas's reconquest, and these Tiwa speakers resettled with the Hopis. Tano speakers from the Galisteo Basin completely evacuated their devastated area and moved to Santa Fe; when Vargas reconquered the region, they resettled in Hopi to form the Tewa village at First Mesa. A number of Pecos and Picuris residents moved in with the Plains Apaches while some of the Jemez went to live among the Navajos. Former Santa Clara residents split up among the Hopis and Navajos for a short time. Since many of the Keresan villages were destroyed, their inhabitants moved first to Acoma and later founded the pueblo of Laguna.[1]

The results of this vast demographic displacement were devastating to the Pueblo farm program. Fields and irrigation ditches constructed and maintained over the centuries of precolonial expansion lay in waste. Many could never be used to grow Indian crops again. Some would be taken over by Spanish settlers, who moved into New Mexico in large numbers in the eighteenth century. Only the western pueblos of Acoma and Zuni remained unscathed during this period of turmoil. From this point on, however, a certain degree of stability had been achieved; with the exception of Pecos, which would later be abandoned, all of the modern pueblos were firmly established.

Following the Reconquest, Pueblo agricultural production increased significantly. Considering some of their responsibilities, this renewal is a tribute to their farming expertise. For one thing their obligations to Spanish officials had not entirely disappeared. While the abuses of the encomienda system ended, forced Pueblo labor in the fields of the alcaldes continued to conflict with native agricultural pursuits.[2] Raids by the surrounding nomadic tribes also diverted Pueblo attention from the normal farm chores. Ironically, the rise in agricultural productivity in eighteenth-century New Mexico made the Pueblos and their neighbors ideal targets. So intense were these sporadic intrusions that the Spanish authorities employed the Pueblos as military auxiliaries. This activity also deterred the Indians from working in their fields, but Pueblo participation was necessary to halt Ute, Comanche, Navajo, and Apache depredations, which endangered Pueblo food supplies as well as Spanish towns.[3]

In spite of attacks by their nomadic neighbors, both Spaniards and Pueblos observed intermittent periods of peace for the purpose of holding trade fairs. The practice of conducting trade with other Indian groups was nothing new for the Pueblos. Even their prehistoric ancestors acquired items from outside the region through barter. In the eighteenth century, however, this activity became more regimented. For one thing, the fairs were held at specific times—generally late summer and fall. They were also held at the same sites. Taos was the most popular, but other pueblos conducted smaller fairs with nearby nomadic groups. Given the potential for violence when Indians who were normally at war got together for the trade fairs, Spanish officials required the presence of soldiers at these events.

Undoubtedly the trade fairs were important for the overall economy of the region. Each group—Pueblos, Spaniards, Apaches, Comanches, Navajos, and Utes—had something to offer that others did not have and something to gain that they did not manufacture. The Pueblos had pottery and other homemade items, but their most important trade item was agricultural produce. On a good harvest year, they could sell some excess crops and store others for less productive times. In exchange they would receive slaves, meat products, hides, and horses.[4]

In one area, pressure on the Pueblos decreased as the demands of the Franciscans diminished in the post-revolt period. Realizing that proselytizing and interference with Pueblo religious practices were major causes of the Spanish expulsion in 1680, the civil authorities in Santa Fe decided to curb ecclesiastic abuses. Although the friars' use of Indian labor was restricted and the number of priests in the area was reduced drastically, religious conversion remained a major concern for both religious and civil authorities but not at the cost of alienating a potential ally.

LATE SPANISH OBSERVATIONS ON PUEBLO AGRICULTURE

Late in the 1700s, members of the religious community received instructions to observe conditions in New Mexico, including the various Pueblo communities. The accounts of Bishop Pedro Tamarón y Romeral in 1760, Fray Francisco Atanasio Domínquez in 1776, and Fray Juan Agustín de Morfi in 1782 provide information on Pueblo agriculture and irrigation, as well as the missions, in the late Spanish period. Since Fray Domínguez was specifically requested to take note of the arable lands, waters, and fruits of each pueblo, his report contains the most thorough observations during this period.

Irrigation ditches of the pueblos impressed the Spanish priests, but their successful application varied from one Indian village to the next. At Tesuque, the Indians farmed a large area entirely under irrigation. Domínguez praised the native farmers for their ingenious canal construction and the productivity of their fields. Some of the Pueblo villages, like San Ildelfonso, San Juan, Santo Domingo, Cochiti, and San Felipe, had farmlands on both sides of the Rio Grande. Others, including Pojoaque and San Ildefonso, shared irrigation ditches with surrounding Spanish settlers. Domínguez's report also pointed out that the acequias of Cochiti, Isleta, and Santo Domingo were very deep and wide.

Some of the Indian villages utilized other means to water their crops. A small spring watered the orchards of San Ildefonso, and Fray Morfi mentioned a large pond north of Taos fed by some three hundred springs. Although some of the priest's fields at Santo Domingo were near the river and required no irrigation, this acreage was decreasing because of erosion. Morfi noted that the farmlands of Santa Ana, located on the Jemez River, were not possible to irrigate because of the lay of the land, and Domínguez stated that farming there depended totally on rain. The same situation existed at Zia. The Indians farmed several small plots, all of which required rain, located in various canyons. They also cultivated areas up and down the narrow streambed of the Jemez River for about a league. These regions received water from the river only during times of great rainfall. Evidently, irrigation canals were not feasible there. In this unpredictable environment, the Pueblos had only sporadic success in crop production.

As the Spanish priests traveled to the western pueblos, they noted fewer irrigation ditches. The pueblo of Laguna, on the Rio San Jose, utilized a mixture of acequias and rainfall. Both Domínguez and Morfi mentioned that the lake upstream from the village, from which the village got its name, occasionally went dry and therefore its waters never reached the Rio Grande. Unlike the other pueblos Domínguez visited, Acoma did not employ extensive irrigation. The Indians farmed in canyons surrounding the mesa wherever the ground was arable. During times of drought, they suffered deprivation. They did, however, irrigate some land west of Laguna in a place called Cubero. In the westernmost pueblo of Zuni, Indians' fields surrounding the village depended totally on rain, and only the convent's land could be irrigated. They still produced sufficient crops for a very large population.

Most of New Mexico's pueblos contained areas that produced food for the local mission. Here the crops introduced by the Spaniards had their greatest impact. At Tesuque, the native farmers planted corn and wheat for the local priest in fields irrigated by Indian ditches, but they did not harvest his crops. In Nambe, the local priest had three irrigated plots planted in corn, green vegetables, and wheat. The priest of San Ildefonso supplied his own food, including vegetables grown in a garden surrounded by an adobe wall. The Indians of San Juan used the community acequia to irrigate the fields of the priest, whose crops they were responsible for sowing and harvesting. The priest's fields at Picuris, located mainly in a damp hollow, were not so bountiful because the site made it difficult for the crops to mature. While the natives of Picuris offered little support in terms of agricultural chores, those of Sandia tended the fields of the local priest. In addition to fields of corn and wheat, they cultivated a small vegetable garden on the west side of the pueblo. The priest at Cochiti had a small vegetable garden where the Indians raised all of his crops. The natives of San Felipe grew corn and wheat for the priest and grew a kitchen garden that included beans, chickpeas, and vetch. Both the priest and the Indians of Santa Ana previously had some large garden plots, but they were destroyed by floods. At Jemez, the native farmers tended four small plots, on which they raised corn, wheat, chiles, onions, and cabbage for the local priest. When Fray Domínguez reached Zuni, the lands originally assigned to the priest had turned to clay. Small plots in the eastern part of the pueblo, watered by a spring, had replaced those fields. At Laguna, the local convent possessed an excellent kitchen garden where Indians harvested a variety of green vegetables.

A number of the pueblos produced a variety of Spanish fruits. Tesuque, Santa Clara, and San Ildefonso all grew apricots and peaches. The latter had ten orchards of fruit trees. Santo Domingo, Isleta, and Sandia had grapevines. San Ildefonso and Santo Domingo produced melons, while

Santa Clara grew plums. A variety of fruit trees were found at Jemez and Sandia. Nambe had six apricot trees, while a small amount of fruit was grown at San Juan. At Laguna the people raised peaches for themselves and the local priest.

The Spanish priests rated each pueblo according to the productivity of its soil. Tesuque, Nambe, Pojoaque, San Juan, Santo Domingo, Cochiti, San Felipe, Jemez, and Laguna were all considered to have fertile fields that produced abundant crops for their people. Lacking oxen, the natives of Zuni had to sow their own land by hand, but they still produced a high yield. Picuris, Sandia, and Isleta grew an adequate supply of food. The pueblo that usually received the highest praise for its agricultural lands was Taos. Fray Morfi remarked that the beautiful valley was watered by four streams, which came from the mountains to the west. Domínguez stated that this large supply of water made the surrounding area extremely fertile. When the other Pueblos suffered from drought, they could always travel to Taos for food supplies. In spite of their abundant harvest, however, Taos farmers could not raise any amounts of chiles and beans because of the high altitude.

Some of the pueblos lacked enough fertile land for their people. While much of San Ildefonso's land watered by the Rio Grande was productive, the fields along the Rio Nambe could not always be irrigated. Upstream users often took all of its water and the river dried up. Santa Clara, located at a high elevation, suffered low yields during times of insufficient moisture. Santa Ana was the only pueblo to buy farmlands separate from its village grant. Fray Morfi reported that the Indian fields within the pueblo were unproductive because floods had engulfed the adjacent areas with sand. The lands of the Indians and the priest at Zia were sterile and unproductive. Agricultural production at Pecos and Galisteo had fallen off dramatically due to drought or Comanche raids. Because of this economic deprivation, both pueblos were later abandoned.[5]

The reconnaissance missions by these Spanish priests and especially those of Fray Domínguez provide an in-depth view of agriculture and irrigation among the Pueblo Indians of New Mexico in the late Spanish period. In many respects, the situation for the native farmers had changed very little from the initial contact with the Spaniards. The introduction of new crops diversified the Indian diet, but of these crops only wheat and certain fruits were used extensively by the Pueblos. Green vegetables were usually found in the local priest's garden, not Pueblo fields, and wheat never replaced corn as the main staple in their food supply.

Unfortunately, none of the priests had much to say about the impact of Spanish tools or draft animals on Pueblo agriculture. Since farming and stock raising were two of the major occupations among the non-Indian residents of New Mexico, these people had introduced various items that made farming chores easier. Mules and oxen were employed as draft animals throughout the region. Spaniards also introduced plows, hoes, and two-wheeled carts. Given the geographic isolation of the colony, acquisition of these items was difficult. In the latter part of the 1700s, this burden was somewhat eased by trade with Chihuahua. Annual caravans made the forty-day journey that became a lifeline for the northern frontier populace.

Since the Pueblos had become working partners with the Spanish in the eighteenth century, most of them also benefited from the acquisition of draft animals and agricultural implements. Domínguez did mention that the pueblo of Galisteo had horses and oxen. He noted that the Comanche traded horses and mules at the Taos trade fairs. His report also stated that while the Indians were required to work on church construction, they did not have to lend their oxen for this purpose. He mentions the priest giving oxen to Picuris farmers and the use of those animals at Taos and Isleta. Draft animals were also used to pull carts full of wheat to the threshing floor. Here, animals would typically walk on the harvest to separate the grain from the chaff. While tools were never mentioned specifically, the priest stated that the Indians cultivated their lands, which implies that hoes and plows were utilized.[6]

PUEBLOS AND SPANISH LAW IN THE EARLY NINETEENTH CENTURY

During the early nineteenth century, the Pueblo Indians continued to seek protection of their land and water rights from encroaching neighbors. This problem was accentuated by two basic phenomenons that occurred throughout the Spanish and Mexican periods—the increase in the Hispano population and the accompanying decrease in the size and number of Pueblo villages. The Indian numbers declined because of disease, increased attacks by nomadic raiders, drought, loss of land base, and battles with the Spanish armies. In addition, some of the Pueblos gradually became assimilated into the Hispanic population. By the mid-eighteenth century the Pueblo population had fallen behind their non-Indian counterparts, and by 1810 Hispanics were twice as numerous.[7]

This demographic shift had periods of rapid growth and those of general stability. Much of the vast decrease in the Pueblo population occurred in the period of initial colonization, as it had with many aboriginal peoples. The non-Indian population made steady gains during the eighteenth century, but it tripled in the first half of the next century to about sixty thousand. Accompanying rapid growth was a tremendous expansion in the Hispanic settlement area.[8] Under these circumstances, competition for the best land and water rights became a keen issue between the Pueblos and their neighbors.

In addition to the negative shift in demographic trends, the Pueblo land base also suffered losses because of weakening power of the Spanish government in the late colonial period. During the waning stage of Spain's empire in the New World, its administration gradually broke down. Laws designed to protect Indian property were enforced only when a conscientious official took their side. When this failed to happen, encroachment resulted.

During these struggles the Spanish authorities were not always willing to see the land base of the Pueblos erode. In 1808, José Manuel Aragón was denied his request that part of the Laguna grant be given to Spanish settlers. However, in 1809, Aragón again petitioned the local authorities for some land within the boundaries of the pueblo for his personal use. He had used this area in the past and wanted title to farm the land permanently. Nemesio Salcedo, the commandant general at Chihuahua, felt that Aragón should have some rights because he was the local alcalde, had a large family, and needed the area to support them. Salcedo ruled that Aragón should be given the right to occupy an area large enough to feed his family as long as he was the alcalde, but that he should not be given title to the land. Occupation did not mean ownership, but it was typically a step in that direction.

Laguna, in particular, seemed beset by encroachment problems at this time. In 1800, thirty Spaniards from Albuquerque were granted a region known as Cebolleta, which the Indians had previously cultivated. This occupation threatened the Paquate tract, which the Lagunas had owned for at least thirty years. When encroachment did occur in 1820, the Indians sent a delegation to the protector for New Mexico—Igracio María Sánchez Vergara. Although he held in favor of Laguna and specified the boundaries for Paquate, the grant would be contested in the future.[9]

Before the Pecos pueblo was abandoned, a question arose concerning its boundaries. In 1818, Juan de Aguilar petitioned the governor of New Mexico. The original measurements, Aguilar claimed, had not been made according to the accepted rules, and, as a result, the pueblo overlapped his property. Instead of following the usual custom of measuring the pueblo from the cross in the cemetery with a cord of a given length, the grounds had been laid off from the edge of the town with a cord twice the normal size. Governor Facundo Melgares stated that the petitioner and his son had witnessed the original measurements and that Aguilar's claim that the pueblo overlapped his property was false. Don Facundo ruled that measuring from the cross in the cemetery was merely a custom, not a fixed rule. The pueblo boundaries had been measured so as not to harm either the Indians or the surrounding citizens.[10]

Cases concerning land did not always center around Spanish-Indian conflict. In 1813, a dispute arose between the pueblos of Santa Ana and San Felipe over lands allegedly bought by the former and taken by the latter. In addition, members of San Felipe had sold the land to various Spanish citizens who cut down timber in the region. Don Felipe Sandoval, the protector of the Indians, petitioned acting governor Manrique, who directed Don José Pino, an alcalde from Albuquerque, to investigate and settle the matter. He assembled representatives of both pueblos and examined their documents. Since neither of these deeds was explicit on the eastern boundary of the land in dispute, Pino established a compromise line agreed to by both parties. However, this decision resulted in the Santa Ana pueblo holding land that had been sold to Spanish citizens by Indians at San Felipe.

At first, the natives of San Felipe agreed to the terms, but later they also petitioned Sandoval about the settlement. The governor sent in Don José María de Arze to investigate the matter. He found one document that indicated that San Felipe was trespassing on Santa Ana land. In another document drawn up by the chief alcalde, Arze discovered a compromise agreed to by the two pueblos. He also suggested that the Indians establish visible permanent monuments that would prevent a recurrence of the problem in the future. Crops planted since Pino's decision were to be harvested by those who planted them, but eventually the Spaniards who bought the disputed land were to be repaid by San Felipe or given land elsewhere.[11]

In 1815, a peculiar land case came about because of claims by former members of the pueblo of Santa Clara. These people held a strip of land within the limits of the pueblo that was acquired by their grandfather, Roque Canjuebe, in 1744 when he severed his relations with the tribe and obtained Spanish citizenship. Seventy-three years later, the Pueblos objected to non-Indians living within their grant. The case came before Governor Alberto Maynez, who eventually decided that Canjuebe's descendants would have to surrender their lands but would be allowed to take up residence in a different area that was to their liking because they had occupied the land long enough to claim title.[12]

The findings of the Indian protector, who was employed to safeguard Indian lands, sometimes worked to the detriment of Spanish citizens. Given his job description, this situation should be expected because often he had a conflict of interest. In 1814, the natives of Sandia had loaned land to a group of Spaniards on the east side of the Rio Grande for five years. Some of these farmers constructed huts during cultivation so that they could live near their fields. Perhaps the Indians felt threatened by this action, for they petitioned the protector, Don Felipe Sandoval. In February 1816, the governor handed down a decision in favor of the Indians. One of the Spaniards, José Gutiérrez, wrote Governor Maynez but apparently to no avail.[13]

Another conflict erupted at Taos in 1815. José Francisco Luján, the Taos governor, petitioned the chief alcalde concerning the protection of his people's grant against trespass. Luján was concerned with usurpation of Taos land by a group of Spanish citizens in an area known as the Arroyo Hondo. Acknowledging that the Indians lacked the legal capacity to protect their claim, Luján requested that Fray Benito Pereyro represent them. The alcalde suggested that the matter be referred to the governor of New Mexico. In April, Governor Alberto Maynez issued a decree that strongly supported the protection of native rights. He stated that the royal grant, as an entailed estate, could not be alienated. If, as was the case at Arroyo Hondo, Spanish citizens intruded on Indian property for a lengthy period, these trespassers must leave the area and lose the work that they had completed. The chief alcalde was instructed to placate the citizens with a compromise that would not cause injury to the Indians. Felipe Sandoval, protector of the Indians, was to see that such action was taken.

Pedro Martín, alcalde of Taos, reported to Governor Maynez that, according to his measurements, 190 families lived in three villages. The citizens had improved all of this land and built a church. In addition, the residents claimed that, when the Comanches attacked the area, the Taos natives encouraged friendship with the neighboring Spaniards in order to gain their support. Martín proposed that the Indians allow the citizens to stay in exchange for a payment of forty-five horses and cattle. He also felt that if the Spaniards lost their land, they should be compensated with parcels that had been previously purchased by the Pueblos. In spite of these strong arguments by the Spanish citizens, Governor Maynez would not reverse his decision. He claimed that his decree was just and well founded and asked the citizens to show respect for his decision by ending all appeals on the matter. The governor claimed that if the matter were to go to court, the Indians would win because the "rights to the league which His Majesty granted to them are incontestable."[14]

It is clear from these cases in the late Spanish period that, as wards of the Crown, the Pueblos were entitled to protection of their grants and the waters within and often received it. However, it should be noted that this guardianship was not always sought. Unaccustomed to the foreign adjudication procedures and, in the case of western pueblos, removed from the mainstream of Spanish jurisprudence, the Indians did not always seek redress of grievances in the courts. Also shown by the previous cases, the Indian protectors could play an important role in Pueblo litigation. However, they often ignored their duty. Equal protection of Pueblo rights then depended on obtaining unbiased judgment from authorities in New Mexico and the unwavering vigilance of the government in Mexico City. When these essential components failed in the Mexican period, the Pueblo land base deteriorated.

PUEBLOS AND MEXICAN LAW

By 1821, the Pueblo Indians had experienced over two centuries of Spanish rule. With the creation of the independent nation of Mexico in that year, the natives of New Mexico came under the jurisdiction of the second of three governments in the nineteenth century. In the Plan of Iguala, the new government declared that "all the inhabitants of New Spain, without distinction, whether European, Africans or Indians, are citizens of this monarchy."[15] Although granting Indians citizenship may have seemed to be an equitable solution to the Indian problem, its results were devastating for the Pueblos. This abrupt measure, designed to be a panacea for vexations that had plagued Indians in New Spain for three centuries, ended by aggravating those problems.

Although the egalitarian attitude of the Mexican government had its roots in the American and French revolutions, it was also a function of the Spanish policy. The goal of eventual Indian citizenship formed the basis of missionary work on the northern frontier and was reflected in the use of native governors among the Pueblos. The Spanish government would never have considered making any Indians equal partners in society unless they adopted Catholicism and became a part of the colonial government. Spain, however, never implemented the individual distribution of Indian land and the taxation of this property in New Mexico. The Pueblos remained outside of the mainstream of Spanish society. Thus, while the new Mexican laws did not embody entirely revolutionary ideas from the colonial administration, they were applied to a people who had never been required to adhere to similar Spanish policies.[16]

It can generally be stated that Mexican law concerning the Pueblo Indians differed little from that of Spain. Although the Indians had been granted full citizenship, they were not able to alienate their property without the consent of government officials. Some authorities have even argued that the Mexican government intended no change in the status of the Pueblos, who, they felt, continued to be considered wards of the government.[17] However, the Mexican period witnessed an evolution in the status of the Pueblo Indians. One of the reasons for this change was the constant attacks by Apaches, Navajos, Utes, and Comanches. The Pueblos continued to aid their Mexican neighbors in thwarting these raids as they had done under Spain. Hence two groups were formed in the minds of many settlers—Pueblos and "*naciones bárbaras*." The first group were considered to be citizens; the others were not.[18]

A more important factor behind Pueblos' citizenship concerned their property rights. Under Spain they had the special status of wards. When

Agustín de Iturbide gained control of the Mexican government, he adopted liberal causes that originated in Spain and this included racial equality. The Mexican Constitution of 1824 did not mention Indians specifically, but it did recognize that all Mexicans were equal before the law. Public documents were no longer to contain the word "Indian." While the Pueblos were thus gaining the title of citizens, they were also losing a special protective status as Indians. They could now alienate their land, and the office of "protector" no longer existed.[19]

Whatever legal status the Pueblos held and the impact this had on their property rights in New Mexico, the federal government could not have cared less. It is important to keep in mind that this was a new national government with little experience in handling its international and internal affairs. Mexico was simply overwhelmed with economic shortcomings, the problems of life after the war of independence, unsophisticated army officers, the age-old problem of the role of the church in state politics, local leaders who were more interested in controlling their affairs, and the threat of invasion by various foreign countries. Some of its best leaders understood that problems in the northern frontier needed attention, but they would have to wait their turn.[20] It cannot be denied that because of the laxity on the part of Mexican officials, illegal alienation of Pueblo land often occurred during the Mexican period. The major offenders were the local alcaldes who colluded with settlers to obtain Pueblo property.[21]

New Mexico remained a province until 1824; then under the new constitution, it became a territory. Although it might seem that the overthrow of the constitutional monarchy of Iturbide and the establishment of a republican form of government would have afforded New Mexico more representation, the territory in fact received less than it had enjoyed earlier. Territorial status during this time was somewhat lower than that of states; thus New Mexico received less attention because of its relegation to a lower position in the Mexican nation. In spite of the new constitution, New Mexico was not provided with a comprehensive form of government, and conditions in this region were ignored.[22]

Because of its preoccupation with various concerns, the federal government did not lead the way in establishing an innovative Indian policy. Instead, each state or territory determined its own system of handling Indian affairs. In New Mexico the governor was primarily responsible for executing Indian policy. Therefore the protection that the Pueblos received vacillated according to the whim of various administrators.[23] Whatever federal or local laws were passed regarding land or water rights, the new Pueblo legal status determined how they lost property. The liberal program that brought about this change also emphasized the need to break up communal land and redistribute it to private individuals. Legislation passed in the late Spanish period laid the foundation for the acquisition of surplus Indian communal property, and when the Mexican government was formed, there were numerous challenges to Pueblo land holdings.

The local *diputación*, under the control of the governor, controlled the breaking up of communal property, and these officials were committed to the liberal philosophy throughout the 1820s. However, the Pueblos asserted their rights as citizens and gained the right to sell excess land as they saw fit. In the decade of the 1830s they did sell a substantial number of land parcels, but there was little choice. There were so many problems with squatters, they were faced with the choice of selling it or losing it. But they did not sell all of their communal lands and often defended their property rights by appealing to local authorities, as any Mexican citizen could do.[24]

Concerning water rights, the Mexican government continued to implement the laws, including the concept of prior appropriations for beneficial use, that originated under Spanish rule. Accordingly, agricultural settlements and pueblos retained the rights to water needed for their development; riparian rights were not considered in New Mexico.[25] Pueblos could distribute waters to lands held in common along unnavigable streams, including lands not adjacent to the water. This Pueblo right superseded the rights of individual appropriations and riparian owners along the stream.[26]

The government in New Mexico realized the necessity for protecting rights of agricultural people. In 1826, some revised statutes were promulgated under the territorial governor, Antonio Narbona. According to this legislation, damage to cultivated fields by wandering animals would cost their owners a fine for each head. The law was so strict, the act of trespassing did not even have to be witnessed. Accordingly, penalty would even be imposed on owners whose animals were on the borders of agricultural fields at night since authorities suspected that herders used this period to sneak their cattle into areas where crops were grown. The statutes also contained laws concerned with irrigation ditches. Because some irrigators did not construct drainage ditches, fields were often ruined during floods. Those who were at fault in these cases were to pay a fine and the cost of labor to repair the damages. Also, a law was passed to govern the misuse of water during times of drought. Greedy farmers who dammed waters for their own use and to the detriment of others would be charged a heavy fine, part of which would go to the injured parties. Finally, the statutes declared that any farmer who did not contribute to the work on the acequia madre would be subject to a fine.[27]

Unfortunately, the new laws did not end the old problem of property loss. Pueblo land problems, a typical vexation during the Spanish reign, actually became more pronounced under Mexico. In February 1824, three petitions, concerning apparently unoccupied agricultural lands of the pueblos of San Felipe and Santo Domingo, were circulated by eighteen people who wanted to secure title to the land. Don José Francisco Ortíz was appointed to examine the area and warn the native people that their property could be alienated. The Indians claimed that the land had been given to them for grazing purposes. The governor decided to distribute to both pueblos the land they held in common so that they could sell what land they wanted in a manner similar to other citizens. Local authorities seemed to be confused over the new status of Pueblo Indians. In 1825, the question of whether the Pecos Indians could sell their lands arose. In other petitions at this time, however, various Pueblo Indians are listed as citizens.

Many of the Pueblos were involved in cases concerning non-Indians who wanted to take over unused portions of Pueblo grants. In the previous Pecos case, the Indians were concerned as to whether they could stop local authorities from donating their unused plots to settlers. Another petition concerned the granting of surplus land belonging to the citizens of San Juan Pueblo. Surplus land at Nambe was also petitioned for at this time. Unwilling or unable to decide these matters, local leaders often referred them to the federal government. At other times, the case was left up to the councils that governed various pueblos.[28]

As the non-Indian population grew, these people continued to set their sights on land that was not being cultivated. In 1831, Agustín Durán and others petitioned the Territorial Deputation of New Mexico concerning lands between Santo Domingo and San Felipe. The matter was referred to the town councils of nearby Cochiti and Sandia. Perhaps Spanish authorities felt that the question of unused land between the two pueblos could best be solved by other Indians. Although the Cochiti council knew of an old Spanish document that gave the pueblos in question title to the land, it claimed that the natives had sufficient lands within the pueblo and were not cultivating the tracts in question. The Sandia council, also aware of the Spanish title, argued that the Indians farmed some portions of the area. Although there is no evidence to show that the petitioners received any grant, in both councils the main question seemed to be the practical use of the land rather than the legal ownership.[29]

While the central government in Mexico City continued to ignore events taking place on the northern frontier, the jefe político in New Mexico wielded considerable power. If he had any concern for Indian rights, lands could be protected. In 1826, for example, Governor Antonio Narbona heard complaints from a Laguna delegation concerning the Paquate section, which had also been contested during the Spanish rule. Although he ruled in favor of the Lagunas, they returned the next year with more complaints about trespass. This time, however, they had to contend with Manuel Armijo, who had replaced Narbona. Unconcerned with Indian land rights, Armijo ignored Hispanic encroachment in the Paquate area throughout the Mexican period.

With the federal government ignoring New Mexico and the governor not always enforcing justice for Indians, the local alcaldes became virtual dictators. Under Spain, Indians could redress their grievances through the protector or perhaps the resident priest. But Mexican rule did not allow for a protector, and the Franciscan missions were secularized in 1826, which meant that most of the Indian villages were without priests. These conditions allowed certain powerful families to run the local affairs. For the Laguna area, the Bacas and Pinos were the dominate rulers. Marcos Baca, who held the office of alcalde in 1830, so abused the Indians that they took their case to Mexico City. While this case may have forced his resignation, his father led a group of settlers to Cubero about three years later. Eight years passed before a boundary line was established between the pueblo and the new settlement.[30] Evidently, the alleged new legal status of the Pueblos, government inefficiency, and competition for arable land in New Mexico at this time was proving to be detrimental to the Indians. These conditions allowed authorities to disregard old regulations in dealing with them. Officials were trying to use new conditions to justify the alienation of Pueblo land on the basis of beneficial use.

Pueblo water rights were also contested during the Mexican period. In 1823, another confrontation between Taos and its Spanish neighbors occurred. In a suit brought before the Town Council of Taos, Hispanic settlers claimed to have water rights to some land acquired by a grant made in 1745. The land remained unused until 1815 when settlers built houses and cultivated their fields. Part of this land was irrigated by the Lucero River, a stream that the Taos Indians used to water their farmlands. In general, the final decision in this matter favored the priority of the native farmers.[31]

In an 1826 case involving the water rights of Laguna Pueblo, Governor Narbona wrote that the stream flowing from the Ojo del Gallo through Cubero was the only water available to the Indians and that he felt further settlement in the area would be detrimental to Laguna's food supply. The Indians were entitled to the water because they had used it for over a century. During the next year, three Indians from Laguna petitioned the authorities for relief from the taking of this water by the pueblo of Acoma, which had enlarged farmland at Cubero and also used water from the Gallo spring. The Lagunas claimed that their neighbors had expanded their agricultural fields and used the water of this spring to irrigate the area, although prior rights gave Laguna a claim to the stream. They also reminded authorities that previous governors had warned the Acomas not to shut off the water supply. The Territorial Deputation, unwilling to become involved in this intertribal dispute, referred the matter to the governor in order that justice be served.[32]

In 1845, two Isleta Indians, speaking for the pueblo, protested against a grant, located at Cabra Spring, made to Don Juan Otero. The natives claimed that this area had always been a part of their pueblo and that their rights had been violated. They asked the governor to take their protest to the Departmental Assembly, the legislative body in New Mexico. Following an investigation, Francisco Sarracion reported that the spring was located outside of the pueblo and did not lie on anybody's property. He claimed that the spring was on an area that had been considered common ground and therefore could be used by everyone. Accordingly, the assembly repealed the grant to Don Juan Otero.[33]

The previous water cases and laws cited during the Mexican period demonstrate that for Pueblos and non-Indians beneficial use of waters over a long period of time was essential in establishing the right to their use. However, while Indian rights were often preserved, the support of these rights was also flexible. If the farmers in the area required water rights for their existence, local authorities saw that these needs were met. The government continued to emphasize the importance of cleaning the ditches, protecting fields from cattle, and controlling floods. As the number of settlers increased in New Mexico, authorities became more concerned with the actual use of land rather than just ownership.

OBSERVATIONS ON PUEBLO AGRICULTURE IN THE MEXICAN PERIOD

As previously mentioned, the number of non-Indians in New Mexico rose sharply in the first half of the nineteenth century. Not all of these people, however, were permanent Hispanic residents. Anglo investigators, traders, and trappers also ventured into the region when the Santa Fe Trail was opened to Americans in 1821. Even before the fall of the Spanish government, a United States reconnaissance mission found its way into the area. This group, under the leadership of Zebulon Montgomery Pike, was arrested and brought into Santa Fe. His observations were the first by an American concerning Pueblo agriculture. On his journey, Pike noted the irrigation practices at Pojoaque and Tesuque. He also observed that corn, wheat, rye barley, rice, tobacco, and grapes were successfully grown in the region.[34]

Mexican authorities were also aware of the Pueblo agricultural program. The New Mexicans used Pueblo Indians as military auxiliaries during the Mexican period as in the Spanish era. While this activity was necessary for protection against raids, it proved to be detrimental to the Pueblo agricultural program as it had in the past. Manuel de Jesús Rada, reporting on the deficiencies of New Mexico's economy in 1827, claimed that most of the people lived in impoverished conditions. The days spent protecting the frontiers from attack by nomadic Indians left little time for land cultivation. In addition, he pointed out that water shortages in the region also hurt agricultural production.[35]

Another holdover from the colonial era involved the economic monopolies the Spanish government had placed on certain trade items. With the advent of the Santa Fe Trail, the Pueblos, as well as the Mexicans, could acquire these items at a cheaper price. Trade flourished at both Santa Fe and Taos, where annual fairs were held every July or August. Comanches and other nomadic tribes came to Taos to trade deer and buffalo skins, slaves, and blankets to the Pueblos and Mexicans for agricultural items and American products.[36] Before agricultural products can be used for trading purposes, however, a surplus of crops must exist. The practice of storing food supplies was a Pueblo trait that extended back to prehistoric times. It had become an essential routine that could protect the people from starvation in times of drought. When droughts came, as they did in 1802 and again from 1820 until 1822, Pueblos could rely on this stored food. Their Spanish neighbors, however, sometimes seem to have lacked food in times of shortage.[37]

It is difficult to determine if the Pueblos consistently produced a surplus amount of food that could have been used as a trade item with Mexico City. Don Pedro Bautista Pino's account, written in the early 1800s, claimed that federal neglect, the general isolation of the province, and Indian depredations made it impossible to export agricultural products from New Mexico. However, he felt that New Mexico had enough good land to allow residents to feed themselves.[38] A few years later, the question of higher productivity came up again. In the early part of the Mexican period Don Antonio Barriero, a legal adviser sent from Mexico City, provided some insight into the farming conditions in New Mexico. He claimed that subsistence agriculture generally dominated in New Mexico and that no large-scale farming existed. Barreiro felt that agricultural expansion could not be attained because of the lack of military protection against raiding Indians.[39]

The first official census, taken in 1827, shows that a large percentage of the Pueblos were engaged in agriculture. Sometimes that survey combined figures for villages, making it impossible to look at each individual group. Table 1 provides a condensed version of the original census.[40] These statistics show that the Pueblos made their living mostly from agricultural pursuits. Even at this early date, however, a significant number of Indians were involved in craftwork. Although agriculture remained the mainstay of the Pueblo economy, industries supplemented the village incomes. Indeed, Barriero observed that while the Pueblos were good farmers, hunters, and fishermen, they were also adept in certain crafts, such as making pottery, tanning hides, and building furniture.[41]

TABLE ONE
PUEBLO OCCUPATIONS LISTED IN 1827 CENSUS

NAME OF PUEBLO	FARMERS	CRAFTSMEN	MERCHANTS	TEACHERS	LABORERS
Cochiti & Santo Domingo	416	63	3	1	94
Jemez, Zia, & Santa Ana	285	90	5	1	113
Sandia & San Felipe	265	97	4	1	99
Isleta	291	96	4	0	103
Laguna & Acoma	376	25	1	1	43
Zuni	417	0	0	0	0
Santa Clara, San Ildefonso, Pojoaque, & Nambe	553	145	19	3	239
San Juan	417	83	6	1	196
Taos & Picuris	503	30	5	2	178
TOTAL	***3,523***	***629***	***47***	***10***	***1,165***

Source: H. Dailey Carroll and J. Villasana Haggard, trans., *Three New Mexico Chronicles* (Albuquerque: Quivira Society Publications, 1942), 88.

An American observer of the Pueblos during the Mexican period, Josiah Gregg, was involved in the Santa Fe trade in the 1830s. He described agricultural conditions in New Mexico as extremely primitive compared to those in America. His comments on implements in the region reflect this opinion:

> A great portion of the peasantry cultivate with the hoe alone—their ploughs (when they have any) being only used for mellow grounds, as they are too rudely constructed to

> be fit for another service. Those I have seen in use are mostly fashioned in this manner: a section of the trunk of a tree eight or ten inches in diameter is cut about two feet long, with a small branch left projecting upwards, of convenient length for a handle.... What is equally worthy of remark is that these ploughs are often made exclusively of wood, without one particle of iron or even a nail to increase their durability.[42]

Gregg made numerous observations on all aspects of Pueblo agriculture. For example, he mentioned the lack of fencing in the area. A few farms had fences made of poles placed upon forks, a loose hedge brush, or sometimes adobe. However, most of the farmers had to rely on the watchfulness of herders to keep livestock from their fields. Cattle owners who failed to do this were forced to pay for the damages. As other observers had done in the past, Gregg complimented the irrigation systems in the area. He felt that the necessity of irrigation would limit agriculture to river valleys, but he believed that this system of watering was far more dependable than relying on the capricious nature of the weather. Gregg also mentioned that the people of the community tended these ditches under the jurisdiction of the local alcaldes and that during periods of drought, officials assigned individual farmers a time of day to use their allotted irrigation water.

Concerning crops, Gregg noted that while wheat and corn were the staple crops, cotton was cultivated only in small amounts. He also mentioned that beans and chiles were the main foods and that the potato had been introduced recently, but there is no record that the Pueblos had begun to grow it at the time. He also stated that little tobacco was grown and that contrary to reputation the region's apples, peaches, and apricots were of poor quality. Even if they were good enough to export, Gregg noted that this would be difficult due to the lack of efficient transportation to outside markets.

Gregg also made some observations on Pueblo agriculture specifically. Recognizing that they had farmed the region before the Spanish arrival, he concluded that "they are now considered the best horticulturists in the country." He backed up this claim by noting that they supplied most of the fruits and a good deal of the vegetables for local markets. Until recent times, they were the only residents in the territory who grew grapes. Essentially, Gregg found the Pueblos to be an honest, moral, and industrious group of people.[43]

AMERICAN GOVERNMENT IN NEW MEXICO

Although the Pueblos had a reputation as law-abiding people, their loyalty to the Mexican government was less than total. The Indians continued to practice their religious ceremonies and to live apart from the Hispanic world in New Mexico. Their isolation was demonstrated by a major rebellion in 1837. The background to this outbreak began during the previous year when the Centralists took over the government of Mexico from the Federalists and reclassified the states into departments, which included New Mexico. Since the Centralists emphasized a strong central government, the new departments had less local autonomy than states had. They became subject to taxes that would be collected by the departmental governments. Rumors spread in New Mexico that the local governor would tax a variety of items, including irrigation ditches. Unaccustomed to a direct tax and fearful that the central government would deprive them of the profit from their canal systems, some of the Pueblos, accompanied by some lower-class Hispanos and *genízaros* (non-Pueblo Indians who had adopted the Spanish lifestyle), staged a major revolt.[44]

Although short lived, the rebellion of 1837 demonstrated once again that the Pueblos could unite to defend their ancient traditions against a repressive government. Unlike the revolt of 1680, the rebellion during the Mexican period did not involve all of the Pueblos. In addition, this latest revolt was not designed to drive the ruling government out of the territory. Instead, the Pueblos and their counterparts were more concerned about establishing a more responsive and understanding government. In a region where a few *ricos* controlled the wealth and the vast majority of residents were merely getting by on limited resources, a tax on irrigation canals (the lifeline of the farming communities) seemed repugnant. Brief though it was, the rebellion demonstrated the instability of conditions in New Mexico and may have helped to encourage later American intervention in the Southwest.

If the authorities in Mexico City sometimes repressed the Pueblos, they more often neglected them. This disregard posed the greatest threat to native farming areas. The competition for arable land became more acute during Mexican rule because of the bestowal of generous land grants during this period. The government made more than three hundred grants, nineteen of which exceeded one hundred thousand acres. In addition, thousands of smaller claims existed within these large community grants.[45] As a result of laxity on the part of Mexican officials, these grants were not surveyed and often vaguely described. The vague boundaries meant that Hispano and Pueblo claims would overlap. Undoubtedly, the loss of land to aggressive colonists was the greatest cause of concern for the Pueblos during the Mexican period. The question of land ownership, complicated by these conflicting claims, would plague the Pueblos for almost a century after the Anglos gained control of the Southwest.

The Mexican period was one of total chaos from which the new government was still reeling after a quarter century. The northern provinces, ignored for most of this time, even resorted to rebellion but to no avail. By the 1840s, aggression that once came from the south now changed direction. It was a decade of expansion for the still-young American nation, and the Southwest lay directly in the westward path of Manifest Destiny. Instead of turning its head from the calamitous situation in New Mexico, the United States faced it head on. With a confidence bordering on self-righteousness, Americans sought to bring the western portion of the continent under their rule. Having failed to purchase the region, the leaders of the country now sought any excuse to engage in warfare to accomplish their mission. For the Pueblos this meant that the ever-changing role of dominant power would soon fall into the hands of a permanent government.

In 1846, when the United States declared war on Mexico, President James K. Polk ordered Colonel Stephen Watts Kearny and his Army of the West to gain control of Mexico's northern territories. Lieutenant Colonel W. H. Emory, a member of the topographical engineers, joined the advance guard with instructions to collect data on the region. He observed the narrow valley of the Rio Grande and visited the pueblos of Santo Domingo and San Felipe. South of the latter village, he noted that irrigation was practiced. Emory felt that because of the sandy condition of the soil, Indian corn grew better than wheat. At Sandia, the colonel observed the native farmers treading wheat with animals in a circular corral. After this, they used a handbarrow, with a bottom made of perforated bull's hide, to separate the wheat from the chaff. He estimated that it took them an hour to winnow a bushel.[46]

The next year, Lieutenant J. W. Abert, another member of the topographical corps, also carried out a reconnaissance mission. He noted the arid conditions of the region and saw that the small supply of water limited the amount of tillable land. He found that corn and wheat were the major crops, but beans, pumpkins, melons, and red peppers also did well. He also commented on the fine orchards of peach and plum trees at San Juan and claimed that the Indians cultivated most of the fruit in the area. Abert ascended the Jemez River valley and found that, in places, the river contained no water at all. At Santa Ana, he found the village almost deserted as the native farmers were gathering corn at a nearby rancho. He was told that Indians farmed on the top of surrounding mesa, but he doubted that anything could be grown there because of the lack of irrigation. Visiting Acoma, Abert learned that the Indians used the lowest story of their buildings to store corn, pumpkins, and melons. In addition, they hung strings of chile peppers; pumpkins and muskmelons were cut into ropes and twisted together to dry for use in the winter.[47]

In 1849, a third military reconnaissance mission under Lieutenant J. H. Simpson entered New Mexico on an expedition into Navajo country. Simpson noted approximately a dozen acres of apricot and peach trees at Jemez Pueblo. He was received well at Laguna and Acoma, where the native people supplied his party with large quantities of muskmelons, which they stored in great abundance. They had peeled them, removed their seeds, and hung them out to dry. In addition to noting the productivity of the region, Simpson also witnessed a corn dance at Jemez and described it in great detail.[48]

These initial American reactions to Pueblo agricultural programs resembled those of Spaniards in previous centuries. The Anglo observers were impressed with the output of the native farms, particularly their corn and fruit trees. However, the Americans felt that the Pueblos still used tools that Anglos considered inadequate or primitive compared to their own. They also felt that the Indians lacked modern methods to increase agricultural production, as exemplified by their outdated threshing technique. These observers also admired Pueblo irrigation canals, commenting that crops would be impossible without them. Lieutenant Abert, reflecting the bias of a person from the wetter regions of the country, did not have any confidence that the dry-farming techniques at Santa Ana would produce crops, but the Pueblos had utilized this method for centuries. Finally, these early American military missions witnessed the ancient Pueblo tradition of drying and storing food for future use.

As previously noted, the Pueblos had suffered from a lack of food during times of drought and also when heavy demands were made on their reserves. In 1846, both of these circumstances dealt a blow to the Indian food supplies. In addition to the lack of rain that year, Kearny's troops needed to be fed. The need for food not only increased the price of grain that year, but also strained the local food supply.[49] This situation was reminiscent of the early Spanish intrusion into New Mexico when Coronado's men depleted Pueblo provisions during their lengthy stay.

Like the Spaniards and Mexicans before him, Kearny also brought a new system of law into New Mexico. In 1846, he instructed Colonel Alexander W. Doniphan and Willard P. Hall to prepare a code of laws for the new territory, which was printed in English and Spanish. The codification, published as the *Laws of the Territory of New Mexico*, included the following section:

> All laws heretofore in force in this territory, which are not repugnant to, or inconsistent with the Constitution of the United States and the laws thereof, or the statute laws in force for the time being, shall be the rule of action and decision in this Territory.[50]

In addition, the new law stated that:

> The laws heretofore in force concerning water courses, stock marks, and brands, horses, enclosures, commons, and arbitrations shall continue in force except so much of said laws as require the ayutementos (ayuntamientos) of the different villages to regulate these subjects, which duties and powers are transferred to and enjoined upon the Alcaldes and Prefects of the several counties.[51]

The so-called Kearny Code had the effect of transmitting parts of the Mexican law and the Spanish law that had survived since 1821 to the United States Territory of New Mexico. This legislation, which included regulations governing "water courses," would have effect until changed by the government in the new territory.

In 1849, James S. Calhoun became the first superintendent of Indian affairs in New Mexico, an office he held for two years before becoming the first territorial governor. During the next year, he commented on the situation that plagued the Pueblos throughout the Mexican period. Calhoun complained that he was constantly being annoyed by protests from the pueblos of Santa Ana and San Juan concerning encroachment by Mexicans. Santa Ana took their case to the circuit court of the territory to defend the land claims that they said had previously been granted to them.[52] This was the first of many land cases that would be presented to the American government.

In the first half of the nineteenth century, the Pueblos had experienced the rule of three different governments. For most people in New Mexico, the change from Spanish to Mexican control in 1821 brought an increase in the economy with the opening of the Santa Fe Trail, and the Pueblos benefited also. However, the local economy, based on subsistence agriculture and some craftwork, remained basically unchanged from the colonial period. Indeed, the Pueblo farm program was constantly threatened by raids and increasing encroachment on their land and water rights as had happened during the Spanish rule. Many of their implements and methods remained tied to an ancient tradition. Pueblo religious ceremonies, many of which revolved around crop raising, continued to be practiced and helped to enhance their traditional attitude toward agriculture.

By mid-century, the United States had officially supplanted the Mexican government. At first there was little change as far as the Pueblos were concerned. Laws regarding their agricultural program remained unaltered. Trespassing by Mexican farmers continued to be a problem. However, the advent of American interest in New Mexico would have profound effects on the farm programs of the Pueblos. Further encroachment, conflicts over water rights, and land alienation continued to cause problems for these natives of New Mexico. Moreover, a new Anglo attitude toward farming had supplanted the Hispanic and

Indian traditions. No longer would the ruling government be satisfied with subsistence production. The emphasis would soon be placed on modern equipment, engineering skills, and soil erosion, with the hope that the Pueblos could eventually become part of the American capitalistic system.

CHAPTER FOUR

Pueblo Agriculture in the Territory of New Mexico (1850–1890)

AMERICAN BEGINNINGS

From the very beginning of the United States, federal policy makers sought to assimilate the country's native people into the cultural mainstream through yeoman farming. Even the strongest supporters of Native Americans felt that the only way to save them from extinction involved their transformation into farm families. This program was all a part of the early American ideal that saw farming as the foundation of civilized society, and it remained the major occupation into the late nineteenth century. For many tribes the transformation from the hunt to the field found native people stuck somewhere in the middle of these two extremes and therefore outside of the standard American cultural milieu and dependent on the federal government.[1]

Because of their unique historical background, the Pueblos represented a challenge for which the federal government was ill prepared. Although they practiced some hunting, fishing, and gathering, agriculture remained the basis of their economy like it did for many Americans. While outsiders appreciated this fact, they sought ways to increase the production of Pueblo fields beyond the subsistence level. This ignored the fact, however, that these Indian farmers had been producing beyond their needs for centuries. Indeed in the 1850s, the Zunis were selling thousands of bushels of excess corn to American troops who came into their region to establish forts to control the Navajos.[2] The Pueblo situation was also unique in terms of possible citizenship and protection of land claims, both of which were interrelated. Confusion over this matter delayed the initiation of an agency for the Pueblos until 1872.[3] It also revealed that while territorial officials rejected the eligibility of the Pueblos as federal wards, authorities in the nation's capitol felt otherwise. So the real problem, as both the Pueblos and federal officials gradually saw it, was the preservation of native land claims.

With the occupation of New Mexico by American forces, Anglo immigration into the region was imminent. Initially, the influx of Americans was hampered by the Civil War, Indian wars, and transportation difficulties. With the defeat of the South, retaliatory military action against raiding tribes, and the construction of rail and stage lines into the Southwest, the number of American settlers increased dramatically. While the Pueblo population remained relatively unchanged (from about 9,000 to 10,000) in the last half of the nineteenth century, the number of Anglos quadrupled from 20,000 to 80,000. In addition, the Hispanic population, which had grown rapidly in the first half of the century, doubled in the last half to approximately 120,000.[4] It was only natural that the new arrivals would desire good land in order to raise crops and cattle. However, homesteading in New Mexico was an arduous task, and many Americans were unfamiliar with irrigation techniques. Since the Pueblos usually possessed the best agricultural areas, once again the large increase in non-Indian population presented a threat to their land and water rights.

According to the Treaty of Guadalupe Hidalgo, which ended the war with Mexico in 1848, the U.S. government agreed to honor all valid land claims by former Mexican citizens who chose to remain in the territory.

Specifically, the treaty held that Mexicans who owned property "shall enjoy with respect to it, guarantees equally ample, as if the same belonged to citizens of the United States." The agreement also stipulated that Mexicans who became American citizens "shall be maintained and protected in the free enjoyment of their liberty, their property, and the civil rights now vested in them according to the Mexican laws."[5] Like the Kearny Code that preceded it, the treaty respected laws that had governed the area under Mexican rule.

Since the treaty was designed to protect the property rights of former Mexican citizens, most authorities were confused as to how this affected the Pueblos. Were they really citizens during the Mexican period? If they were treated as such at that time, how would the American government perceive them? How could the United States accept the status of the Pueblos as citizens when they had never been accustomed to treating Native Americans in this manner? The argument could be made that they were not really citizens under Mexico, and therefore they should not be given that status by the new government in power. The point is that they were either citizens whose property rights should be protected or Indians, and therefore wards of the American government, whose lands could not be alienated. Either way, Pueblo land should have been free from trespass.

Still the Pueblos were unsure of the future actions to be taken by the Americans. In 1849, they expressed their concerns to James S. Calhoun, New Mexico's first territorial governor and head of Indian affairs. Evidently, a number of Hispanos and a few Americans told them that the new government would soon take their land away and move them to some other region. Fearful over the possible loss of ancestral lands and apprehensive about the policies of a recently arrived and unknown ruling body, they needed some assurance about their future well-being. Calhoun realized that these rumors had been started by unscrupulous land grabbers. He quelled their fears by stating that the president would be more inclined to increase their landholdings rather than take them away if their population grew.[6]

Calhoun was keenly aware of Pueblo land problems, especially Hispanic encroachment. As early as 1850, he encouraged the federal government to set up a commission to clear title to Pueblo land. Calhoun said that, in addition to their original grants, some of the Indians had purchased other lands near their villages. Also, some farmed on unappropriated plots. The commissioner acknowledged the encroachment problems with Mexicans and stated that numerous lawsuits were pending between these people and the Indians. Calhoun felt that the Pueblos should be given additional ground outside their grants if they desired it. In this manner, he hoped to avoid bloodshed over land problems.[7]

Calhoun was justified in his fear of another Pueblo rebellion. During the summer of 1851, rumors spread of a possible revolt among the Indians because false accusations had been made suggesting a possible seizure of their land. The new governor even dispatched troops to Taos and other possible danger points to avert conflict. While this situation was finally resolved without open battle, Calhoun was determined to avoid hostility in the future.

In order to insure Pueblo support, he asked some of them to accompany him on a visit to Washington in 1852. Calhoun died en route, but the Pueblos continued their journey with his secretary, D. V. Whiting. Eventually they met with President Millard Fillmore and discussed their problems. The Indians complained about the high price of food, raids on their villages by nomadic tribes, illegal construction of Mexican irrigation canals, and the lack of farming implements. The president was sympathetic and said that he would ask the new territorial governor to look into these matters and to protect their interest.[8]

Calhoun's successor, William Carr Lane, faced similar problems during his brief administration (1852–1853). For example, a contingent of native people from San Felipe, San Juan, and Santo Domingo made complaints about Mexican trespassers on their grants. In addition, Indians from Santa Ana, San Ildefonso, and Santo Domingo complained to the governor about Mexican stock trespassing on their property. Sometimes Lane was able to placate the Pueblos through equitable solutions, but more often, his efforts were in vain.[9]

Obviously, the federal government found something unique about the Pueblo Indians that caused an alteration in the method of dealing with indigenous people. Typically, Indian agriculture, in spite of many success stories, had been viewed for centuries by many Anglo observers as being almost nonexistent, and this was used as a reason to take over native peoples' property and put it into what Americans considered more productive use. In New Mexico, however, the Americans found an exception in the Pueblos; their observations revealed a group of people whom they found to be honest, intelligent, and hard-working farmers. The question was how to keep them in this state by protecting their land from trespass. Early

American authorities realized that part of the confusion over encroachment on Pueblo lands was caused by poorly defined boundaries and that to avoid continuing strife, the Indian land claims would have to be cleared. Agents felt that because of their peaceful demeanor and agricultural tendencies, the Pueblos deserved the protection of the federal government.[10]

The efforts of the Indians and the local administrators finally paid off. In 1854, the United States established the Office of Surveyor General for the New Mexico Territory in order to quiet land claims in New Mexico, including those of the Pueblos. Four years later, Congress confirmed the original Spanish grants for all of the native villages except Santa Ana, Laguna, and Zuni. President Abraham Lincoln issued patents to all of the original confirmations in 1864. Eventually, the three remaining pueblos received patents for their land grants.[11] By this action, the federal government relinquished all claims within the boundaries of the Pueblo villages.

Since the Indian agency dealt with old Spanish documents that proclaimed the Pueblo grants, a translator was hired to deal with this material. Agent Ben Thomas felt that the person who held this position should be well qualified since the work entailed translating numerous old letters and land titles. The translator's efforts were not only needed for verification of the Pueblo grants but also to avoid intrusions on their lands after the titles were confirmed.[12]

It would seem that with the resolution of most of the Indian land titles, the question of the extent of Pueblo lands was finally solved. However, additional quarrels surfaced in the late 1800s. In 1882, Agent Ben M. Thomas had contacted the surveyor general concerning a grant of land adjoining Santa Clara that had been deeded to the pueblo by Spanish governor Juan Bautista de Anza. The next year Agent Pedro Sanchez discovered a Spanish document of 1766 confirming land that belonged to the Indians of Zia, Jemez, and Santa Ana. He referred his findings to the U.S. Indian inspector in Fort Garland, Colorado, in order to get his cooperation in confirming their land claims.[13] Seven years later, Santa Clara, Santo Domingo, Zia, Santa Ana, and Jemez were still trying to establish their titles.[14]

Agents persisted in their support of Pueblo land claims with mixed results. In 1883, Agent Thomas had complained that the Zunis might lose their farmlands at Nutria because of contradictory descriptions of their reservation's boundaries. Because of this error, Nutria was located outside of the reservation. Thomas wanted to establish Zuni's claim to the land so that certain army officers who desired to settle there would not usurp Indian property. Five years later, the surveyor general ruled against a land claim of the Isleta pueblo. The Indians claimed to have purchased the area in question a long time before and had used the land for cultivation and grazing for many years. Agent M. C. Williams felt that they would suffer great hardship if the land was taken away.[15] The land problem, which had plagued the Pueblos for centuries, continued to remain a vexation during the late nineteenth century.

AMERICAN OBSERVATIONS ON PUEBLO AGRICULTURAL METHODS

As the American population grew in New Mexico, its interest in everyday events in that region also increased. Anglos were often critical of what they saw when it did not meet their standards. Observers constantly remarked that the need for irrigation in the area would always restrict the amount of land that could be farmed. To them, the arid climate and infertile soil made the land far more suitable for ranching. It should be remembered that these attitudes were being expressed by a group of people who were raised in the humid and fertile regions of the eastern part of the United States. The very idea that fields had to be irrigated to produce was foreign to them, and dry farming was practically unknown. Many of these people were repulsed by the arid, high-desert areas of New Mexico. They were also prone to look down on racial minorities, and their attitudes toward Indians and Hispanics, and in this case their agricultural practices, were typically disparaging.

William W. H. Davis, appointed as the U.S. attorney for the region, responded in a way typical of Anglos who saw New Mexico for the first time. Many of the conditions he observed during his stay in the 1850s shocked him. He described agriculture as the most neglected industry in the region, claiming that it had not improved since the Spaniards first settled there. He learned that the territory's irrigable land comprised less than 1 percent of the total area. However, Davis also noted that, while the soil appeared to be infertile, it contained decomposed feldspar, which made it very productive. Unaccustomed to irrigation canals, Davis described them accurately because he felt that other American farmers would be curious about their size and use. The acequia madre was three to five yards across, two to six feet deep, and often several miles long. Cross ditches were connected with the main artery, and smaller canals led from these to the actual fields. Floodgates controlled water from the mother ditch to the secondary canals. The irrigable land was divided into beds that measured approximately forty by sixty feet. Farmers heaped dirt around the edge of each plot, which would be individually irrigated. Acres set aside for corn were watered a week before planting; maize and wheat fields were usually watered three or four times a season.[16]

In 1866, James F. Meline conducted a tour of some of the western states including New Mexico. His remarks, published the next year, gave Americans some ideas about agriculture in the Southwest. His descriptions left the impression that farming in this region continued to be linked with archaic traditions. Tools, implements, and methods in the area, he felt, resembled those of ancient civilizations. In fact he thought they would make interesting display items for agricultural fairs back East. Meline noted the absence of metal in the construction of plows, carts, and yokes. Even though Hispanic farmers did not use fertilizers, they produced abundant crops. However, he felt that the number of acres under cultivation in New Mexico could be tripled. Meline marveled at the ability of the "ignorant herdsmen" to irrigate their lands and cited some of the local laws governing the practice.

Concerning the Pueblos, Meline felt that they were equally good farmers as the Mexicans and performed their chores "with more order and industry." He observed that they always stored a year's supply of food and raised every kind of vegetable and fruit known in the region. According to his count, the nineteen New Mexico pueblos had 671 horses, 64 mules, 818 donkeys, 2,143 cows, and 783 oxen. Meline also witnessed the harvesting of the wheat fields near San Felipe. Almost fifty people of all ages and genders gathered, thrashed, winnowed, and cleaned the crop. They used a "smooth piece of ground, surrounded by a circle of small trees . . . with strips of thong all around" for a thrashing floor. Five horses were driven in a circle to thrash the wheat, which was then gathered by the women in flat baskets. It was then cleaned by tossing it in the air. The grain was then "tied up neatly in little bundles of some thirty heads" and given to girls who "pounded out the grain."[17]

Adolph F. Bandelier, an early scholar of the Southwest, also made careful notations on agricultural procedures in the 1880s. He described the typical planting and harvesting events in the San Felipe and Cochiti growing season. Wheat was sown from the middle of February through March; the ground was irrigated before planting only during very dry years. The Indians planted corn toward the end of April; they irrigated the fields a week previous to sowing. Bandelier stated that the soil was broken with plows and oxen; there were almost forty plows in Cochiti Pueblo, but only three or four belonged to the Indians. He also observed that during June and July, the native farmers broke up the ground in the cornfields with iron shovels and hoes; all weeding was done with hoes. Chile was planted from the last week in April to the middle of May, and it was frequently irrigated throughout the season. Melons and watermelons were also planted in late April. They planted beans and tobacco in May and frequently irrigated the latter. The first cut of wheat normally occurred in mid- or late July; the second cut took place in the second week of August. Melons were harvested in early October. Bandelier claimed that wheat- and corn-growing areas, like those for other crops, were often rotated.[18] Other observers felt that many of the Indians did not practice crop rotation.

Another Anglo observer, Frank Hamilton Cushing, lived at Zuni in the early 1880s and left some of the most thorough observations of nineteenth-century Pueblo life. Befriended by the Zunis and inducted into their ceremonial life, he was allowed to observe every facet of their world in minute detail. His observations on agriculture covered everything from clearing the fields to preparing the food. His major emphasis was the growing of maize for, as he stated, the Zunis "make corn the standard measurement for time."[19]

Fields that were cleared with a hoe and axe were located below an arroyo since they could be watered by melting snow in the spring and the monsoon rains in the summer. The Zunis constructed a dam across the bed of the arroyo in order to divert water over a wider area. Other earth banks were added to hold water and silt in certain spots and to make sure that most of the moisture stayed within the confines of the area chosen to plant corn. Cushing also gives a detailed account of the religious activity required to consecrate the field before planting. Once this procedure was completed, the area was untouched for a year so that the soil could become enriched. Older fields were fertilized at this time by planting sagebrush in barren areas so that the wind could deposit dust and sand in this section.

When the fields were ready for planting, Zuni farmers sharpened their planting sticks. This ancient tool was not utilized for planting the whole field but only the seeds that would be sown that day. After four days of fasting they would return with a new digging stick for more planting. Between twelve and twenty kernels were planted from four to seven inches in depth; this was necessary because the seed would continue to receive moisture when the topsoil dried out. Cushing described the methods employed by the natives to keep birds from attacking their crops. Cedar poles, with prickly leaves coming out of the top, were set up at intervals of a few yards. Cords of split yucca leaves were then strung between the poles, forming a network throughout the fields. Old rags and pieces of animal skin were hung across the wires to ward off the crows. In addition, the Zunis made scarecrows using hair from a dead horse's tail, rolled-up cornhusks for the eyes, and cornstalks for the mouth. Cushing claimed that these creations "would astound the most talented scarecrow makers of New England."[20]

Cushing pointed out that the main village was surrounded by small farming towns where fields were irrigated for wheat growing. Viaducts made of large hollow logs directed water into ditches some two miles long, and they conducted the flow into gardens surrounded by walls of earth. Poorer farmers might have fifteen or twenty of these plots while wealthier ones could afford the labor to plant a great deal more. At the village of Nutria forty families farmed an area that was almost a mile-and-a-half square.

The Zunis utilized modern tools to weed and harvest their crops. In older times the instruments used for weeding corn were made of wood and were operated more like short scythes. By the time Cushing lived in the village, they used iron hoes. Weeding usually took place about three times during the summer. The corn was picked shortly after the first frost and hauled with burros or in carts to the village. When the wheat was ready for harvest, the entire pueblo would turn out. They used knives or short sickles for the harvesting. American sickles, which were hard to come by, were broken in half to make two small ones. Like the other Pueblos, the Zunis used the ancient threshing floors to clean their wheat. The finished product was placed in bags and hauled to the village by wagon, cart, or burro.[21]

With an insider's view that others did not enjoy, Cushing did not speak in condescending terms concerning Indian farmers. He genuinely admired their understanding of nature's forces and how to use them to their best advantage. He also noticed that the Zunis had remained fairly traditional in their agricultural practices, including the religious ceremonies that accompanied them. For example, he did not think that the plants, animals, and tools introduced by the Spanish had changed farming procedures at the pueblo. The crops may have been different, but the basic procedure remained the same. No one could doubt the results. The Zunis not only fed their large population, but they also supplied corn to feed the animals at Fort Defiance.[22]

In the first forty years of American occupation of the Southwest, Anglos had gained a great deal of specific information concerning Pueblo agriculture. Procedures for planting, irrigating, and harvesting Indian crops had been outlined by a variety of observers. Some, like Davis and Meline, viewed Pueblo methods with a great deal of sarcasm. Bandelier and Cushing were far more professional in their approach and showed more respect for Pueblo culture. Generally these observers demonstrated that while the Indian farmers continued many of their ancient traditions, the end result was productive and often ingenious. While some modern metal tools could be found in New Mexico, they were still rather scarce. Americans back in the East must have been fascinated by the description of irrigation systems since the practice was foreign to them. Most importantly, these reports were probably read by the American officials and may have helped to form future programs in Pueblo agriculture.

THE AGENTS' VIEW

From the very beginnings of American history, federal government policy makers determined that Indian problems could be resolved by teaching Indians how to farm. Once Native Americans adopted an agricultural way of life, they would give up the hunt, accept smaller tracts of land (thus opening up large areas for white settlement), and appreciate the security of raising crops. Some tribes were offered tools and instruction, and missionaries set up demonstration farms. During George Washington's first administration, Congress set aside twenty thousand dollars for farming necessities and provided that Indian agents take on the responsibility of instructing the native farmers on American agricultural techniques.

No American president had as much faith that farming would be the foundation of America's future than Thomas Jefferson. The country's third president also wanted to extend the benefits of living off the soil to Indians. He felt that through the instructions of missionaries and agents, aboriginal inhabitants would accept the idea of private property, become economically self-sufficient, and be more apt to practice Christianity. In other words, agriculture would make Native Americans more "civilized" and thus more readily assimilable into the mainstream of American culture. Early in his first term of office, Jefferson met with various Indian representatives to discuss the benefits of the plow over the chase.[23]

Throughout the nineteenth century, the philosophy of granting lands in severalty was seen by many Americans as the method to mainstream Native Americans. Federal lawmakers, officials of the Indian bureau, and various friends of the Indians (including the Board of Indian Commissioners) felt that allotment of land to individual farmers on the reservation was a panacea for all of their problems. Unlike many other Indian groups, the Pueblos were not affected by the General Allotment or Dawes Act of 1887, which would have gradually removed the protective status of Indians. In his response to a questionnaire from the board's secretary before the passage of the bill, Agent Dolores Romero stated the Pueblos neither held allotments nor desired them. He also believed that Pueblo allotment would soon lead to the usurpation of their land by cattle ranchers. Nothing, he felt, could protect the Pueblos if federal protection was withdrawn. Three years after the act became law, Romero's advice was countered by Special Agent Frank D. Lewis, who declared that the date of the signing of the Dawes Act should be celebrated as a national holiday among the Pueblos. Believing that their rejection of it stemmed from ignorance of the law, Lewis recommended the Indian youth should be educated as to its opportunities, privileges, and obligations.[24]

By the time the United States took over the American Southwest, the philosophy of making farmers out of the Indians was well in place. Now, however, federal officials were trying to accomplish this task in the driest region of the country. They were also dealing with a group of Indians in New Mexico whose system of irrigation and agriculture predated the American government by more than three centuries. This unique situation meant that among the Pueblos, the agent's major task was not to convert the native people to tilling the fields but to improve their farm program and protect their land and water rights. In order to accomplish these goals, Indian agents often gave special instructions to individual Pueblos concerning their agricultural programs. New governors usually received directions cautioning them about procedures to take in trespassing cases, land alienation problems, and farm education programs. In his annual instructions to Pueblo governors, Agent Ben M. Thomas emphasized that only the natives were allowed to cut hay within the limits of the grant. He also encouraged the governors to protect Indians' rights to irrigation water. Thomas did not want surrounding neighbors to deprive the Pueblos of water needed for their crops.[25]

Although the Pueblos had provided an adequate food supply for centuries, the agents often made suggestions to improve the farm programs. Agent Dolores Romero lamented the subsistence level of agriculture among them, but he also recognized that they had no inducement to improve conditions. He felt that if the government would construct a mill for grinding corn, the Indians would be stimulated to increase production. In turn, they would be able to sell the corn in the region, and that would help to change the Pueblo economy.[26] In the same vein, James S. Calhoun felt that the creation of trading posts among the Pueblos would also help them to advance in "civilization." Being able to exchange farm goods for other articles would inspire them to sow more grain.[27] This change from subsistence to credit economy was seen by the agents as a necessary step in the assimilation process. In their view, this transformation was more important than any threat it might pose to Pueblo cultural institutions.

The idea of increasing the Pueblos' crop production for profit actually started in the late 1870s, but it developed slowly and erratically. Pablo Pobijua of San Juan sold excess hay grown on the pueblo's land. Pueblo governor José Ortíz rescinded the transaction, but Agent Thomas told the governor to allow the sale.[28] Also, Thomas encouraged the governor of San Felipe to allow the local stage line to build a stopping point on an area that the pueblo did not cultivate. The Indians would profit since horses could feed on native corn. He also encouraged the Pueblos to trade ponies that he considered worthless for workhorses, oxen, and wagons. In this manner, Thomas hoped to induce them to cultivate new ground and increase agricultural production.[29] About a decade later, agents Romero and M. C. Williams suggested that the Indian bureau should purchase a pair of horses and a buggy so that they could visit the villages more often, particularly during planting season. Romero explained that in the spring the Indians needed instruction in farming, but paying for transportation was too costly. Williams also said that he needed to spend more time with the Indians to show them how to farm.[30]

Sometimes the agents felt that more drastic measures should be taken to improve the agricultural situation. In 1883, Pedro Sánchez made a report to the commissioner concerning Acoma Pueblo. He explained that the geographic setting caused great difficulty for the Indians. Their mesa-top pueblo, lacking an accessible road and located a great distance from their agricultural fields, forced them to carry crops on their backs in order to reach their homes. Sánchez felt that if they should move their pueblo to a more feasible location, they would profit greatly. He persuaded them to locate near a spring about two miles from the mesa, so that they could take advantage of good grazing land and water. The Acomas agreed to the change but requested wagons and harnesses from the government to move building materials.[31]

Following the well-established government policy, the Pueblo agency set up the position of "farmer" to help modernize the Pueblo Indian farm program. His job was to provide advice to improve production. In addition, a farmer was employed at the Abiquiu Agency (later renamed the Jicarilla Agency in 1881), which was under the jurisdiction of the Pueblo agency.[32] This new employee would continue to be central in the American goal of modernizing Pueblo agriculture in the future, but many of those who occupied the position of farmer were unqualified for the job. Attracting better people was difficult given the typically low pay.

The Bureau of Indian Affairs (BIA) wanted agents to measure the results of their modernization program, but production estimates for the Indian villages were difficult to establish. The Indian fields generally followed the topography of the land. Since the plots were rarely rectangular, they were difficult to measure. Agent Thomas insisted that the individuals who made out annual reports for the Pueblos supply all the necessary statistics concerning agricultural production. Since he also required that this information be submitted on time, Thomas suggested that the people estimate crops that had not yet been harvested.[33] Thus the measurements for yields and acres planted were not always accurate. In 1884, Agent Sánchez estimated that the Indians owned about 1,090,000 acres and cultivated approximately 25,000 acres. He also stated that the Pueblo population was 8,835. This would mean that they were farming approximately 2.83 acres per person, but Sánchez's estimate on the number of acres being farmed was much higher than other reports. José Segura figured that in 1890 there were 8,285 Pueblos living on 992,845 acres, and that they farmed about 5,500 acres. Accordingly, the average number of acres per person would have been 0.66. In his annual report for 1875, Ben Thomas estimated that the Pueblos farmed about 13,000 acres. Special Agent Henry R. Poore reported the same figure in 1890. Poore also estimated that 8,287 Indians lived on 906,845 acres.[34] His figures, which seem to be the most accurate, reveal that the Pueblos farmed 1.56 acres per person.

Throughout the West, Indian agents faced the difficult responsibility of assimilating Native Americans into the mainstream of American life, and farming was seen by federal officials as the vehicle that would bring this about. The acculturation process was often thwarted by insufficient funds, corrupt agents, and conflicts of interest. Pueblo agents in the late nineteenth century, however, remained diligent in their efforts to protect Indian property even though there was some confusion as to the Indians' legal status. As commendable as this attitude was, the agents also hoped to change Pueblo disposition concerning productivity. Numerous suggestions were made to bring about this transformation. Virtually all of the agents saw these actions in a positive light even though they may have posed a threat to Pueblo identity. Always conservative in their outlook on life, these New Mexican natives traditionally viewed change of any kind very cautiously. If the agent's proposals could prove to be beneficial, the Pueblos would probably accept it. However, the goal of agents to transform their economy through modernization was still a long way off.

PUEBLO AGRICULTURAL EDUCATION

As early as 1819 the federal government had set aside funds to teach Indians how to farm, but instruction was left to churches or philanthropic organizations. About twenty years later Indian Commissioner T. Hartley Crawford proposed the creation of nearby manual labor schools complete with model farms. Agricultural education for Indians, however, remained in the hands of agents even though the results were disappointing. As the first New Mexico Indian agent, Calhoun quickly espoused the economic benefits that would flow from the educational process. Moreover, he claimed that the Pueblos were "anxious to have schools established amongst them and to receive agricultural information."[35]

Educational facilities for Indians were slow to develop. With the formation of the Carlisle Indian School in 1879 in Pennsylvania by Captain Richard Henry Pratt, a new era of agricultural training commenced. Schooling was combined with practical experience in order to teach manual skills. Vocational training thus became the model for off-reservation boarding schools. The Pueblos were quick to respond to the opening of Carlisle. The year after it opened Agent Ben Thomas requested permission to take not more than 50 students to the school. Within the next seven years, 129 Pueblo youths had attended. However, Agent M. C. Williams felt that the long absence from their parents was detrimental to the family bond. When students returned home after the schooling, they often felt rejected by their peers and went "back to the blanket." To rectify this situation, Williams stated that they should be educated at territorial boarding schools.[36] So the New Mexico agency focused its efforts on establishing schools for Pueblo youths in order to reach the goal of modernizing their agricultural program.

In the late 1870s, the agency's goal was to establish as many day schools at the various pueblos as funds would allow, but the results were not entirely satisfactory. Low pay (six hundred dollars per year) and lack of job security (the result of funding problems) did not enable the schools to hire and retain teachers of high quality. Agent Thomas tried to rectify this situation by getting the Presbyterian church to supplement the teachers' income. Laguna's school was the first one to be set up in this manner in 1876, and Thomas was pleased with the results. In addition to regular instruction, the Laguna youths learned to dig wells, run ditches, build dams, and plant trees. The Pueblos, of course, had been doing most of these things for centuries, but now the children were learning the white man's way.

Other schools, however, did not share the success of the Laguna school, which averaged 147 students in attendance. As a result of the dearth of supplies, the Zuni day school was failing. Lack of good teachers hampered the progress of schools at San Ildefonso, Isleta, and Jemez. Since all of these Indians were already successful farmers, most schools did not even provide agricultural instruction. However, Thomas believed that the Indians could benefit from a farm education program, but he felt that such instruction could only be successful in industrial boarding schools, where children would be removed from the influence of their homes and placed in an atmosphere more conducive to learning American techniques. Eventually, he hoped to set up a centrally located manual school where all Pueblo children could be housed for training.[37]

The concept of using schools to enhance Pueblo agriculture grew slowly in the late 1870s. Thomas instructed the teacher at Laguna to assist the

Indians in the construction of an irrigation ditch for the school farm and requested that the Zuni governor set aside an area for a garden for its school. Agricultural instruction was hurt by a lack of good teachers and language barriers. Children refused to cultivate land under the teacher's direction and still learned farming from their experienced parents.[38] In the early 1880s, industrial instruction became more widespread in the day schools, but the government did not provide funding for implements or fruit trees. Thomas gave orders to the new teacher at Zuni to train the Indians in both classroom and field work. He also suggested that teachers at Laguna and Jemez concentrate on adult education, as well as training children. During the spring and summer, the teacher would give instructions on improving irrigation, cultivating vegetables, and caring for fruit trees. This program was also carried out at Zuni.[39]

During the 1880s, the call for vocational training grew louder and often ran counter to native traditions. Agent Pedro Sánchez was insistent in his plea for industrial education for the Indian children. He felt that the day school program was not conducive to changing Pueblo practices. Sánchez wanted the children removed from reservation schools to the boarding schools so that their education would not be left to "indolent parents" who were inhibited by their own "superstitions." Since the family unit perpetuated Indian religious beliefs, he felt that it should not be "the basis of the future society."[40] Agent W. C. Williams also encouraged education at industrial schools but asked that parents be allowed to visit often. Frequently, Indian children educated away from their reservation were made fun of when they returned home. Through parental visitation, he reasoned, the child would not be ridiculed on his return to the pueblo, and the influence of the school would be more gradual. Williams believed that the goal of federal programs should be to educate the Indians so that eventually they could do the job without government aid. He believed that the only way to achieve this was to teach the Indians how to use modern equipment.[41]

The problem with encouraging parents to visit was that Indian leaders did not always welcome government education programs. The governor of Acoma was accused of whipping children who attended school and refusing to allow others to attend. Agent W. P. McClure felt that the action taken by an "old ignorant superstitious governor" hindered the Pueblos' progress toward civilization. School attendance problems on reservations were so widespread that President Benjamin Harrison recommended legislation to require Indian children to attend schools.[42] The fact that some students believed they were being exploited may have hurt attendance. A group from Acoma felt that they were not learning anything but rather were simply required to do gardening work.[43] Indeed, Indian schools were often understaffed and underfunded; thus student labor kept the operation going and often took precedence over real academic or vocational education.

Day schools were established at various pueblos, often at the request of the Indians. However, the agents insisted on industrial schools, and the Interior Department planned one for Albuquerque in 1880. Agent Thomas had some difficulty finding a location and requested that the city donate sixty acres of tillable land watered by an acequia. In 1882, land was donated by Albuquerque residents for this purpose. Some of it was already under cultivation and therefore suitable for the students to plant. Although Santa Fe had initially been proposed as the site for the school, it did not have an area available with enough water to irrigate a school farm. To insure that the students would receive advanced training in agriculture, the agency supplied the school with seeds, fencing material, implements, and a windmill tower pump.[44]

Plans for another industrial school at Santa Fe began in 1884. A bill appropriating twenty-five thousand dollars for the Santa Fe school passed in the House, and Agent Sánchez wrote to Senator Henry Dawes in 1885 to enlist his aid in the Senate. Sánchez had initially recommended more day schools for the Pueblos but now felt that the schools had failed to "civilize" them. The same year Agent Dolores Romero reported that the Pueblos had displayed increasing interest in the education of their children. The Santa Fe school was already full and more children wanted to apply. Letters had previously been sent to local governors to encourage them to send some boys and girls to the new school.[45]

The education of Pueblo youths was an important part of the agency's agricultural modernization program. Agents felt that the day schools would not serve this purpose because the children living at home remained strongly influenced by their parents, who discouraged Anglo instruction and encouraged their children to cling to ancient traditions. The expectations of the Pueblo communities generally fostered the idea of a semi-subsistence-level economy. The agency hoped that with the introduction of modern methods and equipment in industrial schools, the Indian youths would acquire the ability and the attitude to maximize production. No longer would education be merely the means to acquaint youths with normal classroom activities. By 1890, the major emphasis in education would be to teach Pueblo children "how to live, how to work, how to farm, and how to grow fruit."[46]

YANKEE INGENUITY

For centuries, the Pueblos performed their agricultural chores with fairly crude instruments usually made of wood. The digging stick and the wooden plow sufficed, but they also made the task extremely difficult. During the Spanish period various tools were introduced, but there was only a limited number, and they were hard to replace when broken. Sophisticated machinery had yet to be introduced. The most significant change to occur in the early American period was the widespread introduction of common metal tools.

Still, many of the Pueblo tools and methods remained crude and tied to the past. A number of Pueblos born in the late nineteenth century remembered the way things were. Juan P. Lente of Isleta, born in 1866, recalled that his grandfather and others used wooden shovels to dig irrigation canals and the jawbones of dead sheep and cows to cut their wheat. They leveled their land by filling a cowhide with dirt and using an ox to pull it to the low places. John Rey Olgin, born seven years later, remembered a time when the Isleta used oak tree branches for plows and hayforks. Antonio Callavasa of Santo Domingo, born in 1870, recalled the Indians using cow hip and rib bones to cut wheat and threshing it by hitting the wheat with sticks. Tomasito Tenorio of the same village, born seventeen years later, remembered farmers using wooden plows with rawhide tied around the ends to make them last longer.[47]

In the early 1850s, Lieutenant A. W. Whipple of the U.S. Corps of Topographical Engineers conducted an exploration for a railroad route through New Mexico. He observed various pueblos and commented on the primitive agricultural practices in the region. Whipple stated that the farmers used only wooden implements, including hoes and crude plows. Seeing that agriculture in New Mexico was limited to irrigated regions, he recommended using artesian wells to rectify this situation. However, well drilling was not always successful. In 1858–1859, an experimental well, drilled near Galisteo to a depth of thirteen hundred feet, did not produce water.[48]

Early Americans brought trade items into New Mexico. In this manner, the Pueblos were introduced to the metal spade and hoe. As federal authorities became aware that the Pueblos needed government assistance, implements were shipped to New Mexico during the 1850s. J. K. Graves and many others compiled a report on Indian affairs in New Mexico in 1865. He requested that the federal government provide the Pueblos with new implements, fruit trees, and seeds. However, not all of the Indians profited from this action. Lieutenant John C. Bourke, who visited the Zunis in 1881, commented that the Indians were still using primitive tools.[49] Perhaps their long distance from the agency hindered their acquisition of implements.

Agent Romero reported that some of the Pueblos received implements from the government and also purchased them from local dealers. He lamented, however, that the great majority of the Indians still used tools that had been employed by their ancestors. Agent Thomas felt that modern implements would do more for the civilization of the Pueblos than anything else. With these improvements, he felt that they would be able to compete with their American neighbors.[50] While the Indians received shovels, hoes, and seeds from the government in the early years of American occupation, larger implements arrived later. In the late 1870s, the agents began to distribute plows to various pueblos. Two kinds of plows were issued—a breaking plow for turning over new soil and a stirring plow designed to prepare old land for planting. Laguna, Jemez, and Zuni were among those who received plows at this time.[51]

Seeds were also important to the Pueblo farm program. When Agent Thomas wanted to purchase ten thousand pounds of wheat seed in 1879, the Indian bureau provided financing. Often, materials arrived too late in the season to be used. Therefore, Thomas decided to construct a storehouse at the agency for different kinds of seeds.[52] In this way, he could offset bureaucratic inefficiency that could be detrimental to Pueblo agriculture.

As previously mentioned, Americans considered wells to be important in this arid region. They could be used to bring water to an area that was normally dry or to combat droughts that sporadically caused trouble for New Mexican farmers. Sometimes, though, well digging could be an exercise in frustration. In 1881, floods ruined a well that the Presbyterian Board of Missions had dug for the school at Zuni. Agent Thomas requested that the federal government dig another well for them.[53]

As the American population began to increase dramatically in New Mexico, mostly due to railroad construction during the 1880s, people brought a variety of new tools into the region. The Pueblos benefited as agents sought to supply them with a number of Anglo implements. In addition to plows, shovels, and hoes already mentioned, the Indians acquired rakes, axes, scythes, spades, hayforks, picks, discs, hammers, saws, and hatchets.[54] While more sophisticated implements would eventually help to increase agricultural production among the Pueblos, these elementary farm tools made the job easier for those Indians who acquired them. This meant that fewer men were needed to accomplish various farming tasks and allowed some Pueblos to diversify their economic pursuits.

For centuries after the introduction of cattle into New Mexico by Hispanic settlers, farmers in the region were plagued by continuous encroachment problems. Agents sent instructions to new Pueblo governors about the procedure to follow if cattle trespassed on their lands. When cattle were found on native farms, members of the pueblo were to drive the animals to the nearest alcalde or justice of the peace. After the authorities verified the details of the intrusion, the owners would be fined for their negligence. Generally the damage payments consisted of corn and wheat, as cash in the late nineteenth century was not yet the typical medium of exchange.[55] Indians found that following this procedure did not always result in just compensation. In 1875, natives from Acoma complained that stock belonging to nearby citizens had destroyed their crops. They followed the agent's advice, but the local alcalde did not enforce the law even though he was aware of fines for trespassers. Instead, he merely returned the stock to its owners. Agent Thomas warned that if this negligence were allowed to continue, he would take future cases to the U.S. district court.[56]

In another case two years later, officers from Santa Ana complained that Don José Perea filed suit against the Indians concerning trespassing stock. However, the Indians claimed that the land in question had been purchased by them in 1739. Thomas wanted to make sure that the local justice of the peace was aware of the Indians' rights before the case went to court.[57] As the local judicial authorities continued to neglect their duties in these cases, agents tried to threaten them into action. In a case involving the pueblo of Nambe, Agent Thomas warned that the matter would be taken to the attorney general of the territory. He also instructed the schoolteacher at Jemez to inform the local justice that if he continued to allow this practice, the attorney general would force him to account for his actions.[58]

Sometimes intruders blatantly abused Pueblo farmland. In 1879, the Santa Ana governor complained about stock in Pueblo fields. The animals in question did not simply wander onto their property, but were actually driven across the Indian grant on a daily basis. As a consequence of the lack of justice in these matters, Indians from Laguna decided to take matters into their own hands. They took possession of about two thousand sheep belonging to Frederico Luna that were being herded on their property. Later they returned the animals to their owner, but six Indians were arrested by the sheriff at Los Lunas. Ben Thomas asked U.S. District Attorney S. M. Baird to look into the matter and see that the Indians' rights were protected.[59]

Obviously, the lack of fencing on Pueblo grants was a major reason for this trespassing. In 1879, the agent noted that the only fencing employed by the Zunis was for small gardens.[60] Not only Pueblos' crops, but also their cattle, were open prey for thieves.[61] Oddly enough, the Pueblos were not always amenable to the idea of enclosing their agricultural fields. In 1883, a native from San Juan informed Agent Sánchez that he wanted to fence off an area that he had cultivated but that local authorities at the pueblo rejected his proposal. Since this case was rather unique, Sánchez decided to refer the matter to the Indian bureau for advice.[62] In another case, Agent Segura contacted a man who removed some fencing at the pueblo of Nambe. He urged that the damage be repaid so that native lands could be protected.[63]

Implements, seeds, wells, and fences initiated the American program for modernizing Pueblo agriculture. Virtually all early observers of this group of Indians gave relatively high praise to their farm program, but regretted that they had not expanded their output. In the eyes of the observers, certain steps forward should be taken in order to reach that potential. The early results were mixed. Certainly some of the Pueblos adapted quickly to the American innovations. These posed no threat to their culture (always a major Pueblo concern) and some were even familiar.

Many of these native New Mexicans, however, remained unaffected. For one thing, federal funding was quite limited for Indian affairs—especially in the early period. Also, not all of the implements reached the Pueblo agency. So a number of the Pueblos did not receive any of these items. Some of them may also have been reluctant to accept the new tools. Sure they made the job easier, but what obligations did their acceptance demand? Was this just another pretext for taking Pueblo land? It is impossible to say how much these attitudes affected Indian inaction, but there was some reluctance. For a group of people who had lost land in the past to outsiders, this was understandable. Unfortunately, threats to their property would not only continue, they were on the rise.

ENCROACHMENT ON PUEBLO LAND

While the trespassing of foreign stock remained a continuous vexation for the Pueblos, encroachment by settlers was potentially more dangerous. Cattle could be removed, but people tended to stay even if the law did not allow it. While the encroachers usually were Hispanics, the number of Yankee trespassers increased during the latter part of the century.

Agent Thomas wrote the governor of San Felipe that so as not to tempt the settlers in the region, Indians should not rent their lands. He felt that they had the wherewithal to farm the area by themselves. Thomas believed that inviting others to cultivate Pueblo land would only result in the Indians being defrauded of their rights. He also told the governor that his authority was only concerned with minor problems and that major difficulties like trespassing should be referred to the agency.[64]

Cases involving encroachment problems were numerous; they touched practically every pueblo. Hispanic trespassers caused problems at Picuris in 1879; at Pojoaque in 1886; and at Taos, Picuris, and Laguna in 1887. In the Pojoaque case, Agent Romero requested that the U.S. attorney general be authorized to eject three Hispanos through legal means. The Indians had been given a patent to the land in question by the federal government in 1858.[65] Anglo squatters also harassed the Pueblos. Cases involving American trespassers occurred at Santa Clara in 1882 and 1884, at Isleta in 1884, and at Zuni in 1886.[66] Americans were also involved in a squabble over a tract of land between San Felipe and Santo Domingo that had been donated to the pueblos jointly by the Spanish government. The American involved built a house and grazed his animals on this property. Eventually, the Indians sent their claim to the surveyor general for confirmation. Agent Sánchez believed that their long-time use of the land and an accompanying ditch gave them the right to the property.[67] Trespassing was not always confined to farmers and ranchers. When, in 1882, miners encroached on the lands of Sandia, Agent Thomas made it clear to the governor of the pueblo that their land title gave them the right to "everything connected with or attached to the soil."[68]

Sometimes other Indians trespassed on Pueblo property. In 1879, the Zunis complained that Navajos were driving their herds onto the farming district of Nutria. Agents for both tribes claimed that the natives under their guidance were in the right. The Navajo agent would not reprimand his Indians until the Zunis could prove their accusation. Agent Thomas claimed that he and the teacher at Zuni had witnessed the Navajos camping near the farms in question and hoped that the problem could be solved without calling in the commandant at Fort Wingate.[69] In another case involving Zuni, the lieutenant governor of Isleta allegedly allowed a new road to be built through the fields of a Zuni farmer. This man had planted and irrigated the land and claimed to have title from a former governor. Part of his area had been granted to somebody from Isleta. Agent Segura desired to clear up the matter.[70]

As this last case points out, a basic problem in these cases of encroachment stemmed from the lack of properly defined titles. The United States recognized the original Spanish grants of the Pueblos, but they failed to acknowledge nearby communal areas that the Indians had used for timber and grazing. Agent Thomas warned the Indians that all land claims had to be filed with the surveyor general so that they could be confirmed by Congress. He said that their original grants gave them title to only four square leagues and that additional claims should be filed immediately.[71]

Part of the encroachment issue resulted from bureaucratic inefficiency. In 1887, the natives of Santa Clara complained about settlers on their grant. This area had been approved by the surveyor general and forwarded to Washington for confirmation. However, territorial authorities never withdrew the land from settlement, and various parties encroached on it. Naturally, the Indians were befuddled to see people settling on their reserved property and threatened to use force if the trespassers were not removed.[72]

Sometimes trespassing cases were left over from an earlier period. At the pueblo of Laguna, the Baca family had caused problems in this regard dating back to the Spanish era. Shortly after the Mexican war, Marcos Baca built a house near the Laguna village of Encinal. Here he farmed, raised cattle, and irrigated from a stream that had long been used by the Indians. As time went on, his heirs expanded their holdings. Due to a twenty-four-year gap between the confirmation of the various Laguna tracts (1860) and their acquisition of a patent (1884), some forty-five non-Indian families had moved onto their land. Margarito Baca, Marcos's son, expanded his land holdings in Encinal Canyon in the 1860s, and in 1874, he petitioned the surveyor general regarding his land claim. Fourteen years later (and four years after Laguna had received its patent), Baca's claim was rejected by Surveyor General James K. Proudfit. Regardless of this decision, the federal government failed to remove trespassers or to halt encroachment at Laguna.[73]

Spanish titles to lands often mentioned physical surroundings such as mesas, springs, or rock formations. Sometimes these boundary markers were indiscernible to observers in the next century. As a result, measurements of the grants began to overlap. Zia, Santa Ana, and Jemez had been given joint occupation of 882,849 acres of grazing land in 1766. According to the commissioner of Indian affairs, this grant was confirmed by Congress in 1874. By the late 1880s, approximately forty Hispanic settlers were living on or using the land claimed by the Indians. However, other parties made counterclaims to the same area, and their grants were also confirmed by Congress. Portions of these Hispanic land claims overlapped the Pueblo grazing areas, and Agent Williams requested a survey to settle the matter. These settlers could not simply be removed from the area because they had been there for many years. Williams felt that immediate legal action was impossible since territorial courts could not eject parties from questionable grants.[74]

In another case involving the pueblo of Zuni, an American settler established a ranch on lands that the natives claimed they had used for many years. Agent Williams attempted to determine the boundary lines of the Indians' claim, but he found that they were vague and indefinite. Williams stated that the Zunis, lacking quantitative measurements, relied on mountains and arroyos to delineate their grant. However, Williams could not make a distinction between these geographical configurations. To remedy this situation, he requested that the reservation be resurveyed and that definite monuments be set up to prevent encroachment problems in the future.[75]

In order to prove their legal claims, the Indians tried to demonstrate their continuous use. In 1890, Isleta natives were troubled with a trespasser on the property they had been granted in 1750. They claimed undisturbed use of the land for agricultural and grazing purposes. However, a man named D. A. Hatch felt he owned the land by virtue of homestead rights. Part of his 160-acre tract coincided with the Isleta grant. The Indians threatened to use force to evict him. Agent Segura called for restraint and filed a petition with the surveyor general's office to investigate the situation. However, he lacked authority and finances to hire legal counsel for the Isletas. Segura realized that the vague nature of their Spanish-era claims would hurt the Indians' chances in court, so he requested that the Indian bureau allow him to obtain a lawyer.[76] Since many Pueblos were unacquainted with American jurisprudence, Anglo settlers often held the edge in land claims cases. Agent Williams also complained about his lack of authority to evict trespassers and employ lawyers for the Pueblos. He suggested that the Indian bureau supply him with general instructions to follow in these cases. Williams tried to thwart increasing encroachments by requesting that the BIA print up notices in English and Spanish that warned parties to stay off Indian property.[77]

Eventually, the plan to deal with intruders on Pueblo grants came to fruition. In 1890, Special Indian Agent Frank D. Lewis sent instructions to various Pueblo governors concerning encroachment on their lands. The Indian leaders were to warn all non-Indians (except traders) and all nonresident Indians that they would be liable for damages that resulted from trespassing. If the encroachers refused to leave immediately after verbal notification, Lewis instructed the governor to gather the names and residences of the trespassers before two witnesses and submit them to the agency. This letter was also to be read to the persons involved so that there would be no doubt about the proceedings that would be taken against them.[78]

Encroachment was not restricted to individuals. As American immigrants pushed west, they demanded better means of transportation and supply, and the railroad was the ultimate answer to this need. The two major lines that affected the Pueblos, the Denver & Rio Grande (D&RG) and the Atchison, Topeka, and Santa Fe (AT&SF) railroads, completed service to New Mexico about 1880. Naturally, these lines wanted to take the most expedient path, which often passed through Pueblo land. Conflicts between the Indians and railroads began almost as soon as the lines entered New Mexico. In 1880, the pueblo of San Ildefonso complained that the D&RG had not paid them for land previously appropriated. They also claimed that, because the railroad had not constructed culverts under the roadbed, some of their crops could not be irrigated. During the original grading process, crops were destroyed at San Ildefonso and at Santa Clara. The Indians, who had not been compensated, asked for fifteen hundred and twenty-five hundred dollars, respectively. The D&RG did not respond quickly; in 1884, they had still not given compensation to the Indians.[79]

The problems with railroads affected a number of different pueblos. The pueblo of San Felipe had a continuous conflict with the AT&SF line throughout the early 1880s. The complaint stemmed from an alteration of an arroyo that resulted in the flooding of Pueblo farmlands. In addition, the railroad also blocked one of the pueblo's irrigation ditches, creating a drought in part of the fields. The railroad had promised to make an opening for the irrigation ditch, but little had been done to rectify the situation. The Indians feared that without corrective measures, they would lose their crops for the year 1882. Although Agent Thomas called for an equitable solution to the problems caused by the company, railroad officials continued to propose answers that he felt were unfair.[80] Acoma was harassed by the Atlantic and Pacific Railroad (A&P), a network controlled by the AT&SF. The Indians complained that the A&P had made one of their irrigation ditches unserviceable because the grading constructed by the line overflowed into the canal. Previously, the Acomas had lost crops because of the same problem. Agent Sánchez threatened a suit against the railroad if the Indians were not compensated. The company agreed to pay them five hundred dollars for work done on the old ditch and promised to build a new aqueduct that would service them better.[81]

Railroads continued to plague the Pueblo villages in the late 1880s with damages occurring at San Juan, Isleta, and Santa Ana.[82] However, they were able to compensate for some of their crop losses by performing wage work for the lines. This allowed some of the pueblos to diversify their economy and become familiar with cash, but the impact was minimal. For the most part, the intrusion of the railroad increased the encroachment problems that had plagued the Pueblos for decades.

AMERICAN LAW IN PUEBLO LAND

As they had under the Spanish and Mexican governments, the Pueblos experienced the effect of a new legal system following the American occupation. The new authorities, however, sought to retain previous legislation particularly suited to the area's unique problems. In June 1851, the year after New Mexico became a territory, the first legislative assembly met in Santa Fe and passed an act that provided for continuation of all laws previously in force that were not inconsistent with the U.S. Constitution.[83]

The territorial legislators began to pass more specific laws to govern farms and water privileges. In 1861, they enacted a law requiring the Pueblos to perform work on acequias that connected with those of villages where non-Indians lived. The governor of each pueblo had the responsibility of controlling the work on the canals by all those who possessed water rights in the acequia. Governors also had to see that water distribution was based on past customs. Finally, Indians who had not paid their share of ditch construction costs could not vote in the election of acequia officers.[84]

Pueblo water rights were affected by policies adopted throughout the western United States, as well as the laws of New Mexico. Since most of the immigrants were not familiar with irrigation, confusion over the legal aspects of water rights existed. The riparian system in vogue in the eastern part of the country recognized the rights of landowners adjacent to a stream. The riparian owner had full rights to this water, but he was required to return an equivalent amount at the lower end of his land. But this was the arid west and a new system evolved.

Since legal systems were not immediately put in force in the California gold rush areas, miners there were forced to improvise the adjudication of water rights. Scholars disagree about whether the system they developed was spontaneous or evolved from European mining traditions.[85] Regardless of origin, the appropriation doctrine that resulted became fundamental to Western water law. According to the doctrine, the first person to utilize water in a stream could take as much as he needed without being required to return it. Priority was thus established on a first-come, first-served basis as long as the user continued to make reasonable use of his rights. The person who acquired these rights did not have to hold land contiguous to the stream.[86]

The Pueblos never considered riparian rights, since they often employed acequias to transfer water to nonadjacent farmlands.[87] It would appear that the appropriation doctrine adopted in New Mexico should have protected Pueblo water rights. As non-Indians increased, however, conflicts between the two groups over water became a major concern. The basic question concerning the protection of Pueblo land or water rights centered on their legal status. It would seem that previous federal laws concerning other Indians should also be applied to the natives of New Mexico, and in 1851 Congress did just that. By this act, all Indians were granted the status of wards of the

U.S. government; thus their land and water could not be sold. Early agents, however, refused to believe that the Pueblos were included in this legislation. They felt, like the Spanish and Mexican governments before them, that these Indians were civilized citizens who should not be classified with the more nomadic types.

The situation came to a head in the case of *United States v. Lucero* in 1867. This litigation, heard in New Mexico's territorial court, actually concerned trespass on Cochiti land but gradually focused on the Pueblo land status. Stephen B. Elkins, U.S. district attorney, hoped to clear up this matter in a final, sweeping settlement. In his decision, Judge John N. Slough was more concerned with the application of federal law concerning Indians in New Mexico than Pueblo status. He concluded that the 1834 Intercourse Act did not, in fact, apply to the Pueblos since they did not live "beyond the settlements or on the frontiers." By this legal technicality, the Pueblos were denied protection of their property. In his decision, Slough held that while the Pueblos were a settled, agricultural people, they "are Indians only in features and complexion" and "superior to all but a few of the civilized Indian tribes of this country."[88] Unfortunately, his apparent praise for the Pueblos dealt a major blow to their land and water rights.

The *Lucero* decision laid the foundation for the most famous decision in Pueblo history. A Supreme Court case of 1876, *United States v. Joseph*, involved an alleged trespass on the Taos grant. The Indians held that this action violated the Indian Intercourse Act of 1834, which prohibited non-Indian settlement on land granted to the native people by the federal government. The court recognized that the defendant had constructed a house on and lived in part of the lands of Taos Pueblo. It agreed that Congress had extended the act of 1834 to New Mexico in 1851. The major question concerned whether or not the Pueblos were considered to be Indians under the terms of the Intercourse Act.

Accordingly, the court held that:

> They are Indians only in feature, complexion, and a few of their habits. . . . The pueblo Indians, if, indeed, they can be called Indians, had nothing in common with this class (nomadic tribes). The degree of civilization which they had attained centuries before, their willing submission to all the laws of the Mexican government, the full submission by that

> government of all their civil rights, including that of voting and holding office, and their absorption into the general mass of the population (except that they held their lands in common), all forbid that they should be classed with the Indian tribes for whom the intercourse acts were made.[89]

The Supreme Court recognized that the Appropriations Act of 1872 had made provisions for a Pueblo agency, which put the Pueblos on parity with other tribes. However, the court felt that this condition did not change their status. Also, the judges did not acknowledge the 1874 report of Indian Agent Edwin C. Lewis, who, like other agents before him, noted that previous efforts had been made to declare the Pueblos citizens and that the main motivation behind this action was to remove the protection of the agency. Lewis felt that this would make the Pueblos open prey to encroachers, who would eventually deprive the Indians of their property and force them into pauperism.[90]

Encroachment on Pueblo property increased following the *Joseph* decision. In part, this can be explained by the large increase of non-Indians in New Mexico in the late nineteenth century. It was also related to a lack of legal expertise. The Indians were not supplied an attorney by the federal government in spite of the constant requests by agents for legal protection.[91] In addition, agents pointed out that the Pueblos were reluctant to pursue legal channels because in the past the lower Hispanic courts in the territory had dealt unjustly with them. The process fed on itself as many citizens, realizing the natives' lack of faith in the courts, continued to impose on them.[92] The expense of court cases was another factor that played against the Pueblos. In a case involving settlement on lands belonging to Santo Domingo, Agent Thomas suggested that the Indians sell the land in order to avoid court costs that they could not afford.[93]

In addition to encroachment problems, the *Joseph* decision caused the Pueblos to lose protection from alienation. Without federal laws prohibiting the sale of Indian land, they could simply sell their holdings as they pleased. Before the court's ruling, Agent Thomas asked the governor of Cochiti to stop his people from selling land to outsiders and restrict the cultivation of the village's land to the Indians of the pueblo. A year after the *Joseph* case, Thomas proposed to increase the grazing and farmlands of the Zuni reservation. He feared, however, that if the Indians received title to the land, they might later alienate it. In the same year, Thomas issued instructions telling the governor of Nambe not to allow the people of his pueblo to sell their land. He feared that if this process continued, eventually the Indians would lose the basis of their economic support system.[94]

With the removal of federal protection, the Pueblos were also allowed to lease their property. Agent Sánchez protested against an apparent thirty-year lease that the governor of Acoma made to Solomon Bibo, the U.S. trader for the pueblo. Sánchez felt that, under the long-term lease, the governor was virtually selling the land. He believed that the *Joseph* decision hurt the Pueblos because they were not equipped to thwart speculators who tried to take over their holdings.[95]

WATER RIGHTS

Attempts to acquire Pueblo water rights accompanied threats to Pueblo land. Since farming and ranching were the major industries in New Mexico, irrigation water became a necessary adjunct to a successful economy. Although non-Indians posed the greatest threat to Pueblo water rights, intervillage conflicts also existed. In 1857, the ancient water dispute between Acoma and Laguna became the first conflict of its kind to appear in the territorial courts of the United States. This case persisted throughout the last part of the nineteenth century. Eventually, Laguna was allowed access to surplus waters not used by Acoma farmers.[96] Another dispute concerning this matter arose in 1890. Agent Segura told the officers of Acoma to see to it that the Laguna farmers continued to receive irrigation water for their crops. Segura wanted Acoma to allow Laguna the use of the water when Acomas were not watering their own fields. If Acoma did not comply, the Lagunas would lose their crops. He warned that continued usurpation of the stream by Acoma would result in a lengthy and costly lawsuit. Finally, after Solomon Bibo was asked to intercede, the Acomas agreed to allow Laguna the use of the water.[97]

Water rights conflicts were particularly bad at Laguna. Shortly after the Americans gained control of the Southwest, acting superintendent John Greiner contacted Indian Commissioner Luke Lea concerning Laguna's water problems with a group of Navajos under the leadership of Sandoval. The Navajos complained to Greiner that the Lagunas were farming upstream and cutting off their water supply. In the process, the Navajo wheat fields were practically destroyed. Greiner told the commissioner that this was but one of many cases of this kind. He felt that the question of water rights was the most difficult problem facing the Pueblos.[98] More than a century later, his remarks would seem prophetic.

One of the major threats to Pueblo water resulted from the interference of the growing number of Hispanic communities in northern New Mexico. For example, the town of Los Padillas constructed an irrigation ditch that apparently conflicted with the ditch at Isleta. The pueblo of Jemez had similar problems with the new town of Torreon, which allocated water from a tank previously constructed by the Indians.[99] Natives at Cochiti shared water with the town of Peña Blanca, whose acequia ran through their pueblo. They agreed to share in the use of water and the maintenance work on the ditch. However, the Indians claimed that the town violated the arrangement and that the Hispanic mayordomo threatened them with fines and imprisonment if they interfered. Four years later, the Hispanic residents constructed a new acequia across Cochiti and denied the Indians use of the canal unless they made a payment to the mayordomo. Eventually the town agreed to let the Indians share in the water if they agreed to do their quota of work on the canal.[100]

Because Taos was situated in the most desirable area, the Indians there were constantly threatened by new arrivals. The natives and the town of Arroyo Seco reached an agreement on the waters of the Rio Lucero. Later, however, the Indians claimed that their neighbors violated the original agreement, and, as a result, they placed a guard at the mouth of the Hispanic acequia. Another conflict arose over the use of the Rio Pueblo. At first the probate judge ruled against the Indians, but Agent Thomas claimed that he did not have jurisdiction in the case. Thomas also believed that the Indians could not sell their surplus waters because they belonged to downstream users. Eventually, to keep peace in the area, Thomas asked the Taos governor to allow local citizens more irrigation time.[101]

Water rights cases abounded during the early American period, affecting nearly every pueblo. Mexican encroachment near Picuris caused Agent Sánchez to call for help from the federal district attorney. Sánchez wanted him to rule on these questions because he feared that the Indians could not win in local courts, which were heavily influenced by non-Indians.[102] American encroachers near Santo Domingo and San Felipe threatened to kill the Indians because of contested claims over water.[103] Conflicts over water also occurred at Zuni,[104] San Juan,[105] Zia,[106] and Nambe.[107] Sometimes the attempts to take water from the Pueblos were rather blatant. In 1884, an American by the name of Charles Lewis dug a well outside of Los Lunas. Unfortunately, his work took place on the Isleta reservation. Agent Sánchez, who wanted to avoid trouble between Indians and whites, requested that Lewis leave the reservation in order to avoid a lawsuit or depredations by the pueblo.[108]

One of the most controversial cases involved the New Mexico Irrigation and Colonization Company, a Boston firm that had received a charter from the territory in 1889. They desired to build a canal on both sides of the Rio Grande from a point six miles above Cochiti down to Texas and Mexico. As part of this grandiose plan, the company also wanted to build reservoirs at various points along the way. In addition, they proposed to dam the river at the beginning point, thus raising the water to a height of sixty-five feet. The canals were designed to run through the pueblos of Cochiti, Santo Domingo, San Felipe, Sandia, and Isleta. The company planned to carry water ten miles from the river so as to increase greatly arable land in the area. Many landowners along the river had already granted the company one-half of the lands to be reclaimed, and the firm sought a similar agreement with the Indians. New Mexican laws restricting interference with the natural flow of the river caused most local residents to doubt the feasibility of such an immense project. Indians protested because their current canals more than sufficed for their agricultural purposes, and they feared a possible loss of water in the future. Special Agent Lewis worried that since local men involved in the scheme were very prominent in the society, they could resort to condemnation proceedings against the Indians.[109] It is doubtful that the Pueblos' protest halted the project, but their water rights in the region could have caused the company many legal complications.

NATURAL DISASTERS

If the Pueblos could not always stop invaders from encroaching on their land and usurping their water rights, they could at least take some measures against these actions. However, there were some matters that they simply could not prevent. Natural disasters sometimes dealt massive blows to their farm program. While the Pueblos tried to cope with these complications in the best way they could, there were times when they needed federal assistance to survive. In the arid climate of New Mexico, droughts posed the biggest problems. The Indians used both practical and theological methods to ward off the dangers that resulted from adverse climatic conditions. Food storage, religious ceremonies, and dry farming were all designed to help the native farmers during lean years.

In the 1850s, W. W. Davis complained about the lack of regular and frequent rains in the area, but his observations were usually based on comparisons with the more humid regions of the eastern United States.[110] Light rainfall in 1873 reduced corn and wheat production to about half its normal yield. Drought about 1880 caused a migration of about 150 Acoma and Laguna Indians to Isleta. In 1886, the agent reported both drought conditions at Zuni and flood damage at Santo Domingo and Isleta.[111]

New Mexican weather can be feast or famine. During extremely wet years, floods caused extensive damage. Flood damage was reported in 1865 and again in 1868.[112] Another flood in 1880 damaged the acequias of San Felipe, as well as endangering the railroad track there. Both the Indians and the company wanted to build a breakwater to avoid similar damage in the future.[113] In 1886, a major flood damaged much of the Pueblo area. Santo Domingo in particular suffered extreme damage to the village as crops and houses were destroyed. Agent Romero reported that the river carried away part of the lands of Santo Domingo, including their old Catholic church. Serious damage also occurred at Isleta, where wheat and cornfields and vineyards were partially destroyed. In part, this overflow at Isleta was caused by the railroad grade and bridge near the pueblo. The Indians were forced to break the grade in order to relieve the flooded fields above it.[114]

Insect infestation also plagued the Pueblos, who were ill-equipped to handle this problem. The main culprit, the grasshopper, often caused crop damage, as in 1865. In a major outbreak in 1877, these insects destroyed wheat and corn crops at Taos, San Juan, Santa Clara, and San Ildefonso. Agent Thomas tried to assist these villages with supplies. He asked that aid be sent to these northern pueblos, and he especially emphasized the need at Taos, which had suffered in the previous year from the same plague. When grasshoppers attacked Laguna in 1878, again Thomas sought federal aid for this destitute pueblo.[115]

The ecological balance in the Pueblo area was always a precarious one that could easily be disturbed by a large population increase. During the early American period, the influx of non-Indians considerably altered New Mexico demographics. This increased immigration occurred not only in the Pueblo area but also in the southern Colorado valleys upstream. As a result of this growth, soil erosion increased. Because of railroad construction and the need for firewood and building material, lumbering became a major industry. Due to a lack of forest conservation, sedimentation in the Rio Grande increased. Also, large growth in the livestock business went unchecked, and the animals were allowed to overgraze. Deforestation and overgrazing, which reached a peak about 1880, led to large silt deposits in the river. This, in turn, raised the water table along the river because of more seepage. Areas that had previously been fertile became regions of alkali, salt grass, and swamp.[116] As the population continued to increase in the next century, this cycle would be repeated with more devastating consequences.

Some of the Pueblos tried to rectify the problems caused by floods and soil erosion by constructing new irrigation canals.[117] The Indians at Jemez decided to run an acequia above their pueblo to a new tract of land.[118] Similar steps were taken at San Ildefonso in 1881,[119] and at Santo Domingo in 1890.[120] Dams and ditches were also built to allow the Zunis more arable land.[121] New irrigation ditches were also suggested for San Juan, but Agent Romero said that a skilled surveyor or engineer was needed to complete the work.[122] While nature had always played havoc with Pueblo farming activities, the new government presented fresh challenges.

The results of the American presence in the Southwest were mixed. In many ways the Pueblos benefited from the American acquisition of New Mexico. Their agents were extremely conscious of the Indians' rights to land and water and sought to protect them from Anglo and Hispanic encroachment. Through American traders or federal government allocations, the Indians acquired small implements to make their agricultural task an easier one. Authorities began to realize the need for engineering skills to improve the Indian farm program. Pueblo land titles were cleared, and Pueblos were given patents for part of their claims. Unlike other Indian tribes, they did not lose major portions of their reservations to Anglos taking advantage of the Dawes Act. During a period of violence and destruction in American Indian history, the life of the Pueblos went on relatively unchanged.

In spite of these benefits, the Pueblos were not always helped by the American presence. For one thing, they did not receive rations like other Native Americans except in time of deprivation. While this action may have fostered independence and the continuation of their agricultural program, it may also have caused them to feel that the federal government's concern for them was less acute than that felt for other Indians. Given the federal neglect of the Pueblos and the tumultuous nature of territorial politics, their agents often lacked the authority and means to meet their needs and redress their grievances. Bureaucratic inefficiency and a lack of legal protection were at least partially responsible, then, for their encroachment problems. This attitude was given further impetus by the *Joseph* decision, which changed the Pueblo legal status and encouraged even more trespassing. More than any other factor, this Supreme Court ruling caused great damage to Pueblo farmers. It meant the loss of protection of land and water that they had enjoyed for centuries.

CHAPTER FIVE

A Change in Status (1890–1920)

AGRICULTURAL CONDITIONS CIRCA 1890

The first years of American rule in the Southwest witnessed a drastic change in the demography of the region. As the non-Indian population gained in numbers, its influence dominated daily activities in the area. Amidst the turmoil wrought by disease, alcohol, relocation, and outright encroachment, American Indian socioeconomic status reached a nadir around the turn of the century. Pueblo reaction to these events reflected more of a traditional approach than an acceptance of outside forces. Lacking the federal protection that many other Indian tribes received, they relied on their usual tenacity to cope with the changes in their physical and social environment.

In spite of the Supreme Court decision that denied the Pueblos the status of other Indians, the federal government continued to treat them like

Figure 2:
Gathering chile, San Juan Pueblo, 1905. Photo by Edward S. Curtis. Courtesy of the Museum of New Mexico, neg. no. 151857.

other Native Americans. An example of this concern was a report of Indian conditions, including the Pueblos, for the 1890 census. This account, written largely by Special Agent Henry R. Poore, contains detailed information on individual villages, including figures on population, occupations, languages spoken, school enrollments, and agriculture. Although Poore did not visit all of the pueblos of New Mexico, two other Anglo observers supplied accounts for four of the villages. Frederick Muller of Taos visited Picuris, and Special Agent Julian Scott went to the western pueblos of Acoma, Laguna, and Zuni. Their observations provided the most thorough account of the Pueblo agricultural situation ever recorded.[1]

Quite naturally, Poore and his associates focused a great deal of attention on crops and their quality. Poore commented on the exceptionally fine crops of the Taos natives, especially corn and wheat, but fruits and vegetables were uncommon because of the elevation. He found San Juan's agriculture to be more like a big garden in comparison to Taos. People of San Juan grew a large variety of vegetables and also harvested apples and plums.

In addition to typical Indian crops, the natives of Tesuque harvested apples and peaches. With the exception of a few vegetables, the crops at Nambe consisted entirely of corn and wheat. Stock owners also grew alfalfa, which they harvested three times per year. Agent Poore remarked on the excellence of the Jemez peaches, which received a higher price in Santa Fe than eastern fruit. He also claimed that the Santa Ana corn crop was the finest in the entire area. The farmers of Santo Domingo had various small orchards in which they grew plums, peaches, apples, and apricots. Although corn and wheat were the staple crops at Isleta and Sandia, grapes were also a major industry at the latter village. Farm products of the Zunis included corn, wheat, oats, beans, chile, onions, pumpkins, and a variety of melons. The Indians of Acoma and Laguna raised garden products consisting of chile, onions, melons, pumpkins, and beans. In addition, they harvested such fruits as apples, peaches, and grapes. Wheat had been a major crop at one time for both pueblos, but these native farmers made corn their chief product because it was more reliable.

The general farming conditions of many pueblos received high praise from the Anglo reporters. As many earlier observers had noted, Taos had a far greater agricultural potential than the others because of its abundance of water. At the time of Poore's visit, it was engaged in a water rights lawsuit, and he commented that it was the only pueblo with enough determination to be aggressive in legal matters. The average farm in Nambe was larger than that of most villages, and it benefited from a constant source of water. Cochiti, surrounded by tillable lands, also received an adequate supply of water. Agriculture at Jemez was far more advanced than at the other pueblos. The Jemez people farmed most of their available acreage; only a few heads of families did not till their own plots. The Indians of Santo Domingo used all of the available farmland within their grant. Additional ground near the riverbed could have been planted, but they feared floods like those that had devastated crops there in the previous decade. Floods had destroyed some of Isleta's farmlands, but these Indians still farmed large sections of their grant. Poore found a big disparity in individuals' economic conditions at San Juan. Some farmed as much as twenty-five acres while others had to work for their neighbors to make a living. Muller found the natives of Picuris cultivated about one-fourth of the arable land within their grant. The families farmed plots that averaged fifteen acres. The fields of Acoma and Laguna were scattered throughout the San Jose and Acoma valleys along

Figure 3:
Waffle gardens and Zuni River, Zuni Pueblo, ca. 1912. Photo by Jesse Nusbaum. Courtesy of the Museum of New Mexico, neg. no. 139172.

small streams and washes where irrigation was feasible. These people also planted gardens wherever they found springs. Within the Zuni village confines, women tended as many as two hundred mud-walled gardens called "waffle gardens" because of their checkerboard appearance.

Farm production at other pueblos was lagging behind. For example, Agent Poore found Santa Clara to be a very poor community. San Ildefonso farming plots were extremely limited because most of the grant was not suitable for agriculture or grazing. Conditions at Zia were not very prosperous either. Although these Indians possessed numerous animals, their agricultural program was haphazard. Much of their grant was sandy and unproductive. Poore felt that the farmers of Zia were less than enthusiastic about their chores and intimated that neighboring Indians called them lazy. In 1876, for example, their neighbors at Jemez had aided them for three days with a force of more than one hundred laborers to dig an acequia, but it had since fallen into decay. The soil around the town site of Santa Ana was unproductive, so the Indians occupied this area only in fall and winter. In March, they moved all of their farming necessities to the summer quarters eight miles away.

Encroachment, a constant vexation over the last two centuries, continued to plague a number of pueblos and often forced them into legal battles they could not afford to pursue. In Santa Clara, two cases involving the pueblo were pending in the local court at the time of the agent's inspection. San Ildefonso was also hampered by settlers on their grant. These Hispanic families had been allowed to purchase land on the pueblo, and they continued to enlarge their boundaries. The sale of nearby Pojoaque property to outsiders had made this pueblo even more destitute than its neighbors. Nambe suffered constant depredations from encroaching Hispanic animals that often destroyed crops and ruined pastures. Jemez was also plagued by Hispanic stock wandering into the corn and grain fields. However, Jemez received fence wire from the agency and constructed a barrier against further encroachment. The pueblo's other crops, such as vegetables, fruits, and grapes, were grown in fields surrounded by high mud walls with locking doors for further protection. Eight Hispanic families lived on the Cochiti grant, but they fell under the jurisdiction of the Indians and were required to perform communal work on the irrigation ditches. A few Hispanic farmers, accepted by the natives of Sandia, raised vegetables on small plots within the grant. Special Agent Scott found that the Zunis made scant use of their large valley, preferring to plant in small side canyons. They had once used the broad valley for agriculture, but by 1890, to discourage white men from encroaching here, they all allowed it to go to waste.

From the beginning of American control of the Southwest, Anglo observers lamented the lack of modern methods and equipment in the Pueblo farm program, and about half of the villages had begun to adopt these by 1890. Poore considered the use of wire fencing at Taos to enclose about five hundred acres of pasture a reflection of their advanced nature. Fencing at San Juan, used only for small plots, consisted of cedar limbs for posts with brush fastened to them by leather strips. Like most of the other pueblos, agriculture at Tesuque was in a transitional stage revealed by the use of both wooden and steel plows. Poore also described Tesuque's ancient method of grinding grain with four different grades of metates and manos. The village of Isleta had once had its own mill for grinding grain, but it had been abandoned.

Figure 4:
Farmland, Laguna Pueblo, ca. 1908–1918. Photo by H. F. Robinson. Courtesy of the Museum of New Mexico, neg. no. 36729.

Modernization at Jemez moved at a faster pace. Agent Poore observed that the native farmers were beginning to fertilize their fields—a practice not employed at other pueblos. A decade before Poore's investigation, the Jemez farmers had used oxen and wooden plows to till their fields, but by 1890, they had switched to horses and metal plows. Evidently, the agency's distribution of equipment was not always governed strictly by a pueblo's needs; the 20 natives of Pojoaque possessed four plows while the 225 people at Santa Clara had only two which they assigned by lot to the Indian farmers.

Throughout their history, some of the Pueblo farmers were able to raise enough excess crops to sell for profit. A large mill located at Ranchos de Taos encouraged the Indians to raise more crops than they needed. Often, to receive maximum profit, they stored their grains until prices were high. Much of the grain produced at Tesuque was exchanged for necessary items in Santa Fe. Santo Domingo natives sold most of their fruits to the railroad station at Wallace. Isleta Indians had planted sixty acres of fruit trees, including peaches, plums, and apricots. They transported much of this crop to Albuquerque and small railroad towns. At Acoma and Laguna, farmers attempted to raise as much corn as possible over their needs. In good years they could raise double the amount required for their people and either store or barter the rest.

As others had stated in the past, Poore and his associates felt that many of the Pueblos could expand the amount of acres they farmed. Poore felt that the natives of San Juan could increase their farmlands by one thousand acres if the government would help them build a main ditch at a higher level. The acequia system at Santa Clara lacked water when he visited the village, but Poore believed that this condition could be alleviated by constructing a reservoir in a nearby canyon. San Felipe had more arable land per person than any other pueblo, but a large part of it was not being used. Agent Poore noted that expansive areas could be farmed if the acequia system of the village could be extended. He estimated that the Indians could easily expand their agricultural area by three thousand acres with additional water. Sandia also had additional land that could be improved. Water flowed through its acequia for six miles before it reached the fields. Poore estimated that the Isletas could farm five hundred more acres through proper engineering skills. Although the village of Picuris possessed sufficient water to cultivate more land, Muller felt that the villagers needed incentive and modern machinery to increase their production.

According to the special agents' observations, farming progressed unevenly in the villages. Pueblos like Taos, Jemez, Santo Domingo, Isleta, and Zuni prospered while others, like Pojoaque, Santa Clara, and Zia, were in poor condition. A few of the villages took advantage of non-Indian businesses to engage in wage work to purchase or trade for food for an insufficient farm program. Some of the Santa Clara Indians supplemented their income through work on the railroad. A few men from San Juan acquired meat through their work on local cattle ranches while others were mainly vegetarians. Generally speaking, the natives of Nambe were also not meat eaters and relied on outside labor to purchase what little meat they had in their diets.[2] Because of their high elevation, the Picuris could not always rely on crop production. Thus, they supplemented their diet with livestock, including cows and pigs. Fishing and hunting for deer, rabbits, and, until the end of the nineteenth century, buffalo also provided sustenance.[3]

While the agency distributed tools to the Pueblos, the Indians did not always make efficient use of them. Indeed, one of the constant complaints of the special agents concerned the natives' lack of engineering expertise. The agents offered numerous suggestions on ways to maximize acreage. While their advice would have led to increased production, many of the Indians were satisfied to raise only enough crops to feed their village. However, the report shows that, as they had done in the past, some villages bartered unneeded farm products. Taos, Tesuque, Santo Domingo, Zuni, Acoma, Laguna, and Isleta made it a practice to sell some of their excess crops for other items or cash. Wages made from ranch, farm, and railroad work, though limited at this time, also helped to supplement Pueblo incomes. Resistance to outside change had been a Pueblo characteristic for centuries, and because of this attitude, the agents' innovations were accepted slowly.

PUEBLO AGRICULTURE STATISTICS

In addition to the general information supplied by the reconnaissance missions, the special agents also provided specific statistics concerning Pueblo agriculture. These are included in table 2.[4]

It is difficult to compare these statistics with the Barriero report of 1827 since neither defines the term "farmer" clearly. However, the agents' reports show that the number of Pueblo farmers decreased by more than half.

Unfortunately, many of these figures for acres cultivated are approximations that do not reflect the Pueblo agriculture program accurately. Within the next decade, agents began to report more exact findings to the commissioner of Indian affairs. For example, the agent's report for 1900 shows that a Pueblo population of 7,883 cultivated 18,379 acres.[5] This means that the average number of acres per person was 2.33. While this figure is significantly higher than that of 1890, the two cannot really be compared since seven of the pueblos did not report cultivated acres in the earlier census.

In addition to the total number of acres cultivated, the agents also began to list a breakdown by crop. The following table reveals the crop production for various items raised by the Pueblos. Again, the 1890 figures are estimates.[6] The crops listed in table 3 were the major items in the Pueblo farm program. Although corn production was lower than wheat in 1899, it rebounded

TABLE TWO

PUEBLO AGRICULTURAL STATISTICS, 1890

PUEBLO	POPULATION	FARMERS	HERDERS & STOCK RAISERS	DAY LABORERS	ACRES CULTIVATED	ACRES PER PERSON
Acoma	566	15	145	60	308	0.54
Cochiti	268	41	—	45	—	—
Isleta	1,059	32	3	13	2600*	2.46
Jemez	428	60	2	—	1400*	3.27
Laguna	1,143	220	25	144	621	0.54
Nambe	79	24	—	—	300*	3.80
Picuris	108	—	—	—	555	5.14
Pojoaque	20	5	—	—	25	1.25
Sandia	140	—	—	3	—	—
Santo Domingo	671	117	—	124	—	—
San Felipe	554	209	—	1	—	—
San Ildefonso	148	26	—	9	—	—
San Juan	406	99	—	50	642*	1.58
Santa Ana	253	117	—	—	750*	2.96
Santa Clara	225	45	—	—	390*	1.73
Taos	401	114	4	33	—	—
Tesuque	91	27	—	—	230	2.52
Zia	106	23	1	—	100*	0.94
Zuni	1,621	342	110	45	—	—
TOTAL	*8,287*	*1,516*	*290*	*527*	*7,921***	*1.77***

* Indicates approximate figures.
** Includes only those pueblos that were measured.

TABLE THREE
PUEBLO CROP PRODUCT

YEAR	BU. OF WHEAT	BU. OF CORN	TONS OF HAY	NO. OF MELONS	BU. OF BEANS	POPULATION
1890	9,000	20,000	20	15,000	300	8,287
1899	65,352	58,801	793	62,000	25,053	8,961
1900	69,184	97,613	1,232	145,368	—	7,883

tremendously the next year. The increase in hay reflects the Pueblo interest in grazing animals. The leading producers of hay were San Felipe, San Juan, Santa Clara, and Santa Ana.[7]

SCIENCE AND ENGINEERING

The major change in the government's attitude toward Pueblo farmers at the turn of the century concerned the use of modern skills and machinery to increase agricultural output. This attitude was peculiar in light of the *Joseph* decision, which legally denied the Pueblos protection and aid from the federal government. In spite of the Supreme Court's ruling, the agency realized that the Indians were not advanced enough to take on the role of a modern farming community. While many of the villages often produced more than enough food to feed their people, they could not always cope with lean years, and only a few were marketing their excess. Through modernization, the federal government hoped to make the Pueblos entirely self-sufficient and removed from the Bureau of Indian Affairs' support. In an ironic twist of fate, Washington bureaucrats were encouraging the Pueblos to play catch-up with a court decision that denied them Indian status and thrust them into a category for which they were ill prepared.

On the other hand, the agency sometimes denied the Pueblos technical assistance. In 1903, the villages of Cochiti and Santo Domingo asked for help following a flood. Superintendent Clinton J. Crandall stated that the Indians were not eligible for federal assistance. He believed that this was probably a wise decision since they sometimes gave their gifts away or refused to heed his advice. The Pueblos, he felt, were self-supporting and should not get government assistance in spite of the precedents that had been established.[8] Superintendent Philip T. Lonergan believed that the Pueblos were denied government support because local non-Indians would not back appropriations to New Mexico's natives. He considered this attitude shortsighted, feeling that development of Pueblo lands could only add to the prosperity of the entire region. Arizona, he pointed out, received almost half a million dollars from the federal government to develop Indian lands. New Mexico's inability to benefit from this legislation meant that the villages would lack roads, cultivated land, and income from leasing property.[9]

The need for engineering expertise, expressed by the agency in the late 1880s, continued through the next decade. In 1892, Agent John H. Robinson recommended the appointment of a competent engineer to oversee irrigation.[10] Five years later, San Ildefonso undertook the construction of an acequia to alleviate the drought that had recently plagued their village. Only part of the ditch had been built, since the pueblo lacked funds to hire a surveyor. Agent C. E. Nordstrom suggested that the government pay for this help so that the project could be completed.[11]

In the late nineteenth century, Congress began to grasp the need for special legislation to develop the arid regions of the West. In 1877, they passed the Desert Land Act, which gave claimants the right to unappropriated state waters for beneficial use.[12] The House and Senate also enacted the Newlands Act of 1902. By this law, which resulted from the earlier efforts of Major John Wesley Powell, the federal government undertook the construction and maintenance of dams and irrigation projects financed by public land sales.[13]

Concern over the need for federal aid for Indian irrigation projects resulted from the increased interest in development of the arid regions. Congress began to make annual appropriations to finance the work of the Irrigation Division of the Bureau of Indian Affairs. For example, in 1903 the Indian Appropriations Act allocated money for ditch construction and irrigation tools. In addition, the bill authorized the employment of two superintendents of irrigation for New Mexico's Indians. John B. Harper held this position for the Pueblos.[14] After Harper's appointment, the Pueblos began to receive more financial aid for irrigation maintenance. Superintendent Crandall reported in 1906 that as much as ten thousand dollars would be spent that year. Two years later the amount of money authorized for Pueblo irrigation tripled, with the greatest percentage going to construction at Zuni. In the following year, irrigation work increased rapidly with projects starting at Taos, Santa Clara, and Zuni. In spite of this benevolent action on the part of federal government, Indians could not always count on assistance. In 1916, for example, no new projects began because Congress did not pass the Indian Appropriations Act for that year.[15]

In addition to the help of the irrigation superintendent, Pueblo villages also needed assistance to modernize their farming methods. Merton L. Miller, who visited Taos in 1896, observed some of the agricultural methods that seemed ancient to Anglos.

> Behind the plow followed an old man or boy (or woman) to drop in the seed. After the field is planted the oxen are hitched to a long pole by rawhide traces fastened to the yoke and to each end of the pole. A man steps on the pole and, holding on by the tails of the oxen, rides around the field to level it off.[16]

To overcome the primitive methods of Pueblo agriculturists, the agency instituted the positions of "farmer" and "additional farmer." These men were assigned to teach the native farmers how to plow and plant their lands properly. In addition, their duties included showing Indians how to care for their crops until harvest time. Then they instructed the Indians on the proper methods of harvesting and storage. These duties included instruction in crop rotation and choosing the best varieties of plants. It was hoped that the agency farmers could enable the Indians to become self-sufficient and thus progress further toward "civilization." Farmers were also required to protect Pueblo land and water from encroaching neighbors. Finally, the farmers took charge of the implements and saw to it that the Indians were taught their proper use.[17]

Numerous Pueblos asked the agency to send out farmers to help with their programs. In 1896, Agent John L. Bullis stated that both Jemez and Zia needed farmers to increase the number of acres that they cultivated. However, he rejected the Jemez request because they refused to deed the government the land used for school buildings and the farmer's housing.[18] In spite of this initial rejection, Jemez acquired a farmer three years later. Agent N. S. Walpole felt that the Indians needed instruction to appreciate the modern methods available to them. He stated that although "the Indians are ready to make use of modern improvements . . . they are not sufficiently cognizant of the advantages offered thereby to provide themselves with them."[19]

By 1900, two additional farmer positions opened up. One of these men would work at Taos while the other was assigned to share duties at Acoma and Laguna.[20] Unfortunately, short funds sometimes eliminated these positions. Pueblos who lost such aid in 1904 included San Juan, Santa Clara, San Ildefonso, Santo Domingo, and Cochiti. Recognizing the need for this job, Superintendent Crandall reported the next year that henceforth the bureau would make it a permanent civil service position.[21]

Government farmers tried to be innovative in their attempt to diversify and modernize the Pueblo farm program. Hiram Jones attempted an experimental garden that included such items as strawberries, raspberries, peaches, and pears. Since the government lacked funds for this undertaking, he asked for donations to initiate the program. In addition, the Indian service set aside land at Acomita for a model farm. Jones had charge of irrigating the field and planting alfalfa. Hispanic trespassers, however, destroyed the fence surrounding the field and tried to farm the land by themselves, thus halting his efforts.[22]

The federal government's employment of farmers to assist Native American agriculture demonstrates concern about changing the subsistence trends for native people as part of its Americanization program. Between 1869 and 1881 the number of people employed in this position doubled. From then until 1905, the number again rose twice as high and covered 80 percent of the agencies in the country. Complaints about the quality of their work, regardless of stricter hiring standards, continued as the advancement of Indian agriculture failed to transpire.[23] Because their agricultural program had been in place for centuries before American takeover of the Southwest, the Pueblos received no training from government farmers until the 1890s, but even then it did little to alleviate their problems.

While the position of farmer seemed like a natural step in the modernization process, it was doomed for failure in its early years. Underfunding, a plague for numerous projects on Western reservations, continued to be a problem for Pueblo farm instructors. The lack of funds affected virtually every tribe in the country. By the turn of the century, the government financed one farmer for every 581 Indians. Inadequate salaries meant that few capable people would be attracted to the job. Those employed at this position were often incompetent, careless, tied to local politics, and encumbered by language barriers. More importantly, farmers lacked direction. The Indian bureau never established an overall plan to direct agricultural policy. They completely underestimated the difficulties of Western farming and the barriers to cultural change.[24]

CONSTRUCTION PROJECTS

Observers of agriculture in New Mexico always remarked about the limitations imposed by the arid climate of the region. Agency personnel attempted various construction projects to increase agricultural production and thwart the dangers of droughts and floods. These works ranged from relatively minor operations like well digging to major undertakings such as dams. Pueblo dams had been built on a small scale before Americans entered the Southwest, and the federal government became involved in minor dam projects in the 1880s. However, major operations were not undertaken until the turn of the century. A step in this direction took place in 1898 when Agent Walpole assigned Charles E. Burton, the supervising teacher at Cochiti, to instruct the Indians in the modern method of dam construction so that they could begin a project on their grant.[25] During the next year, intruders at Jemez attempted to destroy a dam owned by the pueblo. Walpole warned the additional farmer, Albert B. Reagan, of the territorial laws regarding such matters and suggested that the pueblo governor (who also served as mayordomo) take action against the accused.[26]

Dams were especially critical to the western pueblos, which lacked a consistent water source like those enjoyed by villages on the Rio Grande. In 1903, Douglas Graham, the school superintendent at Zuni, recommended a reservoir at Black Rock, along with a boarding school and farm, due east of the village. He felt the location was ideal since it was out of the way of prevalent sandstorms; the school farm, to be located at the foot of a plateau, could easily receive drainage water. Construction on the dam allowed those Indians who could be spared from farming and ranching a chance to earn money from an outside source. Graham, who later became superintendent of the Zuni agency, felt that this alone would help the Indians progress toward civilization. He also stated that when the irrigation system at Black Rock was completed, the model farm would be set up for the new school there.[27]

Floodwater runoff had been controlled by the Zunis for centuries. Innovative check dams allowed them to farm expansive areas and control erosion long before the American arrival. However, conditions in the late nineteenth century exacerbated an already tenuous situation and called for extreme measures. The coming of the railroad in 1881 destroyed large segments of their reservation as lumbermen cut rods for the timber-cutting aspect of the project. The railroad also opened up other Anglo business opportunities for timber and livestock interests over the next few years. Overgrazing and timber cutting heavily damaged the upper Zuni watershed, and channel cutting in the valley floors resulted. Deforestation and overgrazing continued into the first decade of the twentieth century and severely threatened the Zuni agricultural lands.[28]

The construction at Black Rock was the largest project that the agency attempted. While the rest of the pueblos received more than eleven thousand dollars for irrigation and dam construction in 1903, the project at Zuni required almost four times that amount. By 1907, the work on the dam was completed, but canal construction remained. During that year the project received over eighty-five thousand dollars of federal funding. Originally the work was supervised by John B. Harper, but, after his death in 1908, the final canal work was completed under the new superintendent of irrigation, H. F. Robinson. The total cost of the dam alone was almost $263,000. Payment for Indian labor accounted for nearly one-third of this amount.[29]

Even after the Black Rock dam was completed, the federal government financed maintenance and repairs when the dam was damaged. Of the nearly forty-six thousand dollars spent in 1910, over one-third of this sum was expended for Indian labor.[30] By 1918, the reservoir supplied almost five thousand acres with irrigation water. Unfortunately, because the lake was filling up with silt, it had lost 54 percent of its total capacity. Authorities at Zuni predicted that the lake would only last ten more years unless some action was taken. They requested thirteen thousand more dollars in funding in order to extend the life of the reservoir by eleven years. The plan was to increase the height of the dam, but natural surroundings limited this extension. The authorities promised to abandon the project if the silt problem could not be solved.[31]

Whereas Zuni needed a dam to hold water during times of drought, Taos required one to prevent flood damage. Originally officials planned to build a dam a short distance above the pueblo, but the Indians balked at this idea.[32] In addition to flood control, Assistant Engineer Ed Kinney believed a reservoir could increase the amount of arable land in the vicinity. At this time, the Taos Indians irrigated over sixteen hundred acres from the Rio Pueblo and Rio Lucero. Kinney felt that dam construction could increase this amount by almost two thousand acres. He rejected the idea of a masonry dam because of the high cost of transporting cement to the area. Since Kinney thought that the Taos natives could not use all of the land that would be made arable, he suggested making allotments to the Indians and leasing the unused land to non-Indians. He also mentioned the possibility of adding a power site to the construction project.[33]

Superintendent Robinson formulated the proposal for the dam at Taos. The plan contemplated a storage reservoir on the Rio Pueblo that would help to equalize the stream flow. Two sites were considered—one partially on the grant and one entirely on public land. Robinson felt that the Indians could not only use more farmland, but could also strengthen their rights to water in the area. He made a careful outline of construction costs and tried to minimize expense by using local materials. The construction of a power site was a major consideration in the overall plan for the reservoir. Robinson realized that the building cost of the dam could be offset by selling electricity to the surrounding area. He estimated the amount of power that would be made available and the profits that could be reaped from it. Electricity generated at the plant could be sold for household use, manufacturing purposes, and pumping water for irrigation. Rather than asking the Indian office to develop and market the power, Robinson felt that the rights should be sold or leased to a private party. Although there was no immediate demand for electricity in the area, he believed that the town of Taos would eagerly desire power if it were made available.[34]

Power sites and additional land leasing made available by the dam construction were important considerations of the overall project. Congress was still in favor of financing projects on Indian reservations even if those projects could not be reimbursed, but they also hinted that this situation had to change. No method had been developed for procuring payment for maintenance and operating expenses of Indian irrigation projects. The agency felt that individual users, rather than the whole tribe, should pay for the projects in the future. In 1914, Congress concurred, requiring that individual assessments be made on a per-acre basis.[35]

Because the western pueblos of Acoma and Laguna had clashed over water rights in the past, dam construction seemed like a natural resolution. In 1908, three small rock-filled diversion dams were constructed on these two grants to supply existing canals with water.[36] Maintenance on these construction projects was a major concern of the irrigation superintendent. When the Seama Dam on the Rio San Jose required upkeep, Robinson suggested that the Indians attend to it because it was in their best interest. Repairs on the Mesita Dam, six miles below Laguna, cost two thousand dollars. Robinson felt that this expense could largely have been avoided through careful attention by the farmer or Indians in the area.[37] Built without cement, these dams were vulnerable to destruction by heavy floods. In 1912, floodwaters caused considerable damage on both the Mesita and Seama dams. Extreme floods washed out rock and brush at the bottom of the Mesita Dam, allowing part of the construction to sink about four feet and causing leakage. The Seama Dam required rebuilding, and the Indians wanted to know if the Irrigation Division had finances to help. Instead, Superintendent Robinson suggested that a ditch be built through Acoma Pueblo to supply water in place of the Seama Dam.[38]

Another major project was designed to improve Pueblo canal systems. In addition to the typical ditches, the agency attempted to construct flumes to carry water over conflicting canals. These acequias were needed to combat the devastation of floods and loss of water to new settlers in the region. In 1905, the flooding of the Rio Grande caused great damage. Superintendent Crandall suggested that a new ditch be constructed at Santa Clara because the Indians could not get water to their crops after the floods destroyed the original ditch.[39] A ditch construction project took place on the San Ildefonso grant during the next year. Evidently, the canal passed through the Indian grant and property belonging to an American named Hobart. Crandall suggested to assistant U.S. attorney Medler that Hobart contribute to the expense of rebuilding the damaged acequia. The superintendent also wanted Robinson to help him measure the cultivated land of both Hobart and the Indians so that each could be assessed its share of the expenses. In addition to the two miles of ditch work at San Ildefonso, the agency planned similar action at Cochiti and Taos. Crandall also suggested building a breakwater to prevent flooding and erosion at San Juan.[40]

Ditch construction through Pueblo lands was not always the work of the government. In 1898, a court case developed between the Albuquerque Land and Irrigation Company (AL and I) and the pueblos of Sandia, Santa Ana, and San Felipe. The company, made up of local attorneys and businessmen, had proposed to build a major acequia through the Pueblos' grants for the purpose of developing lands for colonization. The Indians opposed this construction because the Rio Grande lacked a sufficient amount of water to supply everyone concerned throughout the year. Because the beginning of the company's project lay above the Pueblos' ditches, they feared that the acequia would cause a scarcity of water on their grants. In the court decision that followed, Judge John R. McFie ruled that the company should be allowed to dig the canal, but that the Indians' prior rights to the river's water should not be disturbed. The decision ignored the traditions of local control and imposed a new shift toward the Anglo water elite.[41]

To compensate the native farmers for the intrusion, the Albuquerque firm was required to pay each village for the right-of-way through its lands. When the company gave the Indians their compensation money, officials assured them that their agricultural lands would not be crossed. Agent Walpole emphasized that the money did not necessarily belong to the pueblo as a whole. If the company's work went through lands appropriated to an individual, then that person was to receive the reimbursement. The amount of payment, which depended on the condition of the land, ranged from twenty-five dollars per acre for tillable land to one dollar and a quarter for arid mesa land.[42]

The Pueblos affected by the court's ruling did not acquiesce in the decision. A meeting of the governors of Isleta, Sandia, Santo Domingo, Santa Ana, and San Felipe was held at the latter pueblo in order to assess the situation. The Indians wanted the company to make an alteration in its construction plans, and evidently the natives of Santa Ana took a hard line against the company. Agent Walpole warned the governor that if he attempted to oppose the AL and I by force, the pueblo would be acting in defiance of the law. He also cautioned that the company was acting within its legal rights and could defend itself with force if necessary.[43]

In spite of the company's legal backing, the Pueblos persisted in their efforts to thwart the AL and I's plan. Walpole stated that if the company did not build a culvert or bridge near San Felipe, the Indians would have considerable transportation problems. More serious was the Indians' accusation that the firm had altered its course from the original survey and that this action caused them great harm. Walpole warned the company that unless the Indians were compensated, serious trouble would result. The company's new route would go through the fields of Santa Ana, and the Indians planned to defend their property if necessary. Walpole tried to pacify the Indians by assuring them that no matter where the AL and I built its canal, the Pueblo prior water rights could not be disturbed. He also stated that if the company crossed the Indian ditches, it must provide flumes so as not to destroy the natives' acequias. On the other hand, Walpole recognized that the court's decision gave the company the legal right to build through Pueblo grants and requested that the Indians respect this settlement and accept the compensation money.[44]

In addition to dams and ditches, the Pueblos received further engineering assistance to enhance their water rights and agricultural program. Because of the possibility of droughts, the agency attempted to drill wells on Pueblo grants. In 1899, Walpole consulted Additional Farmer Albert B. Reagan about the feasibility of digging a well at the Zia pueblo. The agent wanted to erect a windmill so that the Indians could have sufficient water to irrigate lands in the vicinity. Regan aided J. P. Conners in the selection of a site for the well, and Walpole told them to find the spot that would be most beneficial to the native crops.[45]

Wells were a major concern of the western Pueblos. Due to their extremely arid conditions, Acoma and Laguna had a long-running argument over water distribution. In 1915, Hiram Jones, the farmer from Acomita, cautioned Superintendent Lonergan about the need for water at Acoma. Jones claimed that the pueblo had less than an acre per person under cultivation, but he felt that more than five thousand acres near the old village could be farmed if the pueblo had sufficient water. He requested that irrigation superintendent Robinson dig a few artesian wells in the area for this purpose. Jones felt that the Acomas needed more help than their neighbors because Laguna possessed twenty-seven wells while Acoma had only five.[46]

Jones criticized the agricultural methods at Acoma and suggested ways to improve them. He discovered that the Indians divided their fields into small plots and laid them out in such an irregular fashion that they could not be properly leveled for irrigation. Since each farmer was assigned various small plots in different areas, it was difficult to cultivate crops expeditiously. Jones stated that this condition led to useless roads and additional irrigation canals. He suggested that the land be properly surveyed and that each man be assigned a contiguous area of land for his crops. He felt that this action would cause the Indians to build homes near their farms and thereby help to break up old village customs and clans.[47]

When channel cutting began to affect Pueblo farmers toward the end of the nineteenth century, the agency took some measures to cope with it. Superintendent Crandall suggested that a boom of logs be secured across the Galisteo River near Santo Domingo. He wanted the logs to be fastened together and anchored so that, during high water periods, the boom would rise and allow water to pass underneath. At other periods, it would act as an obstruction and cause sand to accumulate, thereby raising the level of the riverbed. Crandall felt that unless this action were taken, adjacent land would become untillable.[48]

The Pueblos had used ditches and small dams in the past, but the new construction was more permanent. It also involved more modern techniques and resulted in fixtures seldom used before, such as flumes, log booms, and wells. The latter would become more prominent in future decades. More importantly, Anglo construction techniques called for the end of small-plot farming and the leveling of large areas that could be irrigated and farmed more effectively. This suggestion, however, posed a threat to traditional land ownership and received little support from the Indians. Thus the clash between the American push to maximize agricultural protection and Pueblo reluctance in this area continued.

MODERN EQUIPMENT

While construction projects generally aided the Indians by increasing their water supply and helping them to make better use of their land, more advanced equipment was also needed. In the 1870s, the Pueblos had been issued implements by the federal government. These supplies continued to be doled out in spite of the apparent change of status that resulted from the *Joseph* decision. Still, some Pueblos felt that the government neglected them because other Indians received goods that they did not. The Zunis, in particular, felt slighted because the Hopis and Navajos got wagons and plows while they were left out. Since Zuni was so distant from the nearest shipping point (Gallup), the village was required to pay shipping charges from that point. Thus, they did not share in goods sent to the Pueblo agency for general distribution.[49]

Although the agency continued to hand out equipment, attitudes about who should receive it changed in the 1890s. For example, Agent John Robertson requested that the government issue him axes, picks, and spades so that he could hand them out to "deserving Pueblo Indians."[50] This phrase, which continued to be used by agents throughout the decade, reflected the agency's opinion that some of the Pueblos had not made significant progress toward "civilization" in spite of the tools and materials given to them by the government. The agency hoped that the offer of tools as a reward for dropping traditional ways of life would induce more conventional Pueblos to adopt Anglo ways.

Sometimes the agency went to extreme measures to insure that assimilation was being achieved. Agent José Segura requested that a mowing machine be sent to Zuni to aid the Indians in their harvesting. However, Segura stated that the machines should not be shipped to Gallup until the Indians agreed to send twelve or fifteen more children to the agency school and keep them there. Furthering this paternalistic attitude, Segura promised that continued good behavior by the Zunis would result in other rewards by the agency, such as having their grant surveyed.[51]

Again in 1903, the agency displayed reluctance to dole out implements. Superintendent Crandall claimed that the Indian office strongly opposed granting assistance to self-supporting native farmers such as those at Taos. Concurring with this opinion, he suggested that the Taos Indians should render some kind of service to the federal government before they received any equipment.[52] Evidently the effect of the *Joseph* decision combined with the self-sufficiency of the Pueblos left some agents with the feeling that this particular group of Indians did not deserve or need government assistance.

Many of the implements that the Pueblos received were not unfamiliar items. For example, in 1892 the agency held plows, axes, picks, and hoes for distribution. However, the same report shows that they also had large amounts of barbed wire and a mowing machine—both of which were new items.[53] The distribution of seeds remained an important item for the Pueblos, but the agency's attitude relative to them reflected the continuing concern over gratuities. Agent John L. Bullis suggested that nothing should be issued to the Indians without their services in return. He added that once seeds had been furnished to the Pueblos, they should start their own seed pools.[54]

As a result of encroachment problems, fencing continued to be a pressing need for Pueblo farmers. The agency took steps to cure this vexation, and by 1892, each village had fenced in large areas of its grant. Not all of the Pueblos, however, used barbed wire judiciously. In 1897, Agent C. E. Nordstrom revealed that during the two previous years, Santa Clara and San Ildefonso had been issued almost seven thousand pounds of wire. While the first village put it to legitimate use, the latter did not. The agent later discovered that the Indians sold their wire at one-fourth of its value to a nearby fruit ranch. Nordstrom decided, in the future, to issue the wire to the local teacher, who would distribute it after notification from the agency. Fearful that the other Pueblos would also sell their wire instead of using it, the agency warned local teachers to use the wire before the Indians disposed of it.[55]

In addition to smaller implements, the agency also started to procure large machinery in order to modernize the Pueblo farm program. Beginning in 1891, the Pueblo agency acquired a mower to aid harvesting. During the same year, the Zunis expressed an interest in this kind of machine. They claimed that they could cut hay for storage and that additional cuttings could be sold at a nearby market. A machine arrived the next year, but Agent John Robertson would not let the Zunis use it because they had not obeyed his orders regarding school matters.[56] In 1895, the agency acquired a hay press and a wire stretcher for the Pueblos. Two years later, the agency purchased more mowing machines and a hay rake. In 1899, the agency bought a hay press to be used at Zuni. Realizing the need for special equipment, Agent Nordstrom requested a breaking plow for making an irrigation ditch. The agency purchased another plow designed to work well in sandy soil in 1913.[57]

As previously mentioned, Pueblo methods for harvesting and threshing grain were, by American standards, very ancient. They still used the old reaping hook to cut the grain and employed animals to tramp it out. Foreign substances had to be picked out by hand after the animals did their chore. This antiquated method did not result in a very clean and

Figure 5:
Threshing with goats, Picuris Pueblo, ca. 1905. Photo by Ed Andrews. Courtesy of the Museum of New Mexico, neg. no. 15123.

marketable crop. Agent Walpole, therefore, asked the commissioner to purchase a number of threshing machines for use by several Rio Grande pueblos. He specifically requested one to be shared by the western pueblos of Acoma and Laguna.[58] As the use of threshing machines slowly spread, each village desired to acquire its own. In 1912, the first one was installed at Jemez. The natives of Santa Clara had high praise for their machine even though the local farmer did not feel that its engine was large enough to carry the load. He stated that with a proper motor, the Indians could put through six hundred bushels a day. Realizing the efficiency of Santa Clara's machine, the people of San Ildefonso asked to use it on their fields.[59]

The distribution of large machinery made a significant contribution to the modernization of Pueblo agriculture. As the Indians came to rely on more efficient technology, they could increase production. The upsurge in output required less manpower as machines performed work formerly done by the native farmers. This in turn allowed Indians who had previously been employed in agriculture to seek other means of support, thereby diversifying the Pueblo economy. However, some of the Pueblo farm plots were so small that large machines could not be used on them.[60] In addition, not all Pueblos sought this aid eagerly nor was the government willing to support the distribution of goods to what they felt were self-supporting Indians. What was needed to change this attitude was a reversal of the *Joseph* decision which would make them legally eligible for increased government support.

NEW LAWS FOR THE PUEBLOS

The Supreme Court decision in the *Joseph* case recognized the Pueblos as members of an advanced Indian civilization who deserved the treatment given to other citizens. In spite of this pronouncement, however, they were not given rights equivalent to Anglos. For example, they possessed no voting rights. Citizenship, even though applied in a half-hearted manner, also brought them responsibilities with which they were ill prepared to cope.

Pueblo agents did not always appreciate the Indian plight in this matter. In 1891, Agent José Segura, in a response to a question from the Lake Mohonk Conference, addressed the issue of American law as it applied to the Pueblos. Segura informed this New York–based Indian rights group that these Indians had "every opportunity that the law affords to any other citizen regardless of race or color." Lacking any government concern or juridical wherewithal, however, the Pueblos had to rely on the dictates of local courts for protection of their rights. But they were not eager to resort to this authority because of a basic lack of faith in non-Indian-dominated tribunals. As Segura pointed out, the Pueblos preferred to arbitrate differences within their villages. Realizing this proclivity, acting agent John Bullis suggested a different approach. Since he considered them to be wards of the government, he thought that Pueblo litigation should be referred to the U.S. district attorney of New Mexico. Here they could get protection that was otherwise unaffordable.[61] As late as 1917, some authorities called for the federal courts to have exclusive jurisdiction in Indian litigation in New Mexico.[62]

In order to protect their rights, the Pueblos required competent legal representation. For years the agents complained that the Pueblos, as private citizens, could not afford to employ the services of an attorney. In 1899,

Congress passed an appropriations act that established the position of special attorney for the Pueblos. This provision was retained in subsequent legislation. The man who was first assigned to this position was A. J. Abbott. Unfortunately, the Pueblos could not always count on adequate help from the man who held this position. In 1917, George Vaux, chairman of the Board of Indian Commissioners, reported to the Indian commissioner that the Pueblos desperately needed an adequate attorney to look after their interests. Francis C. Wilson, who held this position for four years until he resigned in 1914, performed very efficiently, but the agency had failed to find an adequate replacement. Vaux hoped that the new appointee could withstand the local political forces that challenged Indian interests.[63]

Local authorities sometimes questioned the legitimacy of Pueblo citizenship. In 1904, Superintendent C. J. Crandall stated that the Pueblos always depended on outsiders for legal advice, feared the political dominance of non-Indians, and were no more fit for this status than the "wildest Apache." He felt that the Pueblos never attempted to become citizens during the Mexican or American rule. Crandall added that citizenship would further burden them with the responsibility of taxation, which could result in their lands being sold.[64]

In 1904, the Territorial Court of New Mexico ruled that Pueblo lands were subject to taxation. The Pueblos decided to unify and change their status. They asked permission to send one delegate from each pueblo to the nation's capitol in order to deal with the questions of citizenship and taxation before Congress. They desired to ask for special legislation that would temporally exempt them from taxation. Crandall sympathized with this action because he felt that the tax could be made retroactive and thus deal a heavy blow to the very existence of the Pueblos. Realizing the special nature of the case, the superintendent contacted the Indian Rights Association, a Philadelphia-based group, to solicit their cooperation in the matter.[65]

Superintendent Crandall led the Pueblo delegation to Washington and saw the bill pass the lower house. In addition to a tax-exempt status for the Pueblos, it contained a clause that provided that all litigation against Indians be tried before a U.S. district court. The superintendent approved of this legislation because it would remove the Pueblos from the jurisdiction of what he considered to be "petty Mexican" courts. Shortly after his return, Crandall received word that the act relieving the Pueblos from the burden of territorial taxation (including previous taxes) became law. Tax-free items included land, farm equipment given to them by the federal government, and any animals they owned.[66]

This piece of legislation set the stage for a chain of events that eventually led to a reclassification of the Pueblo status. For over a half century New Mexico had pushed Congress for acceptance as a state. Now the legislative branch was willing to cooperate, but one condition was that the territory had to accept a different view of the Pueblos. Bureau officials had created a new atmosphere in Congress by reporting on the harsh economic conditions facing these native farmers. Legislators in the nation's capitol wanted New Mexico to accept the idea that the Pueblos were indeed Indians and thus warranted federal protection to their property.[67]

In 1910, Congress passed an enabling act that would allow the territory to become a state. This act contained a clause that stated that "the terms 'Indian' and 'Indian country' shall include the Pueblo Indians of New Mexico and the lands owned and occupied by them."[68] In essence Congress was going against decisions by the territorial court and the Supreme Court, both of which had decided that the Pueblos did not qualify as Indians. Now something had to be done to realign these judicial interpretations with the new political reality. The case that brought this about involved the selling of alcohol to Indians—something that the federal government had always prohibited. When José Sandoval, a non-Indian, sold liquor at the San Juan village, he was simply doing something that had been approved by the territorial court as late as 1905. According to the enabling act, however, this transaction was illegal.[69]

In 1913, the Supreme Court upheld this new attitude in the decision *United States v. Sandoval.* Justice Van Devanter, who delivered the Court's opinion, cited the enabling act, which stated that the pueblos should be treated as other Indian tribes. He also claimed that, while the Pueblos were sedentary, peaceful, and industrious, they were "nevertheless Indians in race, custom, and domestic government." After all, they lived separate from non-Indians, had "primitive modes of life," and were "influenced by superstition" and "governed according to crude customs." They were, according to the Court, "a simple, uninformed, and inferior people." The justice recognized that the federal government treated the pueblos like other Indian communities by presenting them with farming implements, instructions, agents, training schools, a special attorney, and dam and irrigation works. Van Devanter realized that some people felt that the Pueblos were citizens, but he believed this had no bearing on the power of Congress, in the enabling act, to pass laws to protect them "as a dependent people." The Court also believed that Pueblos held fee simple title to their land, but claimed that it was a communal title since individuals did not own separate tracts. Finally, the Court recognized the previous *Joseph* decision, but felt that it was based on false information that had been updated by recent investigation.[70] Thus the *Sandoval* case reversed a decision that had plagued the Pueblos for almost four decades. Legally, they were now Indians. Thus authority over the Pueblos shifted from Santa Fe back to Washington.

As momentous as the *Sandoval* decision was in Pueblo history, one would think that it would have caused quite a stir in New Mexico. Many non-Indians, however, were undisturbed by it and viewed the case as a status issue and not a land question. The Pueblos, who opposed statehood as another step toward encroachment, now felt that they must take action to halt increased trespassing that would surely accompany their new status. Perhaps many people did not understand all of the implications of the ruling. The question of Pueblo land and water was still hazy at this time. If they legally owned both of these elements, unlike other Indians, could they not sell them at will? The ominous consequence of the *Sandoval* case for non-Indians was, however, that the federal government now controlled Pueblo affairs. Not only could it remove squatters but also any other settlers who had purchased Pueblo property.[71]

Even before New Mexico became a state in 1912, the Interior Department, working with Pueblo Special Attorney Francis C. Wilson, sought to clear up Pueblo land titles. The idea, approved by the Pueblos, was to deed all of their lands to the federal government in trust. Although the trusteeship failed, the plan initiated the idea of determining what lands the Pueblos had lost through previous court decisions. Inspector Dorr of the department also suggested, for the first time, that a legitimate body should be established to determine what non-Indian claims were valid through land surveys and title research. Thus Pueblo titles could be quieted. Surveys began in 1914. Within five years these investigations revealed that approximately three thousand non-Indian families representing some twelve thousand people held claims on Pueblo grants. Irrigation statistics, however, demonstrated that the problem was more acute than originally believed. While the disputed acres represented a relatively small portion of Indian property, they involved the best irrigable land.[72]

In addition to the question of land titles, the Pueblos and the new state government also had to begin clearing up water rights. The Territory of New Mexico solidified the water laws of the region with the passage of the water code in 1907. Some of the old Spanish and Mexican laws concerning water were abrogated, but the general principles remained. The statutes of 1907 were incorporated in the New Mexico state constitution of 1912; both accepted the doctrine of prior appropriation. The constitution confirmed all existing rights for beneficial use. Unappropriated water belonged to the public and could also be appropriated beneficially with priority given to the better right.[73]

In order to make Pueblo water rights a matter of record, the secretary of interior instructed Herbert F. Robinson, the superintendent of Pueblo irrigation, to file the Indian claims with the territorial engineer's office. In 1910, Robinson appointed Howard E. Wanner to expand the Indian canals to protect their water rights. Robinson then drew up documents describing local water sources for different grants and submitted them the next year.[74]

More important for the adjudication of Indian water rights was the Winters Doctrine, which resulted from a Supreme Court decision in 1908. By this landmark case, Native Americans were entitled to all prior and paramount rights to waters that bordered, originated in, or passed through a reservation. Indians were allowed access to all waters necessary to develop fully their lands.[75] More importantly, this decision created a right to water that could not be appropriated under the laws of any state. The federal government held that these native water rights were superior to non-Indians' even if they were not exercised until after the reservation was created and regardless of their continuous use. While the Winters Doctrine did not cause an immediate reaction by non-Indians, its long-term effects would lead to contentious litigation in the future. Confusion arose, however, among Pueblo administrators about how the Winters Doctrine affected them. Prior to the decision, irrigation superintendent Robinson felt that while the Indians had prior rights to water they had been using, they would have to secure title to their rights for additional water in the same way whites did. He assumed that the Winters Doctrine would not apply to the Pueblos because of their unique status. However, Special Attorney Francis C. Wilson, writing after the case, claimed that the Indians had a vested right to water that passed through lands that they had possessed for many years.[76] The confusion over this issue would continue and present major problems for future Pueblo water rights cases.

More than any other event in the period under consideration, the change of the legal status of the Pueblos had a dynamic impact. After years of suffering loss of land and water privileges to encroachers, the Pueblos finally gained the same standing as other Indians. The legal ramifications of the *Sandoval* decision were detrimental, however, to all trespassers on Pueblo lands; a solution to the encroachment problem would wait until the next decade, but the attitude of the federal government concerning the restoration of Pueblo land had changed dramatically. Eventually, however, the idea of clearing up Pueblo titles would bring new actors into the drama who led the way to a revolution in the federal government's policy for all Native Americans.

CONTINUED PROBLEMS OVER PUEBLO LAND, WATER, AND AGRICULTURE

While the Pueblos received more technical advice, construction projects, advanced mechanical innovations, and a new legal status, they continued to deal with problems at the turn of the century that had haunted them since the beginning of American rule. The increase in the region's non-Indian population meant that trespassing and its associated vexations would remain. Education programs, necessary for the Indians' agricultural advancement, had as yet minimal effects.

The most persistent problem, however, was that of water rights. Numerous cases involving virtually every village surfaced during this period. Encroachment by neighboring Hispanics and resulting water conflicts were the most pressing concerns. This situation presented a problem at Nambe in 1899. The Indians complained that the mayordomo of the Acequia del Llano stopped a native from taking water and fined him. The Indians asked that the Pueblo attorney, George H. Howard, bring a countersuit for trespassing. They hoped to stop the mayordomo from interfering and get Hispanic neighbors to respect their water rights.

In the final decision on the matter, the court ruled that each group was entitled to half of the water, but the Hispanics remained reluctant to comply with the order. Agent Walpole reminded the mayordomo that further complaints would result in prosecution.[77]

In 1891, the ancient conflict between Acoma and Laguna appeared to be near settlement. The Laguna governor desired a ninety-nine-year lease to secure the pueblo's rights, but Acoma was not satisfied with that deal. Agent José Segura suggested that the Acomas submit another proposal in order to avoid a lawsuit, which neither of the pueblos could afford.[78] Although an agreement seemed possible in this case, water problems between the villages continued. In 1919, drought conditions in the region exacerbated an already bad situation concerning the Rio San Jose. Acoma claimed to have first rights to the water and therefore appropriated all of it. R. H. Hanna, special attorney for the Pueblos, felt that the time of initial use of the water by each village could not be determined, and therefore, the doctrine of prior appropriation did not apply in this case. In order to determine the extent of use by each pueblo, Hanna suggested that Robinson survey the area. His findings would be used to determine future water use by each village during periods of drought. Thus the problem could be solved internally. If an agreement could not be reached, Hanna felt that a hydrographic survey should be conducted by the state engineer. This action would result in the appointment of a water commissioner with legal authority to apportion the stream.[79]

Taos, long the trouble spot of water problems, continued to have disputes with its neighbors. A case involving Hispanic encroachers began in 1894, when the outsiders tried to take over a pueblo ditch. In 1903, Superintendent Crandall revealed that the citizens of Taos had previously entered into contract with the Indians concerning the use of water flowing through the reservation. After visiting the pueblo, Crandall and irrigation superintendent Harper felt that they had settled the matter. Judge McLaughlin, however, issued a decree that allowed the citizens to take all of the water of the Rio del Pueblo. Crandall rejected it.[80]

Water problems were seldom solved amicably and sometimes led to violence. About 1900, Cornelius Sandoval took over a piece of land near Zia through which the pueblo's acequia flowed. He told the Indians to make a new ditch of their own, but the place where he wanted them to construct the new acequia would have ruined most of their farmlands. Later

Agent Walpole reprimanded him for his action, but he persisted in his efforts. When José de Cruz went to irrigate his fields, Sandoval attacked him and tore his hair out.[81]

Since water was a valuable asset in New Mexico's economy, conflicts over its use existed throughout the area. Mexican encroachments caused problems at San Juan in 1903, at Santa Clara in 1897, and at Santo Domingo in 1894.[82] Not all of the water problems concerned outsiders. Intratribal conflicts usually concerned work on the acequias. These disputes appeared at San Ildefonso in 1906, at Santa Clara in 1900, and at San Juan in 1906.[83] A dispute between San Ildefonso and Santa Clara in 1919 was reminiscent of the old Acoma and Laguna conflicts. In this case Santa Clara would not allow San Ildefonso and Hispanic farmers within Santa Clara's grant to use irrigation water. Santa Clara also rejected a proposal that would allow a right-of-way for a new ditch. Farmers on the San Ildefonso grant hoped to adjudicate the matter through Attorney Hanna since they could not afford to go to court over the matter.[84]

Encroachment on Pueblo land by animals and settlers continued to plague the Indians. The agency instructed them on the proper course of action to be taken when foreign cattle intruded on their property. Indians were required to notify the owner immediately. If he could not be found, they were to contact the justice of the peace. If the owner did not pay the damages, the judge could sell the animals for payment.[85] In 1905, C. J. Crandall complained that a lack of fencing at Zia led to stock trespassing. He also stated that, at Cochiti, a Hispanic used fencing to enclose part of the Cochiti grant. Two years later at San Juan, Francisco Trujillo cut down the Indian fence and harvested a wheat crop. Agency workers were not always helpful in these matters. Crandall felt that Felipe Valdez, who worked for the Indian service, was in collusion with a group of Hispanic squatters.[86]

Like water rights cases, those involving squatters occurred on a number of Pueblo grants. Most involved Hispanics, as did the ones at Sandia in 1894, at Taos in 1903, and at Laguna in 1894.[87] At Zuni, however, most complaints involved Navajos who wandered onto the Zuni grant because of a lack of clear boundaries.[88] A peculiar case occurred at Santo Domingo in 1893 when the Indians complained about a French squatter.[89] In 1900 and again in 1910, Laguna went to court in cases concerning the Baca family.[90]

The agency continued to explore the possibility of modernizing Indian agriculture through educational programs at the Pueblo schools. A new facility opened at Zuni, and the children there received instruction in farming. However, the Pueblos constantly battled the agency on the question of boarding schools. Agent C. E. Nordstrom regretted that the Indians would not consent to sending their children to distant schools. Because the Pueblos supported themselves, they were more independent of the agency than other tribes, and Nordstrom believed that the government should force them to send their children to school.[91]

Throughout the first two decades of the twentieth century, the federal government continued to encourage the development of Indian agriculture with the hope that this action would solve their economic hardships. By 1913, the Indian bureau employed 249 farmers, 13 assistants, and 37 stockmen. The Department of Agriculture supplied the Indians with technical brochures, advice on insects, and seeds to use in time of droughts. The bureau also set up thirty-eight agricultural fairs, similar to those of Anglo farmers. Cato Sells, Indian commissioner from 1913 to 1921, strongly emphasized agriculture as a means to make Indians self-sufficient. Unfortunately, the number of acres cultivated by tribesmen increased slowly, and leasing to white farmers and ranchers became commonplace.[92]

World War I caused both Anglo and Indian farmers to increase production. Sells thought this was an opportune time for Indians to gain experience in a highly productive market. The number of tillable acres on reservations increased, but poor soil conditions and cultural adjustments tempered this upswing.[93] The Pueblos were well aware of the need to increase production. President Woodrow Wilson notified Indian farmers of the need to grow more food for the Allied Nations at war. Superintendent Philip T. Lonergan relayed this message to the local agents and encouraged them to get the largest possible production for the international crisis. He also asked the pueblos of Acoma and Laguna to put aside their differences concerning water rights. Lonergan felt that these were "petty differences" that should not stand in the way of the war effort.[94]

It would take more time to cure problems that had long vexed native farmers, but the agency had taken major steps around the turn of the century in order to improve Pueblo agricultural conditions. The Poore Report of 1890 demonstrated that considerable government intervention was necessary to alter ancient customs among the Indians. The agency responded to this challenge by introducing modern machines, scientific methods, and major construction projects. The Supreme Court and the Congress insured that federal aid would continue when they helped to reclassify the Pueblos. In spite of these efforts, progress was slow. Modernization entailed the introduction of revolutionary ideas and the destruction of old concepts. While the Pueblos readily accepted items that would assure them protection from encroachers or a constant water supply, they were still unwilling to do away with all of their ancient traditions. Agriculture remained the basis of the economy, but most of the Pueblos preferred to raise only enough food to suit their needs. Before any further progress could be made, the question of trespassers had to be solved. The effort to bring this about would eventually cause a complete transformation of federal Indian policy.

CHAPTER SIX

Prelude to Modernization (1920–1938)

THE PUEBLO LANDS ACT

Unquestionably, the Supreme Court's decision in the *Sandoval* case represented a major turning point in Pueblo history. With the Court's sanction, the Indians gained protection from the federal government concerning the alienation of their land. While this decision marked the beginning of the restitution of Pueblo lands that had been lost to encroachers during New Mexico's territorial period, it also caused strife for non-Indians living on native grants. These settlers had either purchased their lands in good faith or simply squatted on them. Those who had paid felt that because of the apparent emancipation of the Pueblos, the Indians had the right to sell their property. The *Sandoval* decision changed all of that. Now, about twelve thousand people found themselves unlawfully on Pueblo land.[1]

Slowly, the federal government began to react to the encroachment problem. In 1919, authorities in the Interior Department assigned Richard H. Hanna as special attorney to the Pueblos, and he filed suits to quiet titles. Court battles ensued between the Indians and settlers on their grants. A year later Congress intervened. The House Committee on Indian Affairs began hearings at Tesuque. José Archuleta of San Juan, representing seven of the pueblos, stated that they wanted the federal government to remove non-Indians from their grants. A. D. Renehan, attorney for the latter group, claimed that the Pueblos were not really Indians. Ignoring the ruling of the *Sandoval* decision, a case that he had lost in court, Renehan stated that the Pueblos had excellent livestock and "the most modern implements of farming." Influential New Mexican legislators also maintained that the legal status of the Pueblos was the same as that of the non-Indians. These people hoped that congressional legislation, not court decisions, would clear up land titles in their favor.[2]

In 1922, Senator Holm O. Bursum of New Mexico introduced a bill that was designed to remedy the situation. However, a close examination of the Bursum Bill reveals that it would have put the burden of proof of disproving the rights of the settlers on the federal government. This procedure went against the usual tradition regarding land claims, in which the claimant was forced to prove his case. Leo Crane, who had been connected with the Indian bureau for a long time, felt that this action was the combined scheme of Senator Bursum and Secretary of the Interior Albert Fall, a former U.S. senator from New Mexico.

Although the Bursum Bill received the backing of the Harding administration, opposition to it surfaced immediately. Two groups, including a New Mexican association concerned with Indian affairs and a collection of women's clubs, led the fight against this legislation.[3] Opposition centered on the fact that the bill would lawfully confirm a large majority of non-Indian claims. This constituted a major threat to Pueblo agriculture since most of these claims involved valuable irrigation lands. Furthermore, one section of the bill would have allowed Pueblo water and land rights to be adjudicated in unfriendly state courts.[4]

In 1921, the women's clubs hired John Collier, future commissioner of Indian affairs, as field representative and employed Francis C. Wilson of

Santa Fe for legal services. Encouraged by the action taken by non-Indians, the Pueblos also organized the following year to defeat the bill. The All Indian Pueblo Council (AIPC) met at Santo Domingo in a concerted effort to thwart the act. Because of these efforts, the measures were defeated and new legislation was proposed. When it was finally passed, the Pueblo Lands Act of 1924 provided for restoration of native lands which were lost to encroachers or financial compensation for areas that could not be legally restored.[5]

The aforementioned act also created the Pueblo Lands Board in order to clear up the controversy created by the *Sandoval* decision. The board, composed of the secretary of the interior, the attorney general, and a presidential appointee, was to investigate land values and water rights that resulted from non-Indian claims. Located in Santa Fe, the board could subpoena witnesses and reported on each pueblo separately. The act further allowed the Indians to receive compensation money for any loss of land, water rights, or improvements on lost areas where title was extinguished from negligence by the U.S. government. Although financial compensation was not immediately appropriated, it eventually had a profound effect on the native economy. While most of the Pueblos relied on farming and ranching to establish their subsistence economy, others traded their excess products. Because of the Pueblo Lands Act, cash eventually became an important asset for the Indians.[6]

The act prevented non-Indians from acquiring further Pueblo land. In addition, it provided that:

> all non-Indian claimants to Pueblo land had to prove exclusive and continuous occupancy with a title to the land and taxes paid since 1902 or exclusive and continuous occupancy, but without title since 1889.[7]

The Pueblo Lands Board hoped to clear up land titles that had been disputed for decades, and this included Laguna's battle with the Baca family, which dated back to the Spanish period. The board surveyed that pueblo in 1929 and ruled in favor of the Indians. Following a compromise decision, each side received certain portions of land and compensation money. A decade later, Laguna purchased some of the Baca land.[8] When the Pueblo Lands Board completed its work in 1938, controversy over claims to land within all Pueblo grants was removed for the first time since the coming of the Spaniards.

Congressional action that created the Pueblo Lands Board received further support from the courts. The *Sandoval* case of 1913 had established that the Pueblos were really Indians, but what about the period before that time? In 1851, Congress had extended the benefits of the Indian Intercourse Act of 1834 to the Indians of New Mexico; thus their lands could not be sold. Neither the *Sandoval* decision nor the state's enabling act had clarified this situation for the Pueblos. In 1926, the Supreme Court ruled, in the case of *United States v. Candelaria*, that the Pueblos were, in fact, protected by the 1851 act. Therefore, it was impossible for them to sell their land after that date.[9]

With the passage of the Pueblo Lands Act and the *Candelaria* decision, it would seem that these native New Mexicans would no longer lose land to encroachers. Before either of these actions were taken, however, the Santa Fe Northwestern Railroad sought a right-of-way through the pueblos of Santa Ana, Zia, and Jemez. The latter village, in particular, protested because the route would pass through some of their best farmland. In 1923, their governor contacted Secretary of Interior Fall, but Indian Commissioner Charles H. Burke telegrammed Superintendent H. P. Marble that the railroad builders had acquired the right to proceed with construction. The federal government had given them permission based on a congressional act of 1899 that allowed railroad companies to acquire passage through Indian lands. Jemez filed suit in accordance with the Pueblo Lands Act, but in 1926 Congress passed a bill providing for the condemnation of Pueblo lands. Two years later the railroad company received permission from the Interior Department to proceed with construction. A major flood in 1941 destroyed the railroad line, but the condemnation act continued to haunt the Pueblos and was not repealed until 1976.[10]

Although the Pueblos received money and land as a result of encroachment problems, the confirmation of these awards came slowly. Sometimes this situation led to confusion between natives and non-Indians concerning the use of areas under question. In 1928, Attorney E. P. Davies of Santa Fe warned C. J. Crandall, superintendent of the Northern Pueblos Agency (NPA), that upstream users were taking water that belonged to the pueblo of San Ildefonso. Crandall stated that he knew of no way of solving this problem because he could not regulate the appropriation of water in the vicinity. During the next near, new superintendent Thomas F. McCormick contacted Supervising Engineer H. F. Robinson regarding similar problems at Nambe and San Ildefonso. In both cases, Hispanic farmers diverted water by constructing dams and canals. The ditch at Nambe went through property that the Pueblo Lands Board awarded to the Indians, but this area had not been confirmed by the courts. As a result, both sides laid claim to the tract although neither had title to the area.[11]

A major question connected with compensation concerned how the money would be spent. In 1928, the Northern Pueblos Agency considered repair work at Taos Pueblo. The Indians complained about the need to remodel the headgates of their irrigation system, but the agency lacked sufficient funds to cope with this expensive project. Engineer Robinson suggested that the work be financed with reimbursable money that would be deducted from financial awards from the Pueblo Lands Board, but this compensation, which totaled almost fifty thousand dollars, had still not been appropriated by Congress.[12]

One of the communities hardest hit by encroachment and loss of water was Tesuque Pueblo. Non-Indians had confiscated most of the water designed to flow through their property. In 1928, Governor H. J. Hagerman reported that new landowners in the vicinity had increased their farming operations by transferring land titles. This change of conditions resulted in a decrease of the Pueblo water supply. Acting commissioner C. F. Hauke suggested that Walter C. Cochrane, special attorney for the Pueblos, arrange a meeting of leading authorities so that a protective program for the Pueblos could be arranged. Hauke realized that the Pueblo Lands Board outlined specific procedures to deal with the water and suggested that Cochrane confer with board members to form a plan of action.[13] Dr. Kirk Bryan, part-time employee of the Geological Survey and a Harvard professor, conducted an examination in order to determine how the Indians could obtain more water. Hagerman remained interested in the Tesuque situation because the Pueblo Lands Board awarded some twenty-nine thousand dollars to these Indians based on the fact that they had lost a considerable amount of water. The board also recommended that part or all of the compensation money be used for recovering their water.[14]

Unfortunately, Bryan's geological study revealed the unfeasibility of additional water development in the area. Nevertheless, Superintendent

McCormick continued to make plans for the construction of more canal work. Designs for Tesuque included drainage of marshy grounds, installation of a gate to hold water during nonirrigation season, and a cement-lined ditch to prevent seepage. Engineer Robinson pointed out that while these projects could be beneficial, the deficient-water problem was worsened by the fact that Indians did not use their water economically. For example, instead of using the new cement ditch that the agency provided, they continued to employ their old canal, which was prone to seepage. Commissioner Charles Burke felt that expenditures of large sums of money from the Pueblo Lands Board should be contingent upon the beneficial use of water by the Indians.[15]

While the Pueblo Lands Act allowed for compensation for lost land and water, the Pueblo Lands Board took a conservative approach concerning the amount allowed to Indians. Disregarding its own appraisals, the board set a maximum figure at $35 per acre (about one-third of the actual market value) for Pueblo farmlands. The Pueblo Lands Board, reflecting its non-Indian bias, ignored a ceiling for non-Pueblos and complied with the recommendations of its appraisers. Unsatisfied, the Pueblos began to pressure Congress for equal economic footing, and this action led to the Pueblo Compensation Act of 1933. As a result the Pueblos received additional funds amounting to $761,954 that were applied to the purchase of land and water rights and irrigation-works construction.[16]

By the mid-1930s the agency began to clarify its position on the use of compensation money. It developed programs to help the Pueblos and spent appropriations when the finances finally became a reality. The agency allocated a great deal of this money for the acquisition of additional land for the Indians. In addition, it used compensation funds to acquire farm machinery.[17] It also financed irrigation projects in this manner. In 1935, the agency submitted a list of these undertakings to the Indian bureau, and they totaled almost two million dollars. Unfortunately, the bureau's irrigation service possessed only one-tenth of this amount. This lack of funding meant that a selection process based on need diminished a large number of the projects. Commissioner John Collier felt that the Pueblos could use their compensation money for irrigation plans. Although some of the Indians never received any of these funds, he believed that those who acquired them would be wiser to use the money on irrigation projects than on the acquisition of new land.[18]

THE CONSERVANCY DISTRICT

Another step taken by the authorities in Washington to assist the Pueblo farm program concerned the work of the Middle Rio Grande Conservancy District (MRGCD). Since the Rio Grande is one of the main rivers in the West, its waters were important to settlements from southern Colorado to Mexico. While the population along its banks remained small, conflicts over its use were minimized except during times of extreme drought. Around 1700, there were approximately 73,000 acres under cultivation, and this figure increased to around 125,000 acres in 1880 as immigrants came into the region. While settlement increased along its banks in Colorado's San Luis Valley, southern New Mexico, Texas, and Mexico, grazing lands were misused resulting in a buildup of silt in the river. This, in turn, led to a rising water table that waterlogged adjacent lands. Because of this situation, the cultivated acreage decreased to 40,000 acres by 1925.

The construction of reservoirs along the Rio Grande and its tributaries between 1910 and 1938 partially alleviated some of the loss of water to upstream users. These major projects included the Elephant Butte and Caballo dams on the Rio Grande and the El Vado dam on the Rio Chama. These projects revealed that the federal government had become more aware of the importance of water development in this region and paved the way for increased federal funds for similar action concerning the Pueblos. For example, in 1925 completion of the Isleta drainage system cost forty-five thousand dollars. As a result, the pueblo reclaimed thirty-five hundred acres for almost fifteen dollars per acre.[19]

The Irrigation Division of the Indian bureau became an important agency for water projects throughout the country. This agency controlled the initiation, construction, operation, maintenance, and funding of irrigation and drainage projects on Indian reservations. The federal government created five irrigation districts, and each had a supervising engineer to conduct work under its jurisdiction. District number five, whose headquarters were located in Albuquerque, regulated affairs in New Mexico, northern Arizona, and Colorado. This branch office was responsible for water development in all but six of the Pueblo grants. These latter pueblos, including Santa Ana, Cochiti, Santo Domingo, San Felipe, Isleta, and Sandia, fell under the auspices of the Middle Rio Grande Conservancy District. Formed in 1925 as a political subdivision of the state of New Mexico, this public corporation was designed to provide conservation, drainage, irrigation, and flood control for over 130,000 acres of swampy land along the Rio Grande. The land in question extended 150 miles and included various Indian grants that made up approximately 17 percent of the area.[20]

In 1927, Congress passed legislation to appropriate federal funds for reconnaissance work for Pueblos in the district. After the survey was completed, the federal government authorized the secretary of the interior to enter into an agreement with the district to extend its works to the six Pueblo tribes under its jurisdiction. In 1931, a representative from the Interior Department traveled to Albuquerque to negotiate with the director of the conservancy district concerning Pueblo projects.[21] The district emphasized the modernization of existing irrigable Indian land that consisted of 8,346 acres, but another concern involved developing additional property that was capable of receiving irrigation water. In the first case, the district proposed to update canal systems considered to be antiquated and inefficient. This work was exempted from future operation, maintenance, and improvement changes. In the case of the raw land, Indians were required to repay the government for work in this area, which included approximately 15,000 acres. Eighty percent of this region could be farmed by the Pueblos without applying their proceeds to the cost. The rest was to be leased or farmed as a means of reimbursing the government.[22]

Agency personnel were determined to develop raw land because the standard philosophy held that by increasing the amount of tillable acreage for the Indians, economic growth would follow. In order to bring this land under cultivation, it first had to be cleared of tree growth and then leveled and bordered for proper irrigation. According to the official plan of the district, engineers would usually carry water to within one-half mile of new farms. This meant that the Pueblos had to construct their own primary and secondary laterals. In the case of existing farms, engineers would improve current canals by enlarging them and changing the course of laterals.

Because of the high water level along the Rio Grande, drainage was an important adjunct to the district's irrigation program. By 1930, progress in this area led to the development of a considerable amount of irrigable land. While non-Indians took advantage of this situation by rapidly preparing new land, Pueblo work lagged behind. Supervising Engineer H. C. Neuffer felt that the Pueblos' land preparation should be carried on at the same pace as the district's construction work so that the new areas could be farmed immediately. He also encouraged the development of small tracts sandwiched between existing farms since they could be irrigated by older ditches.

Previous work on land preparation at Isleta demonstrated that the Indians were ill equipped to handle clearing and leveling projects. Neuffer suggested that the government should complete these undertakings with machinery. Similar work had been carried on at the Pima reservation and proved quite successful. He feared that if the land was not properly leveled, it would require 50 percent more water for irrigation. Therefore, he requested that the federal government purchase a tractor, rotary scraper, and leveler so that the land could be planted as soon as possible.[23]

Because of the vast work done by the conservancy district, agency officials expressed concern about the loss of Pueblo water rights to newly developed non-Indian farms. According to the district's original agreement, all present water rights connected to existing Indian farms were prior and paramount to the rights of the district or any property holder within it. However, concerning the remaining fifteen thousand acres of newly reclaimed land, the district recognized the Pueblo water rights as being on an equal basis with other conservancy members. Old and new rights were not subject to loss by nonuse or abandonment as long as the Pueblos held their title. The loss of supreme status for water rights in the newly developed areas reflected some confusion concerning the application of the Winters Doctrine to the Pueblos in the district.[24]

Work of the conservancy district progressed rapidly in the early 1930s. In 1931, it appropriated $325,000 for its construction projects. By the next year, much of this work had been completed. Initial developments emphasized the immediate need for land drainage and river protection to halt land loss due to floods. The district completed only half of the actual irrigation projects by 1933.[25] Eventually, the district increased the amount of irrigable acreage under its jurisdiction by 118,000 acres. The cost of this project was $9.4 million, of which more than half was supplied by the Reconstruction Finance Corporation, a New Deal agency designed to lend money to various companies, including agricultural agencies. The federal government contributed more than $1.3 million for the work on Indian lands. A breakdown of how these undertakings affected the individual pueblos within the district is contained in the following table.[26]

TABLE FOUR

MIDDLE RIO GRANDE PROJECTS FOR THE PUEBLO INDIANS

	POPULATION	PRESENTLY IRRIGATED ACRES	ULTIMATELY IRRIGABLE ACRES
Isleta	989	3,370	6,478
Sandia	74	738	3,513
Santa Ana	226	667	1,005
San Felipe	475	1,168	4,931
Santo Domingo	1,000	1,700	4,547
Cochiti	300	703	1,490
TOTAL	*3,064*	*8,346*	*21,964*

According to these figures, the pueblos in the district would almost be able to triple their quantity of farmlands. However, not all of the tribes were interested in or capable of taking advantage of the situation. Due to a lack of modern machinery and techniques and the unwillingness of traditional pueblos like Santo Domingo to change old practices and expectations, farming potentials in the region remained unfulfilled. Also the conservancy district benefited only six of the pueblos while the thirteen villages outside of its jurisdiction received nothing. The district was not really designed to aid Indian agriculture. Those who did benefit were simply fortunate enough to be located in this region.

REPORTS ON PUEBLO AGRICULTURE AND IRRIGATION

While the work of the conservancy district and the Pueblo Lands Board allowed for a gradual acceptance of federal programs by the more progressive tribes, it also helped to increase interest in Indian affairs in New Mexico and throughout the country. This heightened awareness of Native

Americans' plight led to studies concerning Indian affairs, which were conducted in the late 1920s. For example, in 1928 the Brookings Institution published a document following its exhaustive research that demonstrated the shocking economic and social conditions of Indian people. Commonly dubbed the Meriam Report, this publication did not deal extensively with irrigation and agriculture. However, it emphasized the need to modernize Indian farming methods in order to make the high cost of irrigation projects worthwhile. In addition the Meriam study called for a change in qualifications for Indian farmers, the establishment of a permanent farming education program, and the development of a one-crop farm system in order to bring about a self-sufficient economy for Native Americans.[27]

The Preston-Engle Report of the same year provided a more thorough investigation of irrigation systems on Indian reservations. Previously, Secretary of the Interior Hubert Work instructed engineers from the Bureau of Reclamation and the Bureau of Indian Affairs to examine Indian irrigation projects. They recommended annual crops surveys that included acreage utilization and data on crop yields. Their report suggested that any irrigable land not being utilized by Indians be leased and that complete information on all facets of new projects be furnished before their implementation. The engineers felt that the Irrigation Service should be reorganized and have closer coordination with the Reclamation Bureau concerning project designs, legal work, and feasibility investigations. Finally, the report recommended that the federal government should employ full-time specialists to educate and interest the Indians in farming techniques.[28] This latter concern would represent a major thrust of the New Deal farm program for the Pueblos.

Most of the information in the Preston-Engle Report dealt with irrigation projects on underdeveloped reservations. Specific information on Pueblo irrigation was confined to charts concerning the expenditures of the Irrigation Service, data on reimbursable appropriations, and individual Pueblo expenses for drainage and river protection. According to the statistics of the Preston-Engle Report, the federal government invested over one million dollars for preliminary surveys, engineer charges, and construction, operation, and maintenance charges on Pueblo irrigation projects by the fiscal year 1928. Over 60 percent of these funds were allocated for the Zuni pueblo, which required reservoirs to hold an adequate water supply in its arid region. Of the remaining sum, 20 percent went to the pueblos of the conservancy district. Other pueblos that received large allotments for irrigation included San Juan, San Ildefonso, Santa Clara, Zia, and Laguna.[29]

The increasing awareness of the economic and social hardships on Indian reservations inspired further investigation of these conditions. In 1928, Senator William H. King of Utah introduced a resolution calling for a congressional survey of the Indian situation. When the bill passed, Congress appropriated money to conduct hearings on Indian reservations throughout the country. The first of these was held during the same year, and they continued until the summer of 1943.[30] Congressional hearings for the Pueblos took place in 1931 at Albuquerque, Santa Fe, Taos, Santo Domingo, and Zuni. Ten senators conducted the discussions, including Sam G. Bratton of New Mexico, Robert M. LaFollette Jr. of Wisconsin, and Burton K. Wheeler of Montana. The committee interviewed agency personnel and superintendents, engineers from the Irrigation Service, experts in Indian affairs, and representatives of the various pueblos. Subjects discussed included the work of the Pueblo Lands Board and the conservancy district. Senators were supplied with the acts that created these agencies so that they could be familiar with the responsibilities of each.

One of the major topics of the Pueblo hearing concerned the development of water on the various grants. H. C. Neuffer, the supervising engineer of the Fifth Irrigation District, estimated that it would cost approximately two hundred thousand dollars to complete water development for the New Mexico pueblos. This work included drilling wells for windmills, spring development, and the construction of surface reservoirs. He favored springs over wells because of the higher operation and maintenance costs for the latter. Neuffer admitted that due to the arid conditions, the reservoirs could not always carry sufficient water to last the entire year. He supplied the committee with a list of sixty-eight well and windmill projects on sixteen grants. He also included an inventory of irrigation developments throughout the Pueblo region. Neuffer estimated that when the present projects were completed, the number of acres supplied with water would increase from 36,517 to 77,604. Most of the work in this area involved the diversion of water from the Zuni and Nutria rivers in the west and the Rio Grande and its tributaries in the east.[31]

As representatives were called before the committee, they explained water developments on their grant. For example, Tesuque constructed a diversion dam in 1923, but it required twenty-five hundred dollars worth of repair work. Neuffer felt that it was difficult to get a large quantity of water there because of long droughts in the summertime. He recommended canal maintenance but rejected the idea of a reservoir for the pueblo. He also planned to install a sluiceway in Tesuque's diversion dam so that it would not fill up with sand.[32] Richard Tafaya, an officer at Nambe, told the committee of a new ditch under construction at his pueblo. The natives spent eighty-five thousand dollars of their compensation money on it, and Tafaya felt that it would work well. However, he was more concerned about the protection of water rights in the canal because of past encroachments. Earlier, Neuffer pointed out that the Irrigation Division planned to develop an area known as the Tonoya tract, which had been given to the Indians by the Pueblo Lands Board. However, he predicted that they would have to proceed with litigation to determine water rights in this area.[33]

The ancient conflict between Acoma and Laguna surfaced in the committee hearings. Recently, the federal government had constructed a new dam at Seama, and the Irrigation Service was forced to put a ditch through the Acoma land in order to supply water for Laguna. Although the Acomas were compensated financially for this action, their bitterness about it remained. The bureau proposed a new dam for Acoma, but authorities felt that it would silt rapidly and they wanted an alternate solution.[34]

At Cochiti, the Indians desired to farm an area on a mesa top, but the location was at least twenty-five feet higher than the main irrigation ditch. Neuffer suggested that a pump could be installed to deliver the water to a higher location, but he realized that this would be an expensive undertaking. In addition, the pueblo lacked electricity to make the project feasible. Picuris also wanted to farm more land, but this expansion would require the construction of a reservoir. The Indians felt that this project would allow them to reclaim land presently in sagebrush.[35]

Federal authorities encouraged water development projects in both arid and wet regions in New Mexico. At Taos, the Pueblo Lands Board granted the people an area previously cultivated by non-Indians. They farmed about eight hundred acres of this tract, but Neuffer hoped to triple this amount by using compensation funds. The Zunis wanted to construct another dam above the reservoir at Black Rock in order to have enough water to increase their farmlands. The Indians possessed fertile land at

Ojo Caliente and worked on a dam in this area. However, it failed to hold enough water for them to irrigate all of their crops.[36]

Committee members also discussed the acquisition of modern equipment at the hearings. Ten of the pueblos received a yearly appropriation of seventy-five hundred dollars for farm machinery, but this fund was reimbursable. Senator Wheeler also stated that the committee was informed that the use of this equipment went against the religious beliefs of some of the Indians. He felt that the more progressive and industrious ones should be encouraged to use modern implements.[37] In order to operate this machinery, the Pueblos required adequate instruction by a competent farmer. Albert Paytimo of Acoma stated that the government employee lacked agricultural training and therefore did not know about implements.[38]

All of these government reports were crucial to the agricultural modernization of the Pueblos. While the Meriam and Preston-Engle accounts said little about activities on Pueblo grants, they reflected an interest in improving the general Indian economy. The congressional survey was more specific about Pueblo needs and current Indian projects. For example, the end of the report contained a pueblo-by-pueblo account of land confirmations and farming activity. In conclusion, these reports would pave the way for further government action during the New Deal period.

WATER DEVELOPMENT

As witnessed by the reports on Pueblo agriculture, the development of the Indian water supply became a major concern of the federal government. This activity had always been a part of the agency's farm program, but prior to the *Sandoval* decision, a lack of funds restricted these projects. Even before the Court's ruling, the Black Rock reservoir at Zuni and minor projects along the Rio Grande reflected the agency's tendency to treat the Pueblos like other Indians. However, the bulk of the water development projects occurred after the creation of the Pueblo Lands Board, which helped to increase federal awareness in this field. Agency activities in these areas involved a wide assortment of projects designed to increase the supply of water and the amount of tillable land. For example, in 1920 Professor Fabian Garcia, a recognized expert on southwestern horticulture, lectured Mexican farmers on the advantages of draining waterlogged areas. Superintendent Leo Crane desired to extend this information to the Pueblos, and he called a meeting at Isleta to allow Garcia a chance to speak to the Indians.[39]

Construction work on dams, flumes, and irrigation canals occurred throughout the Pueblo region. In 1923, work commenced on a dam project at Tesuque, and laborers completed the project in time for the next year's planting. Superintendent Crandall stated that it would hold one and a half times the amount of water needed by the village. He also predicted that it would end their water problems forever. Previously, H. F. Robinson received permission to proceed with a new dam near Acomita. On this project, the Indians furnished both logs and labor. They were, however, paid for work on a local flume project. The Indians at Jemez furnished the labor on their flume works while the agency supplied materials.[40]

The Irrigation Service did not always approve water projects. Cochiti had suffered ditch damage because of flood conditions in 1922, and Supervising Engineer H. F. Robinson inspected the situation. He felt that a longer ditch should be dug in order to get a better heading. While he considered the shorter canal impractical, Robinson believed the longer one too costly to build and too difficult to maintain. Since the Indians had lost land to encroachers and floods in the area, they could only farm about 350 acres under the ditch. Therefore, Robinson could not justify a large expenditure for such a small area.[41]

Often Pueblos and non-Indians used the same irrigation ditch—a condition that led to conflicts over repair work. This situation existed at Tesuque where the native farmers shared a mother ditch with E. D. Newman. Having had previous difficulty with the pueblo, Newman refused to allow the Indians to enter his land for the purpose of repairing the canal. Superintendent A. W. Leech finally made an agreement by which Newman furnished certain materials, paid for his share of the work, and received a temporary right to the water.[42]

Non-Indians also sought to develop water, but this work often conflicted with Pueblo interests. Kenneth Heron, civil and irrigation engineer from Chama, proposed the construction of an irrigation canal through the San Juan grant. He felt that the ditch could eliminate the need for numerous small diversion dams in the area that required yearly maintenance. Heron also claimed that an additional six hundred acres of Indian land would become irrigable. C. J. Crandall inspected the area and concluded that the Indians would not benefit from the project because Hispanics owned or claimed almost all of the land under consideration. Furthermore, the superintendent felt that the natives would lose water rights in the process. By giving up the right-of-way for allegedly increasing the value of their land, the San Juans would be forced to buy water rights from the canal owners. Therefore, Crandall rejected the proposal.[43]

The concept that the agency employed in the 1890s to aid only those deserving Pueblos who demonstrated the desire to implement programs still existed prior to the creation of the Pueblo Lands Board. In 1923, Assistant Commissioner E. B. Meritt stated that the San Ildefonso Indians were not agriculturally industrious, and he felt that they probably would not attempt to increase their tillable acreage even if the agency provided them with additional water development.[44] Agency personnel, however, disagreed with this attitude and continued to aid Pueblo water projects toward the end of the decade. While Commissioner Meritt's attitude concerning San Ildefonso might have been applicable in the less progressive villages, it did not apply to Santa Clara. In 1930, Desiderio Naranjo, a native of the pueblo, complained about the lack of irrigation water and the adverse effect it had on additional land development. He felt sure that if sufficient water could be made available, then the Indians would immediately increase their tillable acreage accordingly.[45]

The unwillingness of the Pueblos to adapt to modern times continued to be a major concern of agency personnel. In his 1933 annual report of the Southern Pueblos Agency (SPA), agricultural extension agent Milton A. Johnson commented on the traditional attitudes of various tribes and how these affected farming progress. He felt that the ancient customs and superstitions of the Pueblos made it difficult for extension personnel to gain the confidence of the Indians and that the situation hindered accomplishments. Johnson recognized, however, a vast degree of difference in the Pueblo region concerning the adoption of modern practices. He labeled Laguna and Isleta as progressives, Zia, Santa Ana, Acoma, Jemez, and Sandia as semiprogressives, and San Felipe and Santo Domingo as nonprogressives.[46]

Still, water projects commenced in numerous villages toward the end of the decade of the twenties. San Juan, in particular, required a great deal of work. In 1928, H. F. Robinson reported that the main ditch San Juan shared with neighboring Hispanics required work at the heading since the river was twelve feet lower than the intake of the ditch. Previously, Indians and Hispanics had worked jointly on a temporary dam that continually washed out during flood years. Robinson considered the construction of a permanent dam in order to avoid a lengthy and expensive extension of the main canal. However, this project required the building of four flumes, which greatly increased the cost. Robinson estimated that the entire project would run over one hundred thousand dollars and benefit some three or four hundred acres, of which the Indians held only about half. Since the Hispanics would not finance their share of the work, Robinson felt the project was unfeasible.[47] Unfortunately, the San Juan construction, like the previous work at Cochiti, could not be justified because the cost was too high for the amount of tillable land it would yield.

Agency personnel viewed other projects more favorably. For example, Superintendent T. F. McCormick requested the expenditure of twenty-five hundred dollars to complete water development at Tesuque. Heavy rains at Cochiti in 1929 damaged some farmland and endangered the main irrigation ditch. Assistant Engineer Robert H. Rupkey suggested the building of a retard made of cedar brush and fastened to an anchored cable in order to stop more land from being washed away. Three years later, Laguna's farm agent Sidney R. Miller reported that the agency began floodgate construction and irrigation work at Mesita. The Irrigation Service also completed its development work at Santa Clara in 1930, which allowed sufficient water to irrigate 102 new acres. The land was in a raw state and unfenced, but the Indians had already divided it up and planned to plant it the following spring. Since they lacked implements to break up the new land, Superintendent J. W. Elliott suggested that funds be authorized for this purpose.[48]

One of the more unique suggestions for water development concerned the pueblo of San Ildefonso. In 1929, a group of non-Indians planned to build a wheel with a series of buckets around the rim that would continuously catch water as the current moved the wheel. The water would then be dumped into an old irrigation ditch formerly used by the Indians. In order to make this water available to the group, the agency would have to construct a fifteen-hundred-foot flume and complete repairs on the canal. Since the cost was so high and the possibility of getting a necessary amount of water was so small, Robinson decided the plan was impractical. He did, however, consider other projects, including the renovation of an irrigation ditch and the construction of three flumes to make usage of the canal more feasible.[49]

In 1930, Robinson submitted a report on water development projects for the New Mexico pueblos to the Indian bureau. Funds for these activities from federal appropriations totaled $16,500, but additional finances from the Pueblo Lands Board and unexpended irrigation money from the previous fiscal year were also available. The projected work involved every village except Zuni, Pojoaque, and Picuris, which did not require new development at that time. Although the conservancy district took care of projects for those tribes under its jurisdiction, Robinson considered river protection work at Cochiti and ditch cleaning at Isleta.

Projects for the Northern Pueblos Agency included the construction of a diversion dam on the Rio Lucero, a concrete ditch, and a flume across the river near the northern boundary of Taos. Robinson felt that the river protection at San Juan should be done in cooperation with the state. He also outlined a drainage project that would yield almost three hundred acres and a high line canal to bring water to an area previously considered wasteland. At Santa Clara, the supervising engineer planned river protection work, including replacement of a culvert that carried water from a nearby creek. He also planned to develop further underground water by implementing an infiltration pipe at San Ildefonso. Robinson hoped to extend the high line ditch at Nambe and develop additional water through drainage at Tesuque.

For the pueblos in the Southern Pueblos Agency, Robinson proposed similar work. At Zia, the natives needed a headgate for their main canal, and the Jemez Indians planned to build a diversion dam. Robinson scheduled numerous projects at Laguna, including canal building, dam construction, and flume repair. Robinson also suggested flume work on a canal extension at Acoma. In addition, this pueblo required repair work on small diversion dams.[50]

Water development projects increased in the thirties. For example, in 1934 Nambe required flume construction, and Santa Clara appropriated funds for ditch lining and well digging. The agency designed the ditch project to stop the loss of water in the sand, which amounted to 60 percent of the stream flow. H. C. Neuffer, supervising engineer for the Fifth Irrigation District, stressed that Indians should be given preference in

Figure 6:
Cochiti Diversion Dam, Rio Grande, New Mexico, ca. 1932. Courtesy of the Museum of New Mexico, neg. no. 59142.

hiring for these projects, including white men with Indian wives. In 1935, he also requested funds to construct stations where stream measurements could be taken so that the Pueblos could validate their claims to water in the area.[51]

Dam construction continued to be important work for the Pueblos in the 1930s, but the results were not always satisfactory. A concrete dam north of Cochiti increased the flow in irrigation canals, but the guaranteed water supply led to a decrease in ceremonialism designed to assist natural weather conditions.[52] Continued logging in the mountains around Zuni ruined natural drain-offs. From 1929 to 1937, the federal government constructed seven additional dams, but all of them failed at one time or another. Combined with overgrazing, this situation led to erosion, siltation in reservoirs, and a loss of agricultural land.[53]

Well digging increased among the Pueblos in the late thirties. Sometimes this work consisted of improving old structures, such as the ones at San Juan in 1935. Other projects, like the five wells dug on the

Isleta grant during the next year, were completely new. Outsiders usually performed the labor, but in 1936 a reported case of collusion among well drillers, the drilling inspector, the project manager, and the purchasing officer changed this procedure. The agency decided to purchase its own equipment so it could conduct projects with its own employees and Indian labor.[54]

In 1937, Engineer Robert H. Rupkey reported on unused, cultivated land. A U.S. Supreme Court case by the state of Texas against New Mexico and the Middle Rio Grande Conservancy District necessitated this action. Since this litigation concerned the use of irrigation water from the Rio Grande, Rupkey felt the agency should record its past and present use by Pueblos. Tribal councils sent out old men to accompany representatives of the Indian service in order to determine where the Indians had previously farmed. Rupkey estimated that the Indians of Isleta no longer tilled over fifteen hundred acres because the land had become waterlogged. Floods at Santo Domingo forced the natives to abandon approximately one thousand acres of farmland. Indian population at Sandia had decreased to the point that they farmed between three and five hundred fewer acres.[55]

ATTEMPTS AT MODERNIZATION

While water development programs helped the Pueblos to obtain necessary items to conduct their agricultural program, these did not assure that they would actually increase their farmlands. Two ingredients required for increased production were modern equipment and advanced agricultural methods. These practices would allow the Indians to increase production per acre farmed and to decrease the number of people involved in agriculture. Unfortunately, as in the past, modern methods often collided with native traditions.

As in the case of improvements in water development, none of the Pueblos willingly took advantage of modern farm implements. In 1921, Farmer Walter L. Bolander told Superintendent Horace J. Johnson that it did not seem worthwhile to ask for planting and harvesting machinery for the Picuris pueblo because of their low crop production. Bolander blamed the Indians' "primitive methods of planting and harvesting" for their low output. However, he felt that it was impossible to teach them

Figure 7:
Wagon, Acoma Pueblo, 1940. Photo by Ferenz Fedor. Courtesy of the Museum of New Mexico, neg. no. 100471.

modern methods without adequate machinery. Two years later, Superintendent A. W. Leech noted that a binder at Picuris had never been used by the Indians, and he doubted that they would ever employ it. He thought that another village like Taos might be able to use the binder and that Picuris could be aided by means other than machinery.[56]

Throughout the 1920s and 1930s, agency personnel ordered a variety of equipment to be distributed throughout the region. Included in this list were hay bailers, scrapers, plows, hay rakes, barbed wire, and mowing machines. The Indian bureau also approved the purchase of a cultivator, harrows, and a tractor. In 1933, Superintendent C. E. Faris ordered two mowers for Taos and also wanted to purchase two bunchers to be used with them. He considered these latter items to be more practical in harvesting grain on small plots because they were smaller and more pliable than heavy machinery.[57]

While the more conservative Pueblos rejected or ignored the mechanical advancements, progressive farmers adopted them into their agricultural program. In 1920, Santa Clara and San Juan shared the government threshing machine. While the former produced 250 bushels of grain, the latter had more than ten times this amount. Picuris acquired a mowing machine in 1926, and Tesuque purchased a hay press on the reimbursable plan two years later.[58] On the other hand, a report on Zia showed that in 1924 the pueblo possessed only a plow and spades. The Indian bureau authorized the purchase of a thresher for Jemez in 1928, but by the middle of the next decade, conservatives continued to oppose the use of the cultivator and the harrow. The Acomas employed few implements. These Indians used mowing machines and a rake, but they did not possess cultivators, corn grinders, or shellers.[59]

Other villages were in a transition stage concerning the use of modern machinery. While the principal implements at Santa Ana included the hoe and the shovel, these Indians also employed a cultivator, mowing machine, plow, hay rake, tractor, and threshing machine. Cochitis demonstrated an early appreciation for machinery when they purchased a thresher in 1916. They used it until 1934 when the tribe bought a second machine. Traditional attitudes hindered progress at the pueblo of Santo Domingo. A report in 1938 revealed that the farmers felt machines wasted the wheat. Another decade would pass before they would purchase a thresher. San Ildefonso also purchased one of these machines in the mid-1930s.[60] Toward the end of the 1930s, Pueblo farmers adopted more sophisticated agricultural implements. In 1936, the Irrigation Division granted Laguna the use of a power shovel to help in dam construction. The next year Nambe purchased a hay press from the J. Korber Company of Albuquerque, and the agency acquired a tractor and thresher for San Felipe. Zuni received a power-lift rake from the agency in the same year.[61]

In order to take advantage of this equipment, the agency tried to instruct the Indians in modern techniques. In 1933, the Southern Pueblos Agency stationed a farm extension agent at Laguna, and another one handled the pueblos of Sandia, Santa Ana, Zia, Jemez, San Felipe, Santo Domingo, and Cochiti. The latter person covered an area seventy-five miles long with a total population of 2,939. Conflicting opinions existed between the Indians and the farmers concerning the benefits of the government employees. While the natives of Santa Ana complained of the farmer's laziness or incompetence, the Indians' persistence in following their ancient ways

Figure 8:
Winnowing wheat, Picuris Pueblo, ca. 1935. Photo by T. Harmon Parkhurst. Courtesy of the Museum of New Mexico, neg. no. 22697.

upset the agency personnel. In spite of these differences, the trend toward modernization slowly gained momentum.[62]

While the Indians generally accepted modern machinery, they tended to cling to some old methods. At Cochiti, they used a combine harvester, but after threshing, the grain was brought into the pueblo where they employed the ancient means of letting the wind separate the chaff and other foreign particles. At Santa Ana, the Indians practiced crop rotation to a certain extent. They alternated corn, wheat, and alfalfa, but it was not done as a systematic matter of principle. An idea that received numerous criticisms concerned the use of fertilizer. Apparently, the natives of Taos did not like Farmer Rolander, and they refused to follow his advice on the fertilization of their crops. In 1923, Superintendent Crandall remarked that the Tesuque wheat crop also suffered from a lack of fertilizer.[63]

To counter this trend, District Superintendent Chester H. Faris arranged agricultural education programs for agency farmers in 1928, and representatives from the northern, southern, and Zuni agencies and the Albuquerque Indian School attended these sessions at the New Mexico School of Agriculture. In 1932, Farm Agent Sidney Miller reported that he initiated a farmers' club at Laguna to discuss farm methods and to try to increase yields. The Indian schools also demonstrated a trend toward modernization. A representative of the J. Korber Company conducted a disc-repair work session at the Albuquerque Indian School, and the Santa Fe Indian School expressed interest in purchasing a tractor. By the early 1930s, the Indian service began to encourage vocational programs at these schools applicable to reservation conditions with an emphasis on basic land-use training.[64]

To stimulate Pueblo agriculture further, the agency tried to increase garden production. In 1925, Superintendent Crandall reported to the commissioner that the agency planned a number of large gardens for Taos, San Juan, Santa Clara, Nambe, and San Ildefonso. School principals managed some of these while government farmers took care of the others. Four years later, two district farmers planned to increase garden acreage. They constructed hotbeds for plants with the idea of distributing them to individual Indians for use in their home gardens. Local 4-H clubs sponsored some of the garden activity. Agricultural extension agent Lorin F. Jones noted that gardening received the most emphasis in 4-H activities.[65]

Toward the end of the 1930s, the agency increased the emphasis on agricultural education programs. In 1936, it stressed closer cooperation between extension agents and educational facilities. The concept emphasized agricultural education in the schools and introduced adult education in the villages through the schools. Since the Pueblos were somewhat isolated, agency personnel realized that adults had few opportunities to learn by observation. They felt the need for adult education was greater than before because of new problems such as overgrazing, inadequate water, and the desire for more irrigable land.

One of the initial problems of Pueblo agriculture involved the consolidation of farm holdings for the sake of a more economical farming unit and a more efficient use of available water. The agency's educational program for 1938, therefore, stressed land holdings, irrigation engineering principles, economical use of water, land leveling, and improved crops. Concerning this final goal, the Extension Division planned to set up demonstration farms to show how certain vegetables could be grown on small acreage. The educational planners stressed that their program was based on a careful study of the Pueblos rather than just the white man's way of thinking.[66]

To implement these programs, the Indian service decided to upgrade the position of agricultural advisor in 1938. For one thing, the bureau changed the job title from "farmer" to "farm agent." More importantly, the bureau increased requirements so that an applicant had to have a degree from an accredited college with a major in agriculture. In addition, the federal government required that the person applying for the position have three years of practical farming experience.[67] Years of untutored and incompetent assistants were finally being addressed.

PUEBLO AGRICULTURE STATISTICS

It is difficult to estimate how the amounts of land farmed and individual crops raised differed over a period of time. Early measurements of these figures were often approximations and their reliability is questionable. Indian farm plots, typically divided into small areas for individual families, made exact measurements concerning acres planted difficult to achieve. Therefore, it is hard to compare statistics for figures over a given time span. However, table 5 provides a look at each of the pueblos' dependence on agriculture and a comparison of agricultural figures over the period from 1900 to 1936.[68]

While the population figures for 1900 appear to be accurate, the land statistics for Santa Ana, Zuni, Tesuque, Jemez, and Isleta in that year seem exceedingly high.[69] If they are subtracted from the totals at the turn of the century, the resulting acres per person ratio for that time period would be 1.15 as opposed to 1.30 for 1936. This means that while the number of cultivated acres increased among Pueblo farmers (disallowing for high estimates of 1900), their ability to feed themselves through agricultural increased slightly.

Sometimes Pueblo farmers had low crop production figures because of infestation, destruction by animals, and natural disasters. The most common complaint had to do with grasshoppers. In 1937, Jemez, Isleta, and Nambe reported the heaviest infestations. Zuni, Acoma, Laguna, Cochiti, San Ildefonso, Santa Clara, Picuris, and Taos also had the pests in varying degrees. Three years before, farmers in the northern Pueblo

TABLE FIVE
PUEBLO AGRICULTURE STATISTICS 1900–1936

1900				1936			
CULTIVATED				CULTIVATED			
PUEBLO	POPULATION	ACRES FARMED	ACRES/ PERSON	POPULATION	ACRES FARMED	ACRES/ PERSON	CHANGE
Acoma	492	342	0.70	1,139	868.9	0.76	+0.06
Cochiti	198	260	1.31	307	627.6	2.04	+0.73
Isleta	1,021	510	3.44	1,137	2,300.5	2.02	–1.42
Jemez	449	2,557	5.70	648	1,052.5	1.62	–4.08
Laguna	1,077	664	0.62	2,300	1,336.6	0.58	–0.04
Nambe	80	64	0.80	141	379.5	2.69	+1.89
Picuris	96	103	1.07	97	187.9	1.94	+0.87
Pojoaque	12	12	1.00	20	29.5	1.47	+0.47
Sandia	76	383	5.04	125	547.3	4.3	+0.66
San Felipe	514	1,200	2.33	624	1,273.3	2.04	–0.29
San Ildefonso	138	99	0.72	127	253.9	2.00	+1.28
San Juan	379	506	1.33	550	637.2	1.16	–0.17
Santa Ana	228	1,182	5.18	251	558.8	2.23	–2.95
Santa Clara	221	440	1.99	433	361.4	0.83	–1.16
Santo Domingo	771	918	1.19	923	1,407.1	1.52	+0.33
Taos	414	142	0.34	773	1,338.0	1.73	+1.39
Tesuque	80	655	8.19	133	188.9	1.42	–6.77
Zia	114	30	0.26	208	225.5	1.08	+0.82
Zuni	1,523	5,201	3.41	2,080	2,070.5	0.99	–2.42
TOTAL	***7,883***	***18,379***	***2.33***	***12,005***	***15,644.9***	***1.30***	***–1.03***

area feared that prairie dogs would destroy crops. Santa Clara reported troubles with caterpillars, and a researcher in Santa Fe tried to determine how to control the problem. In 1936, beavers were destroying irrigation ditches at Jemez, and in 1922 chipmunks destroyed wheat on the Taos reservation.[70] Flood damage caused additional loss in crop production. Considerable destruction occurred at Laguna in 1932 when floods wrecked the irrigation system. The Indians did some temporary work in the area, but the Irrigation Division furnished material for more permanent construction. Floods also engulfed land in San Felipe in 1937 because weeds clogged the culvert built for the railroad tracks.[71]

Events in the 1920s and 1930s were so monumental that the Pueblo farm program made its greatest shifts ever. The threat of the Bursum Bill led to the creation of the Pueblo Lands Board, and the impact of this situation resulted in reports concerning Indians throughout the country. The conservancy district, actually set up for non-Indians, helped to increase the available agricultural land for certain pueblos. The major activities designed to help Pueblo farmers involved the development of water projects. In spite of these activities, the Indians were still having a difficult time supplying food for their expanding population. While future expansion required vast expenditures for land leveling and water products, the agency also had to overcome the Indians' reluctance to change. To bring about this change in attitude, the Pueblos would have to be pressured by drastic circumstances. The Depression and World War II would provide the necessary atmosphere.

CHAPTER SEVEN

Transitions in Pueblo Agriculture (1938–1948)

NEW DEAL BEGINNINGS

During the 1930s, economic depression engulfed the country, and many Americans faced a jobless future with an increasing sense of panic. A growing number of the unemployed felt that the crisis had to be faced with a strong sense of urgency. Never before had the federal government had such an open commitment from its electorate to revamp the economic structure of the nation. To combat the hardships brought on by the Depression, the administration of Franklin D. Roosevelt enacted a plethora of legislative measures known as the New Deal. The effects of such a broad, sweeping program carried over to every segment of the population, including Native Americans.

During their earlier periods of wardship, the Pueblo Indians accepted and incorporated certain agricultural innovations while rejecting others. There was nothing unusual about this. Anthropologists have noted that the Pueblos had a proclivity to resist dramatic changes in their culture while adopting certain foreign modifications. This practice, dubbed "compartmentalization," entailed integration of only those innovations that did not cause major alterations in the Pueblos' ancient cultural traditions.[1] Their basic patterns would, in the main, continue. While not all the Pueblos shared this view, it was a concept that shaped federal policy. The Bureau of Indian Affairs recognized that any farm program it initiated that ignored this Pueblo trait would be doomed to partial acceptance at best.

In the twentieth century, however, the rapid increase of population after decades of stability necessitated alterations in ancient farming methods. At the turn of the century, the Pueblo Indians of New Mexico numbered a little more than 9,000. By the early 1930s, this figure increased 20 percent to 11,346. The Pueblo population increased almost 18 percent during the Depression and an additional 10 percent, to 15,309, through the World War II era.[2]

Fortunately, land acquisition programs during this time increased the number of acres that Pueblo Indians owned. The Pueblo Lands Act of 1924 afforded them compensation money for land, water rights, and reclamation. These funds not only introduced the Pueblos to a cash economy but also enabled them to finance additional land purchases. As a result, fifteen of the Pueblo tribes bought 47,054 acres. While the acquisition of much-needed acreage may have appeared to solve the problem of an increasing population, the benefits of the new land were limited, since the Pueblos irrigated only 2,869 acres or 6 percent of the land purchased through compensation funds.[3]

Like most people in the United States, the Pueblos felt the impact of the Depression. Crafts, tourism, and off-reservation work, which had helped to bolster their economy before 1930, suffered in the decade that followed. As an adjunct to the New Deal legislation, Congress passed the Indian Reorganization Act (IRA) in 1934. In part, this act provided for the establishment of sound Indian economies. The Indian New Deal, however, posed a threat to the established traditions among the Pueblo agriculturists, and, as a result, the IRA had no immediate impact on the Pueblos.

NEW DEAL IMPACT

The Bureau of Indian Affairs, however, sought to implement New Deal programs on Pueblo reservations. In 1935, as a measure to increase bureaucratic efficiency, Commissioner of Indian Affairs John Collier formed the United Pueblos Agency (UPA) with headquarters in Albuquerque and named Dr. Sophie Aberle to replace the three superintendents of the northern, southern, and Zuni jurisdictions. Given the precarious climate of the arid regions of New Mexico, the Pueblo Indians required modern agricultural expertise to increase their output and feed their expanding population. Therefore, Collier decided to seek aid from the Soil Conservation Service (SCS) to rehabilitate eroded land.[4] Dr. Aberle contacted the director of extension and industry, A. C. Cooley, in 1936 concerning Pueblo eligibility for the SCS program. Cooley felt that the Pueblos should be allowed to receive benefits in the same manner as non-Indians. The responsibilities of the SCS on Indian lands included the formulation of land-use management plans, which had to be approved by the Soil Conservation Service district director and Dr. Aberle.[5]

Soil Conservation Service projects for 1936 included bank protection work at Nambe, ditch cleaning at Santo Domingo, work on spreader dams at Sandia, and jetty construction at Jemez. Senior Agronomist H. C. Stewart also desired to establish a demonstration on winter irrigation at the Laguna reservation in the hope that it would lessen the need for spring and summer watering. In 1937, E. R. Smith, district manager for the SCS, proposed a survey to establish a priority of projects within the Rio Grande watershed with special emphasis on areas where erosion control was critical. In order to accomplish UPA projects, the conservation service agreed to loan workmen to the Irrigation Service.[6]

Unfortunately the Pueblos did not universally accept the SCS. Superintendent Aberle hoped that land-use plans could be extended to all of the Pueblo farms, but the SCS only wanted to perform services for those Indians who had signed an agreement with them. Another hindrance to the successful completion of SCS projects involved Indian workers who were sometimes required to do community ditch work or participate in religious rites. Realizing the necessity of these activities, Director of Land Use Activities H. R. Fryer suggested that the SCS work should not supersede normal Indian activities.[7] This change of attitude contrasted sharply with the views of earlier Indian service personnel who chastised the Pueblos for their involvement with ancient traditions.

Other New Deal programs had limited effects on the Pueblo region. As a special unit within the SCS, the government established the Technical Cooperation Bureau in 1935. The bureau designed this organization to make technical and economic surveys on Indian land in order to prepare feasible plans for controlling soil erosion and insuring proper management of Indian farms and grazing areas. In 1937, the Pueblos also received limited funds from the Public Works Administration for construction of diversion dams, headgates, sluiceways, and siphons. Two years earlier, the Emergency Conservation Work (ECW), a forerunner to the Civilian Conservation Corps (CCC), was scheduled to govern Indian workers on projects for the Soil Erosion Service (SES). In 1937, Dr. Aberle asked W. A. Wunsch, assistant director of extension at New Mexico State College, if the Pueblos could apply for the Agricultural Adjustment Administration's (AAA) programs. Later that year Wunsch, who became executive secretary for the AAA, stated that the Indians were eligible for the program, but they had to wait for the next farming season (1938) before they could apply.[8]

New Deal programs, like other steps toward modernization, developed slowly among the Pueblo farmers. Understandably, Indian reluctance to adopt changes in their traditional programs hindered progress in this area. Agency personnel were also confused about the eligibility of Indians in these activities. Even when they became available to the Native Americans, the New Deal plans had to be adjusted for reservations and this further delayed their impact. Another concern involved the lack of uniform acceptance of these innovations among the various reservations. Although federal officials encouraged modern equipment and agricultural education programs in the early 1930s, their acceptance varied throughout the Pueblo region. However, Indian attitudes started to change as they began to feel the pressure brought on by the drastic economic circumstances of the Depression.

The Indian Division of the Civilian Conservation Corps (CCC-ID) was the first New Deal agency to exert a profound impact on Pueblo agriculture. Commissioner John Collier and the leaders of the Indian bureau argued that Indians should have their own branch of the CCC. In April of 1933, President Roosevelt approved this idea and allocated almost six million dollars for work on land restoration. The commissioner wished to help the Indians maintain their subsistence-level agriculture and improve their farming techniques through educational programs. The CCC maintenance work camps established for the Indians, unlike non-Indian work centers, resembled the reservation community. Since the Pueblos were reluctant to accept government innovations, however, the instructive efforts of the CCC-ID progressed slowly, making little headway until 1938.[9]

CCC activities were handled differently for the Pueblos than for most tribes throughout the country. Superintendent Sophie Aberle, who exercised almost complete independence in her administration, controlled the activities of the CCC with little interference from the Indian bureau. The duties of the CCC leaders among the Pueblos, unlike those on other reservations, were carried out by officials from the SCS and the Soil Erosion Service. Therefore, the work of these organizations had a strong influence on the Pueblo CCC program.[10]

Petty jealousies between the SCS and the Indian bureau, however, soon led to conflict between the two groups. Although the conservation service maintained that the Pueblos should rapidly adopt its programs, the conservative Indians were not willing to accept passively radical government innovations. Bureau officials, more aware of this tendency among the Pueblos than personnel of the SCS, were stymied by the aggressiveness of the conservation service. The Department of Agriculture, which funded the SCS, thought that its methods should not be questioned by the bureau and, as a result of the conflict, tried to secure drastic cuts in the funds for the Pueblo CCC program in 1938.[11]

Despite these conflicts, the CCC moved ahead. One of its activities was well drilling. These wells were designed to replace or supplement existing domestic water supply systems; many of the Pueblos used this water on garden plots located within the villages. In 1939, Engineer Alan Laflin scheduled projects at Isleta, Sandia, San Juan, Santa Ana, Zuni, and Acoma. The Irrigation Division handled work at Zuni and Acoma while maintenance forces constructed the other four projects. Laflin estimated the cost of these projects at $12,400, adding that the Zuni construction work would be delayed if the funds were insufficient.[12] Naturally, this decision would be unpopular at Zuni. It would alienate further a group already resentful of the consolidation efforts of the United Pueblos Agency.

Another function of the CCC-ID was the development of springs. In 1939, Assistant Field Aide William Rapp encouraged the immediate commencement of work on these projects. While springs usually had been used for stock watering, Rapp suggested that if they could be used for domestic purposes, the conservation corps should make this water supply available for community use. In more remote areas, springs might be used to irrigate land otherwise unsuitable for agriculture.[13]

An adequate water supply for the Pueblos also concerned officials. At the turn of the century, the federal government had constructed reservoirs for Pueblo Indians who were without a constant water supply. Unfortunately, as a result of silt collection, many of these reservoirs were no longer capable of holding an adequate water supply. The inadequacy of the reservoirs was a major concern of the Indian bureau and Dr. Aberle. Conservationist T. P. Bixby of the Department of the Interior asked the Pueblo superintendent how much silt removal work needed to be done. After obtaining this information, Bixby hoped to purchase the equipment necessary to restore Pueblo reservoirs to their full carrying capacity.[14]

The Agricultural Adjustment Administration was another New Deal program that influenced practices of the Pueblos. Recognizing that agriculture had suffered greatly throughout the 1920s, New Dealers thought that proper action should be taken to raise farm prices. In 1933, Congress established the AAA to carry out this goal. Because the organization hoped to encourage soil-building methods, American farmers were offered government subsidies for withdrawing acreage from production. Realizing the importance of these measures, the Indian service urged that the AAA include Indian farmlands. Activities for the AAA on Indian reservations differed only in that Indian service extension personnel were employed to administer the program.

In 1939, this New Deal program was extended to the Pueblo Indians, who began to receive money for soil-building programs. Under the AAA, payments were made to the Pueblos for renovation, planting, leveling, manure plowing, and fertilizer use. For renovating alfalfa fields through the elimination of grass and weeds, the Indians received $0.75 per acre. Planting new alfalfa fields paid $3.00 per acre, and leveling land paid $6.50 per acre. This latter process was more lucrative than other innovations because it allowed the land to be irrigated with a smaller amount of water, it controlled erosion, and it avoided the burning of high spots and the drowning of low areas. The Pueblos also received $1.50 per acre for plowing under green manure crops and applying commercial fertilizers.[15]

A survey of eight of the pueblos in 1939 (table 6) shows the impact of the AAA program on alfalfa planting, land renovation, green manure plowing, and parity checks.[16]

TABLE SIX
PUEBLO AAA PAYMENTS—1939

PUEBLO	ALFALFA	RENOVATION	MANURE	PARITY	TOTAL
Taos	$61.80	$4.95	$4.20	$265.73	$336.68
Picuris	12.00			87.72	99.72
San Juan	45.00			180.00	225.00
Santa Clara				58.00	58.00
San Ildefonso				16.28	16.28
Nambe	15.00			41.76	56.76
Pojoaque	3.00			3.85	6.85
Tesuque				22.55	22.55
TOTAL	***$136.80***	***$4.95***	***$4.20***	***$675.89***	***$821.84***

These statistics indicate that most of the Pueblos surveyed were reluctant to become involved in the AAA program. Their natural tendency to reject government innovations remained dominant. For many Indians, concepts of the administration were difficult to understand. Why should the federal government pay a farmer for not growing crops in an area he had used for years? Accustomed to using silt as fertilizer, Indians reacted negatively to the use of manure. These figures also reveal that Taos alone received more than 40 percent of these AAA benefits and along with Picuris and San Juan collected more than 80 percent of the administration's funds. At the same time, although the concept of parity might have seemed alien to the Pueblos, all Indians who were interviewed participated in the program to some extent. However, the practice of compartmentalization that the Pueblos had followed for centuries limited the effectiveness of this New Deal program on their reservations.

Interest in technical assistance, on the other hand, was increasing among the Pueblos. Regulations that governed AAA activities became more stringent each year, and the acting director of the Land-Use Division, Dan T. O'Neill, feared that Indians would lose some benefits if their activities were not approved. O'Neill suggested that the Pueblos needed range assistance from the Soil Conservation Service, but they did not require technical advice for their farm programs.[17] However, Antonio Abeita of Isleta Pueblo requested "an 'honest to goodness' farm agent down here—a white man that [*sic*] would contact the Indian farmers, in view of combating the many farm problems that arise."[18]

By 1940, benefits to the Pueblos from the AAA program were increasing. In April of that year, economist L. B. Liljenquist met with officers and family heads at Cochiti to discuss the AAA farm and range programs. After the meeting, twenty-six Cochiti farmers signed applications allowing them to take part in AAA farm activities during 1940. Undoubtedly the payments made to the ten farmers who participated in the soil-building program in 1939 were an incentive to others. Many of the farmers applied part of their AAA money to payment on debts that had accumulated in their reimbursable accounts.[19]

Information about the administration's payment programs spread unevenly throughout the Pueblo region. Although the United Pueblos Agency had hoped that New Deal legislation would be widely accepted among the Pueblos, even the agency's more moderate efforts were thwarted on the conservative reservations. Dr. Aberle wrote the governor of the Jemez pueblo at the end of 1940 to explain the benefits, purposes, and procedures of the AAA, and to assure the governor that the program was not designed to steal Indian land. She stressed that the government had no claim to the improved land. Aberle also reminded the governor that payments Indians received did not have to be repaid.[20] Despite these assurances, a few Pueblos still feared federal interference in their agricultural programs.

With the assistance of the AAA and the CCC-ID on Pueblo reservations, some Indians took advantage of the land obtained through acquisition programs. In addition, the soil conservation efforts of the AAA allowed Pueblos to make better use of their traditional farmlands. Table 7 shows the relationships among total acres, farmland, crops, rangeland, and population for the Pueblos in 1940. The cropland figures represent that portion of available farmland being cultivated.[21]

These figures reveal that most of the Pueblo land was used for grazing. Rangeland accounted for almost 90 percent of the total acreage, while cropland made up only 1.4 percent. Almost half of the arable acreage was not being cultivated. The number of cultivated acres per person shows the extent to which each community depended on agriculture. For example, while Zunis

TABLE SEVEN

PUEBLO ACREAGE FIGURES—1940

PUEBLO	TOTAL ACREAGE	FARM-LAND	CROP-LAND	RANGE-LAND	POPULATION[22]	CULTIVATED ACRES PER PERSON
Isleta	185,295	4,683	3,006	180,612	1,304	2.31
Sandia	20,974	1,618	765	19,356	139	5.50
Acoma	150,294	1,850	1,113	115,622	1,322	0.84
Laguna	240,773	3,400	1,570	231,993	2,686	0.58
Zuni	334,361	6,300	2,485	299,251	2,319	1.07
Santa Ana	24,860	714	595	20,502	273	2.18
Zia	16,689	650	327	13,584	235	1.39
Jemez	33,013	1,600	1,369	29,995	767	1.78
San Felipe	41,471	3,230	1,035	36,071	697	1.48
Santo Domingo	70,952	3,478	1,329	65,271	1,017	1.31
Cochiti	23,641	1,364	702	21,250	346	2.03
Santa Clara	45,200	700	520	44,500	528	0.98
Tesuque	17,078	290	190	16,298	147	1.29
Nambe	19,282	600	279	18,273	144	1.94
Pojoaque	11,593	50	43	10,794	25	1.72
San Ildefonso	19,989	720	401	14,540	147	2.73
Picuris	14,975	220	186	8,848	115	1.62
Taos	22,798	4,800	2,015	16,134	830	2.43
San Juan	12,329	1,250	922	11,079	702	1.31
TOTALS	***1,305,567***	***37,517***	***18,852***	***1,173,973***	***13,743***	***1.37***

Figure 9:
The wealth of the Zunis, sheep herd, Zuni Pueblo, ca. 1890. Photo by Ben Wittick. Courtesy of the Museum of New Mexico, neg. no. 51828.

devoted more acres to agriculture than any other tribe, this practice did not dominate their economy. On the other hand, Sandia had only one-third as many acres in cultivation as Zuni, but no other pueblo emphasized crop growing to such an extent. For the Pueblos to develop the full potential of their farmland, more government aid and Indian cooperation were necessary.

AAA regulations were explicit and strictly enforced. In 1941, for example, the administration's regulations for the United Pueblos Agency included a stipulation that a deduction would be made from the subsidy checks of those who planted more acres than AAA contracts designated. Every participating farmer was issued a marketing card, which he had to present when he sold his products to the government. All Indians were required to let administration authorities measure their acreage to see that the land planted complied with government standards. The AAA contended that an Indian who usually grew less than ten acres of wheat could benefit from the program. While the agency emphasized the voluntary nature of Pueblo involvement, it did urge Indian participation.

For example, the government would not guarantee the purchase of Pueblo wheat unless the Indians joined the program.[23]

Government personnel felt that the logic behind their actions helped to overcome the traditional Pueblo fear of outside interference in the native agricultural program. The Pueblos who adopted federal agricultural programs became more proficient farmers. Conversely, those who rejected the AAA or did not adhere to its policies would lose benefits that could help feed the expanding population. To help overcome Indian reluctance, the federal government and the UPA pointed to the economic hardships Pueblos had experienced during the Depression.

IRRIGATION PROJECTS

As population grew in the Pueblo region, demands on the Indians' water supply also increased. The usurpation of Pueblo water rights by users upstream led to yearly water shortages for seven of the tribes, and the demands of farmers downstream caused additional loss of water. The overtaxing of the limited water supply in New Mexico affected the pueblos of Santa Clara, Tesuque, Nambe, Pojoaque, San Ildefonso, Acoma, and Laguna.[24]

In January of 1940, the Upper Rio Grande Drainage Basin Committee held its first meeting. The committee, formed under the National Resources Planning Board, represented state and federal agencies concerned with Rio Grande irrigation water. The committee's purpose was to allow various parties to discuss irrigation projects and protest against the possible loss of water rights to new developments along the river. Because projects in Colorado, New Mexico, and Texas utilized almost all of the water of the Rio Grande, the future of new irrigation projects among the Pueblos was in jeopardy.

According to the terms of the Rio Grande Compact, which governed these three states, the Pueblos agreed to work with the state engineer. Generally their projects would have been approved without much conflict. However, because of complaints from users downstream, the United Pueblos Agency was required to submit a list of future irrigation developments for those Indians in the Rio Grande watershed. Acting agency superintendent Alan Laflin feared that the long list of projects, which required approximately twenty-five thousand acre-feet of water, would not receive approval from the basin committee. He thought that if the projects were submitted in a piecemeal fashion, the probability of approval would be considerably better.[25] Laflin was also reluctant to submit a list of projects that might be interpreted as a complete forecast of future Pueblo irrigation undertakings and therefore lead to limitations on the potential of Indian agricultural development. He informed the planning board that his present inventory was temporary and that he would revise it in the future. However, knowing that there would be no funds available after the end of the 1940–1941 fiscal year, Laflin was pessimistic about the future of Pueblo irrigation plans.

All of the Pueblo irrigation projects remained under investigation until March of 1940. It was desirable to start construction during the following summer because vast amounts of labor were needed to maintain older ditches and expand the irrigation system so Pueblos could develop their agricultural program. Laflin estimated the total cost for irrigation construction at $2,200,500.[26]

The federal government, of course, could not anticipate natural disasters, and thus it was not prepared in the late spring of 1941 when the Rio Grande overflowed its banks for almost two months. This catastrophe inundated more than fifty thousand acres of land above the Elephant Butte Reservoir. The damages and emergency measures cost approximately $1,678,800. Since the flooding also involved the entire Rio Grande watershed, small farms along the tributary systems suffered land and crop losses.[27] The results of the flood were devastating to the subsistence economy of the Pueblo Indians in this region. In some areas, many acres of cultivated land were washed away, and additional farm acreage was sufficiently flooded to cause severe crop loss. The United Pueblos Agency applied for federal aid, and Dr. Aberle estimated that $153,700 in additional funds would be required to repair flood damages and to prevent further destruction. Concerned about future flood damage among the Pueblos, the agency urged immediate action on bank protection. Both the Irrigation Division and Soil Moisture Conservation Operation Division (SMCO) would work with the CCC-ID to construct protection projects for the Pueblos. Dr. Aberle estimated that approximately $70,000 of the regular funds appropriated for the CCC and the SMCO could be put toward the total cost of repair work.[28]

The plans to increase the construction projects on irrigation canals and to supply flood protection presented no threat to traditional Pueblo lifestyles. On the contrary, these projects would enhance the agricultural programs of the Indians by increasing arable acreage and repairing faulty aqueducts.

TABLE EIGHT
PROPOSED IRRIGATION COST LIST FOR 1940–1941

Pueblo	Cost	Pueblo	Cost
Acoma	$140,000	San Ildefonso	69,500
Cochiti	107,000	San Juan	96,500
Isleta	127,000	Santa Ana	$34,500
Jemez	83,000	Santa Clara	150,000
Laguna	160,000	Santo Domingo	147,000
Nambe	83,000	Taos	325,000
Picuris	$9,500	Tesuque	89,000
Pojoaque	15,000	Zia	68,500
Sandia	111,500	Zuni	294,000
San Felipe	90,000		

For this reason, the Pueblos were willing to accept government assistance with these projects. Such was not the case, however, for agricultural education programs that attempted to reverse traditional concepts.

AGRICULTURAL EDUCATION AND WORLD WAR II

In the summer of 1940, Dr. Aberle held a meeting to discuss better coordination between the Extension Division and the boarding school agricultural programs under the jurisdiction of the UPA. Two large boarding schools at Santa Fe and Albuquerque had approximately 550 and 700 students, respectively. Both schools emphasized vocational training, including instruction in craftwork and agricultural techniques.[29] At the meeting, one member of the agency argued that education should be directed toward a more profitable use of Indian homelands. Others at the conference began to evaluate Pueblo resources and to design specific procedures in the Extension Division and the schools that would help the Pueblos use their land more efficiently. In an attempt to carry out these procedures, officials divided the pueblos into five groups according to similarities in physical resources, industry, and initiative: Taos and Picuris; Tewa pueblos; Middle Rio Grande pueblos; Acoma and Laguna; and Zuni.

First, conference attendees proposed a winter agricultural course for adults. However, because of myriad problems in the different Indian villages, two members suggested that adults be trained at local day schools. A plan to instruct a particular adult in each pueblo and use him as an advisor was rejected because of the long training period involved and the lack of influence that the trainee might have on his community. Mr. Laflin suggested that the increased irrigation construction for the Pueblos was a waste since the canals already exceeded the Indians' ability to utilize land. Noting the Pueblos' dislike for innovations in methods and implements, he concluded that only by educating young Indians at the boarding schools could modern ideas be initiated on the reservations.

UPA personnel generally agreed that Pueblos could increase their subsistence-level economy by cultivating unused irrigable lands. It was imperative that they develop those lands quickly, since pueblos in the Middle Rio Grande Conservancy District would have to pay their operation and maintenance charges after 1944. At this time, the agency was not overly concerned about economic diversification but rather emphasized the need for agricultural expansion to feed the increasing Indian population.

The agency committee suggested that vegetables raised at the Albuquerque Indian School could be used at the boarding schools and the UPA hospital. They also mentioned that, in the event of war, the government might establish farms to cope with food shortages. Similar action had been taken when there was a lack of food in the First World War.[30] In 1940, however, no one could have realized the impact war would have on the Pueblo agricultural economy.

The New Deal programs for the Pueblos laid the basis for future government innovations. Although the federal government had tried previously to impose agricultural programs, the Pueblos had always preserved their traditional concepts. During the Depression, however, conditions were ripe for a change. Because of the Indians' drastic poverty, the federal government and the agency had been able to introduce the AAA and CCC-ID methods for increasing farm production. To help the sagging Pueblo economy through better land utilization, the agency had also begun to concentrate on agricultural education. Aware of the Indian propensity to resist change in cultural traditions, the UPA had not tried to impose educational programs directly on the reservations. Rather, the agency sensed that introducing new farming methods to Pueblo youths in boarding schools would ultimately lead to greater acceptance of the programs.

Although World War II dealt a blow to the New Deal programs, the agency's desire to revolutionize Pueblo agriculture continued. To boost agricultural production, the agency began to implement ideas designed to modernize Pueblo agriculture. To meet these ends, the UPA emphasized vocational agriculture programs in the Santa Fe and Albuquerque schools.

A report on the progress of these schools in January 1942 indicated that more classroom instruction was necessary to make the program efficient. Walter Nations, who was responsible for the report, suggested that half of each day be spent in the classroom and the other part devoted to individual projects.[31] This concept of dealing individually with the students embodied a philosophy that the federal government thought crucial to any change in the Pueblos' subsistence economy. It was the agency's goal to emphasize individualism, through stronger competition among farmers, in order to change the Indians' subsistence economy.

The immediate objective of this vocational program was the utilization of irrigated land to its maximum potential. To correlate educational and reservation activities, schools taught students methods of irrigation and agriculture compatible with their previous experiences. The agency also attempted to make its activities among the Pueblos more pervasive. Since the draft had depleted Indian manpower, the extension agent encouraged girls to obtain vocational training in agriculture at the schools.

In addition to these programs, other changes were introduced. Because of the shortage of certain food items during the war, the federal government emphasized nutrition and asked students to grow crops that would diversify and balance their diets.[32] Although new food items had been introduced as early as the Spanish period, the Pueblo diet had changed very little. If the agency could convince Indians to alter their eating habits, another phase of agricultural assimilation could be accomplished. With the encouragement of the agency, a number of Pueblo women (including the wives of governors) attended nutrition courses in Albuquerque in the fall of 1942.[33]

The federal government also encouraged the Pueblos to grow "gardens for victory." Officials hoped that these gardens would enhance the war spirit on the home front and help assure an adequate food supply. According to Aberle, Pueblo garden production increased by almost 25 percent in 1942. Under the direction of the agency, the extension men, schoolteachers, and children worked together on victory gardens. They constructed hotbeds at the schools so that the children could grow plants, and at the end of the school term the children transferred these plants to their homes. Since the teachers were busy with selective service work and rationing, the agency encouraged village leaders to supervise the growing of victory gardens. In this way the agency hoped to enlist village elders in the process of modernizing Pueblo agriculture.[34]

The Department of the Interior strongly urged the agency to assist the Indian garden program. Aware of the limited amount of garden seeds available, Interior officials stressed the need to start permanent seed pools among Indian groups. If the Pueblos could not grow their seeds, the Bureau of Indian Affairs proposed that officers purchase the seeds with tribal funds. The combined support from local agents and the federal government therefore aided agricultural assimilation within the Pueblo communities. Indeed, an estimated twenty-five hundred families planted gardens in 1942.[35]

Because of the shortage of certain food supplies, the federal government also imposed rationing stamps on the Pueblos immediately after the beginning of American involvement in the war. Under these circumstances, the Office of Price Administration allowed Indian superintendents to serve as rationing administrators, an office usually under state control. The Interior Department encouraged the use of local rationing facilities rather than the employment of Indian service rationing boards.[36] This emphasis on UPA control enlarged the agency's power over Pueblo foodstuffs. Given the drastic measures taken by the government during the wartime to enlarge food supplies, Indians would have to increase agricultural production to help combat unpredicted food shortages.

Agricultural supervisor Will R. Bolen visited the pueblos of Sandia, Santa Ana, San Felipe, and Santo Domingo in the summer of 1943. He observed that the Indians were growing excellent potatoes, a crop that had failed previously. The farm agent at San Felipe showed Bolen the demonstration garden in which improved varieties of fruits and vegetables were grown. Largely because of efforts of farm and extension agents, Pueblos had begun to emphasize crop rotation and fertilization. Bolen predicted that, through the industry of the Pueblos, total acreage for the year would increase, and food production would rise.[37]

In addition to helping the war effort at home, the Pueblos also served well on the battlefields. The percentage of these Indians who served in World War II was higher than that of any other ethnic group in the country.[38] Because so many young Indian men went into the armed forces, adjustments had to be made in Pueblo agricultural programs. Labor-saving machinery was needed, for example, to allow the Indians to harvest their surplus crops for market. In addition, John G. Evans, the new general superintendent, requested a power hay press for Isleta Pueblo in 1944. At that time, these Indians cultivated almost one thousand acres of alfalfa for feed and sale.[39]

Conditions at Sandia were similar. There, too, modern equipment was needed to make up for the loss in manpower to the war. The Indians were cultivating approximately five hundred acres of land in small grain, mostly wheat. The combine they had rented the previous year at a high price was not available, and Evans asked for funds to purchase one.[40]

Altogether, the impact of the war on Pueblo agriculture furthered assimilation. Until the war, modern agricultural equipment had a limited impact on the Indians. Around the turn of the century, the federal government had introduced modern farm equipment, but the Pueblos used these devices sparingly. They were accustomed to traditional methods of crop gathering, which, although ineffective, were part of an ancient lifestyle. But with the need to increase agricultural output during World War II, Indians became more willing to accept innovations in their farming methods.

The program in agricultural education for the Pueblos expanded near the end of the war. For example, the Taos day school taught Indian farmers how to increase their farm output. Taos Indians, who had previously rejected the use of manure as fertilizer, began to use it in their fields after observing its benefits. At Zuni, which lacked a consistent water supply, there was less enthusiasm for the agricultural program. However, the agency felt that if the development of a dependable water supply could be achieved on this reservation, an agricultural education program would be quite successful.

Although the Santa Fe school also suffered from an insufficient water supply, its major concern was the development of educational training. The agency proposed that a student instructor be added to assist the farm teacher. In addition to teaching methods of agricultural production, these men would offer instruction in the maintenance, repair, and operation of farm machinery.[41] By 1945, the agency's educational program had begun to emphasize the use of modern farm equipment.

A comparison of Pueblo agricultural acreage in 1940 and 1944 helps measure the efficiency of the agency's program during this time. Although the total acres cultivated increased by more than 15 percent, nine pueblos showed a decrease in lands farmed. Only nominal gains occurred in three other pueblos. However, many of the Rio Grande pueblos lost crop acreage in the flood of 1941. Yet there were gains in other areas. The major increases in agricultural lands were at Sandia, San Felipe, Santo Domingo, Taos, Zuni, Acoma, and Isleta. In the seven years preceding 1944, production of forage crops, cereal crops, and garden produce increased considerably.[42] In short, judging from the amount of land available for farming, the government's program bore mixed results.

A major part of the agency's modernization plan was the emphasis on individual, competitive agricultural programs. Like other members of American society, Pueblo soldiers had difficulty in adjusting to the postwar world. During the war, they had undergone changes similar to non-Indians. However, in some respects, the impact had been greater on Indians. For the Indian soldier, competitive individualism without community support was a new experience. While in the service, he was not only accepted as an individual but also judged as one.[43] As a result the Indian veterans had trouble readjusting to their traditional communal lifestyles. This was evident in many pueblos and perhaps was confirmed in the increase in craftwork and off-reservation employment in the postwar era. This trend toward nontraditional occupations began during the New Deal programs and increased as Pueblos became involved in wartime industries.

A few Anglo observers in the postwar period reported a decrease in religious ceremonies concerned with crops and weather and an increase in nonagricultural activities. Some persons claimed that this lessening interest in religion was reflected in a separation of church and state and a shift in power from the religious heads to the civil government in the more progressive pueblos. Florence Hawley, who wrote about New Mexican Indians in the late 1940s, contended that participation in the ancient agricultural ceremonies waned because individual wage earners "no longer need[ed] cooperation of deities to achieve economic security."[44]

Since agriculture had been the basis of the Indian subsistence economy for centuries, the task of modernizing methods of Pueblo crop growing appeared monumental in the early 1930s. The Indian bureau officials saw that if this aspect of Pueblo life were to be altered, economic, governmental, and

TABLE NINE

COMPARISON OF AGRICULTURAL LAND, 1940–1944

PUEBLO	1940	1944	% CHANGE
Isleta	3,006	3,465	+15.6
Sandia	765	1,878	+145.5
Acoma	1,113	1,391	+25.0
Laguna	1,570	1,588	+1.1
Zuni	2,485	2,833	+14.0
Santa Ana	714	585	–18.1
Zia	327	312	–4.6
Jemez	1,369	1,345	–1.0
Santo Domingo	1,329	1,644	+24.7
San Felipe	1,035	1,419	+37.1
Cochiti	702	630	–10.3
Santa Clara	502	547	+5.2
Tesuque	190	177	–6.8
Nambe	279	288	+3.2
Pojoaque	43	35	–18.6
San Ildefonso	401	274	–31.7
Picuris	186	176	–5.4
Taos	2,015	2,369	+17.6
San Juan	922	899	–2.6
Total	***18,953***	***21,855***	***+15.3***

religious patterns would have to change to accommodate innovations in farming techniques. The UPA, recognizing the Pueblos' tendency to resist modifications in their ancient cultural patterns, formulated the agency's program accordingly.

The Pueblos did not accept rapidly the New Deal agricultural programs. However, the work of the AAA and the CCC gradually introduced agricultural innovations, and, after the war began, the agency continued to exploit the propensity for change fostered during the brief New Deal period. Rather than directing its programs entirely to the reservations, the UPA decided to emphasize the education of Pueblo youths to encourage the gradual acceptance of modern farming methods.

The federal government and the agency were able to promulgate their programs during the war because the loss of manpower necessitated the use of modern equipment. Wartime nutritional programs also helped to diversify the Indian diet. Although a few observers noted a decline in traditional Pueblo ceremonies and forms of government, conservative Pueblos were reluctant to abandon their traditions.[45] Indeed, the breakdown of ancient agricultural practices that resulted from the pressures of the Depression and war years did not necessitate a decrease in ceremonialism. To believe that adaptation to a modern, diversified economy would automatically result in a loss of this kind would be an underestimation of the strength of the religious character of the Pueblos. It is also clear, however, that while Pueblos retained some of their traditional agricultural practices, they were now utilizing modern methods of farming introduced during the Depression and war years.

CHAPTER EIGHT

Postwar Pueblos (1948–1962)

AGRICULTURAL MODERNIZATION

More than any other event in the twentieth century, World War II made a dramatic impact on cultures throughout the planet. As war-weary veterans returned to their homelands, they sought to put their lives back together the way millions of soldiers had done in the past. But just as the conflict had changed the participants, it had also made readjustment a trying experience. Even during the early stages of the battles, concerned observers questioned whether formerly tradition-bound Pueblo Indian farmers would want to return to the fields that had fed their ancestors. In the fall of 1942, Thomas Nickerson, writing on postwar considerations, questioned whether Pueblo soldiers would want to take up where they had left off. Older soldiers, who had spent many of their younger years embracing an agricultural tradition, might make the adjustment more easily than younger ones who had spent many of their formative years in the military. He felt that it would be up to Pueblo officials to cajole returnees back to the fields.[1]

Overriding the question of future job choices was the threat of modernization to Pueblo agricultural traditions after the war. Although the degree to which the various villages held on to ancient practices varied from one group to another, the Pueblos adopted change only when it posed little or no threat to the ways of the past. The exigencies of the Depression era, however, weakened some of those ties, for the Pueblos had to boost agricultural production to feed their growing populations. The war years only intensified this tendency; the Pueblos, like other tribes throughout the nation, increased production to support the war effort. Indeed, while Native Americans expanded cultivated ground by 150,000 acres nationally, Pueblo land devoted to agriculture rose by almost 3,000 acres. Much of this increase was a consequence of land-acquisition programs that also helped them to expand garden and livestock production. Ironically, this dramatic change occurred at a time when one-third of the Indian population across the country had left reservations to serve in the military or work in war-related industries. The Pueblos sent to the armed forces a higher percentage of men than any other ethnic group in the country, significantly reducing the number of available farmworkers. The increase in Pueblo agricultural production, though uneven throughout the reservations, stemmed from the introduction of modern farm machinery—a trend they had resisted in the past.[2]

Thus, modernization edged its way into Pueblo life in the decade before the end of the Second World War, but their traditional subsistence agricultural program faced a greater long-term threat in the conflict's aftermath. The war opened a different world to the Pueblos and other Native Americans, bringing them new skills, opportunities, and a greater degree of acceptance. Whether they served in the war or worked in domestic industries, Pueblos acquired experience that prepared them for employment in the modern economy. In conjunction with these abilities, some Pueblos utilized the GI Bill after the war to prepare them for a different way of life. For many of these people, the search for employment would lead them to cities distant from their homelands.[3] Those who stayed behind would face an agricultural future far different from the past. For them, a major question surfaced: How could they compete in an agricultural world that was undergoing a technological and economic revolution?

Industrial America underwent dramatic expansion after the war and change in the agricultural sector was even more striking. The dramatic growth in agricultural productivity was the product of rapid mechanization, scientific advancements, increased specialization, modern management, and government subsidies. At an accelerated pace, the traditional family farm began to give way to the new agribusiness that drastically reduced the farm population, greatly increased production, and yielded lower farm prices.[4] Commissioner of Indian Affairs William Brophy noted that future Indian agriculture would likewise require larger units of production, the application of scientific and technological innovations, and large investments of capital. He also acknowledged the supreme challenge this presented to Native Americans, who generally lacked capital, lived on lands in need of rehabilitation, and faced grave challenges to their water rights.[5]

In many ways, the plight of the Pueblo farmers paralleled that of small farmers nationwide—both eked by against great odds. Government subsidies benefited agribusiness, and urban areas offered better wages and job diversification. As a result, farm populations on and off reservations declined dramatically after World War II. While the agricultural revolution created a new kind of rural poverty among poor whites and minorities, its impact was even greater on Native Americans because of their unique historical experience. Indian law and bureaucratic red tape complicated advancement, and, traditionally, reservation development lagged behind non-Indian businesses. Many Pueblos, like other Americans, turned to new occupations—mining, Los Alamos jobs, ranching, recreation—while others profited from the growing interest in Native American arts and crafts in society and the marketplace. However, farming still remained important to the Pueblos, and their transmission of traditional agricultural knowledge to younger generations was rare among western native peoples.[6]

For those who remained on the reservations after the war, the opportunity to succeed in the new economic order dominated by industrial production—even in agriculture—rested on the diversification of the existing reservation economy. Such growth and change presented a tremendous challenge to Pueblo farmers whose traditional methods and religious beliefs hindered economic and agricultural adaptation. But even those willing to utilize the most modern methods available found obstacles to their progress overwhelming. Farmers discovered their small, broken parcels on the pueblos could not compete with the highly mechanized, corporate farms that gradually became the norm in white America. Like other Native Americans, Pueblos lacked adequate machinery, education, money, and, some observers believed, the competitive drive to contend with the forces of modern capitalism that increasingly dominated American agriculture. Like many indigenous people around the globe following World War II, the Pueblos lived in societies generally unstructured and egalitarian but now faced a future shaped by modern corporate capitalism and market agriculture—an economics with which they were unfamiliar.[7]

However, after decades of resisting American innovations to their ancient traditions, the Pueblos had to embrace modern science and technology and take advantage of a change in government policy toward self-determination before both opportunities slipped away. Uncertain at the time was whether the federal government would catalyze the transition or whether Pueblo leaders would reassert themselves after years of resisting change. The farming transition would undoubtedly require substantial federal assistance. If these political and economic changes came about, would the Pueblos remain what they always had been—a culture steeped in an agricultural way of life that had defined the core of their existence for centuries?

TERMINATION

It was not simply an adjustment to civilian life that faced Pueblos and other Native Americans in the postwar setting. There was also a change in federal government policy that would come to be known as "termination." From the very beginning of America's relations with its native people, the United States attempted to assimilate them into the cultural mainstream. Unfortunately, this policy had a limited chance to succeed since few Indians wanted to undergo a drastic departure from their cultural traditions. Under John Collier's New Deal programs, the idea changed to one of cultural pluralism, which sought to integrate Indians without destroying traditional ties. But in 1945, Collier resigned following numerous conflicts with Congress, and a much more conservative mood settled in the nation's capital.

With the end of the war and America's great victory came a strong sense of nationalism and a prevalent desire for national unity. Cultural preservation seemed out of phase to policy makers and to some Indian veterans as well. Almost overnight the idea of Indian progress changed from holding on to past traditions to embracing an economic future in the nation's mainstream. Federal lawmakers emphasized that the new policy would give Indians equality with whites, but the underlying theme was the severing of legal responsibilities between the government and its native people, which had come about through treaty agreements in the previous century and a half. More than freeing Native Americans from an allegedly subservient position, Congressional conservatives hope to erase the expense of being in the Indian business.[8]

Of course severing ties with the federal government meant that Native Americans would be on a lonely road to economic independence. Bureaucrats hoped to ease the burdens for the Indians seeking jobs after the war by initiating a movement to the cities that was known as "relocation." For those who had acquired the proper training and could somehow manage to deal with omnipresent racial discrimination, there was at least a chance, albeit small, to overcome the obstacles toward upward economic mobility. While this change of residence complemented the termination program, it also threatened Pueblo cultural traditions.

Congressional adoption of House Concurrent Resolution 108 in 1953, which initiated termination, sent shock waves throughout the Pueblo community, and Commissioner Glenn Emmons visited with representatives from various villages in the fall of that year to assure them that all was well. An ardent advocate of termination during the Eisenhower administration, Emmons hoped to overcome their fear of the Malone Bill, which was pending in Congress. Pat Toya of Jemez and Clement Vigil of Nambe expressed their concerns over the bill, which they thought would force them to sell their land. Emmons found their statements quite disturbing but told the Indians not to worry because he would take the matter to court if necessary. Trying to console the Pueblos, he claimed that the federal government's withdrawal from the administration of Indian affairs was a long way off. He promised that the reservation land would be turned over to its original owners and that he would "fight having the government sell out your land from under you."

Yet while Emmons's remarks were designed to placate Pueblo reservations about the change in government policy, his real mission was to prepare them for the inevitable. He warned them that they should immediately "start planning and making programs for the time when the federal government steps out." When this time came, he promised to help each tribe get established, but he cajoled them to work toward the future so that they "won't be turned loose until they are ready." Abel Padilla of Tesuque was troubled by talk of abolishing or withdrawing the Bureau of Indian Affairs, which he felt was important for older Pueblos, many of whom did not understand English or how to run a business for themselves. Emmons disdained the term "withdrawal" and felt that the transition should be looked on as a "readjustment." His word games attempted not to leave the impression that the government was backing out of its commitment.[9]

In order to placate Indian concerns about the transformation in government policy, the commissioner's office called for a three-day meeting of bureau officials and tribal leaders of the Southwest in Phoenix during late January of 1958. This conference at the Phoenix Indian School involved numerous sessions covering issues important to Pueblo farmers, including irrigation, water rights, and soil and moisture conservation—all under the major theme of termination. The idea was to set into motion plans and programs designed to assist Indians in taking responsibility for resource management. The commissioner hoped that bureau employees, working with native leaders, could reach an understanding on the methods of developing Indian "self-reliance necessary to conduct their personal affairs with the same degree of independence as other American citizens." Having made little progress toward that goal since termination commenced, superintendents were goaded to set up long-range programs "without delay" so that Indian people could "free themselves and their affairs from federal control." The urgency of the situation in the commissioner's mind was underscored by his admonishment of bureau personnel not to wait for specific instruction from area offices or the initiation of congressional legislation.[10] Indeed haste seemed more important than eventual success.

The movement toward quick implementation of the overall program at the national level was tempered somewhat by planners at the conference. Speaking on the subject of soil and moisture conservation, Homer Gilliland of the Colorado River Agency pointed out that the idea of long-range planning for terminating these activities went back to a report requested by Evan Flory in 1946. The former director of the Soil and Moisture Conservation Program urged his staff to prepare a report on the work and funds necessary to complete soil and moisture conservation activities on Indian reservations. He proposed that construction activities would be completed within a twenty-year period, leaving only maintenance and technical assistance. Eight years passed before the Department of the Interior became interested in the program as the exigencies of termination brought it into focus. The grand scheme called for the construction phase to last from 1956 to 1975 with cost peaking in 1963. Gilliland realized that technical services of this nature could never be taken away from Indian farmers but rather transferred to the Soil Conservation Service of the Department of Agriculture. After all, they all had been made available to all other American farmers since the early 1930s and were now seen as a fixed necessity in modern agriculture. Therefore, he urged that Native American farmers set up their own conservation districts and work closely with county, state, and federal agencies to help bridge the gap when termination arrived.[11]

Most important for the success of such a grandiose concept as termination was the application of the policy to individual Indians. Leroy Horn, field land operations officer from Santa Fe, spoke specifically about how this applied to the Pueblos. He laid out certain steps necessary to assure some reasonable measure of success. They included an explanation of how termination affected Indian farmers, a call for written record of ownership and description of farmlands, a discussion of future services by various agencies, and establishment of a board of review for field personnel to handle special problems. Realizing that many younger Pueblo men had left the farms, Horn cited problems in working with older Indians, including a language barrier, their having little or no education, and a lack of any conception of an economical farming unit. James Simpson added that a great deal of education was necessary in order for Indian farmers to make important decisions in the use and management of their resources. Past experience had proven that unless individual farmers could take these steps, development plans would typically become inactive.[12]

Indian officials felt that their work in the area of irrigation was a necessary adjunct to termination. Charles Corke, United Pueblos Agency land operations officer, believed that this work was essential to bringing Indian people up to parity with non-Indians and getting the government out of the irrigation business. Irrigation standards, he felt, should be based on government obligations, basic engineering requirements, economic feasibility, and most important, what Indians thought would fill their needs. Corke desired standards high enough to minimize operation and maintenance costs and rid the government of any obstacles in the way of relinquishing irrigation services. R. N. Hull, UPA general engineer, recognized the difficulty of assigning payment for the costs because of the different conditions involved in every Indian irrigation project, which included a wide variation in assessment rates. The situation for the Pueblos was rather unique since they had constructed their own irrigation systems before Anglo intervention, and government improvements had only occurred since the turn of the century. Since these systems were always operated by the Pueblos and were included in their religious activities, they were reluctant to make any changes that might cause them to lose control. They generally performed their own maintenance work and only called on the government for heavy equipment during times of emergency. However, a number of them had left their farms for jobs elsewhere, causing their lands to go idle and unleased. Many of those who stayed behind were satisfied with the old ways and had no desire to adjust their farm methods in order to improve their standard of living. Hull felt that only an alteration in land tenure that increased the size of individual holdings would bring about necessary changes for modernization.[13]

Area General Engineer Robert H. Rupkey of Phoenix lectured the audience of the conference on termination on the subject of water rights, laying out the basics of appropriative rights and their various types of classifications, including court degree, state permit, continued use, implied reservation rights, and those based on treaties with foreign nations. He contended that the Pueblo irrigation water came out of the Treaty of Guadalupe Hidalgo, which ended the Mexican war, but this, he felt, was no guarantee to its preservation. G. B. Keesee, area general engineer from the Gallup area office, was more specific about how termination would affect Indian water. Recognizing that the dramatic increase in municipal, industrial, and agricultural water use had forced Indian officials to be increasingly engaged in defense of native rights, he pointed out that outsiders viewed with envy any water that was not being beneficially used. With termination, the responsibility of defending their precious rights to water would fall into the hands of Native Americans. Keesee contended that the best defense would be "continued beneficial use of their water to irrigate all their lands." In order to make a proper defense in the future, bureau officials had to acquaint Indians with the legal status of their water rights; keep accurate and continuous records of water diversion, return flow, and crop production; and assist tribes in setting up organizations to administer water use.[14] In the future the Pueblos would become all too familiar with the necessity of heeding this sage advice.

As the conference drew to a close, bureau officials made some general comments concerning the problems to be expected with termination. Phoenix area director Fredrick Haverland admonished administrators that they were "dealing with human beings . . . and cold, hard, tough agreements just will not survive." He asked them to consider the humanity of their job and not simply "put a price on the Indians." He said they should be patient because such a drastic change could not happen overnight but at the same time to keep pushing the natives toward progress. After all, he stated, it was Indians and not the bureau who must lead termination. However, the transition was not going well with the Pueblos. As UPA Road Engineer Leo Honey pointed out, some Indians within the agency had "a general anti-termination attitude." It appeared to him that they had "adopted the Irish attitude that 'The Devil you know is better than the Devil you don't know.'" As far as UPA Agency Forester Albert Palmer was concerned, the reluctance of the Pueblos was generally irrelevant. Since most of their reservations were increasing in population and their land base was too small to support them adequately, changes were already in progress to bring them "abreast of modern society" and force the majority of them not to depend on a reservation economy for their livelihood.[15]

The glowing prognosis of the meeting was summed up by E. J. Utz, assistant commissioner in charge of resources for the Indian service, who stated that American Indians were headed for ultimate independence and that while it may take some time, "eventually the Indian will be free." But would they? By December of that year Area Director T. B. Hall, reporting on the future maintenance of wells for the Pueblos, stated that of the nineteen tribes, nine of them did not have sufficient funds to do the job and that the others might be able to do it. The worst case was at Zuni, where there was very little tribal income. Other pueblos, especially Taos, made "substantial tribal income" ($15,000) from recreation and tourism, but they were the exception rather than the rule. While the Pueblos were expected to furnish all materials, labor, and supplies beyond the 1962 fiscal year termination date, Hall suggested that a trained technical staff be carried after June 30.[16]

Having sustained themselves for centuries, most of the Pueblos were fairly independent, but the cost of modern agriculture—flood control, crop irrigation, well drilling, and ever-changing equipment—was staggering. With termination, technical assistance would be more difficult to come by. Clearly the progress of Pueblo farmers would be stymied by this change of policy, and some government officials, recognizing the decline of Indian agriculture and the rise of off-reservation employment after the war, were already backing away from agricultural programs on the reservations. Indeed, termination was a strong signal to all tribes that the U.S. government had given up the idea of assimilation through agriculture.[17]

FLOODS AND DAMS

One of the first major problems to attract the attention of the Pueblo administrators was the flooding that devastated the Rio Grande valley at irregular intervals. At a meeting of the All Indian Pueblo Council in 1947, the governors discussed a general outline for the middle Rio Grande valley concerning flood control, possible dam locations, and the cost of their construction for the tribes involved. Recognizing that current land holdings of the Indians were insufficient for their needs, tribal leaders wanted to protect every irrigable acre from the ravages of flood damage. Chairman Abel Paisano contacted the Bureau of Reclamation to let them know that Pueblo officials were ready to meet with their engineers to push the program forward.[18]

By the early 1950s individual Pueblos began to develop their own flood-control plans. The Santa Ana Tribal Council proposed the construction of detention-type structures to control the disposal of arroyo floodwater that was impacting some 265 acres of prime irrigation land between US Highway 85 and the Albuquerque Main Canal. An additional 79 acres of mixed farmland above the pueblo were also affected, and 15 acres had already been abandoned because of siltation due to flood damage in the arroyos. The tribal council estimated the value of the land to be $103,200, which yielded crops worth $25,000. They submitted a proposal to the United Pueblos Agency requiring land leveling and canal reorganization that they claimed would raise the value of the land to $127,000.[19]

It was not always Indian lands that were damaged by faulty irrigation canals. In a rather unique case, seventy-eight-year-old H. P. McKenzie complained to the Interior Department in 1953 that his lands within the San Juan pueblo were flooded after the department had deepened the village ditch without widening his own. McKenzie's private land, previously granted him by the Indian Claims Commission, was overrun by a torrent of water that ran through the new Indian canal and broke his smaller ditch. Having farmed the area all his life and too proud to accept welfare, he hoped that the Indian bureau would accept the responsibility of fixing the ditch, of which he was a joint water user. Eventually the Branch of Irrigation straightened a sharp turn in the canal, which had caused the problem, to the satisfaction of everyone concerned.[20]

While this situation at San Juan was relatively minor and had more to do with negligence on the part of the main canal operator, it did point out the need for foresight in dealing with flood problems. Two years later Guy Williams, superintendent of the United Pueblos Agency, contacted Area Director Wade Head on possible flood-control construction in the San Juan area. Williams was specifically concerned with the stopgap measures that were typically employed to handle flood problems. Rather than attacking a problem before it happened, the BIA relied on short-term solutions that occurred after the fact. Funding from their office typically came during a crisis stage when emergency work was needed to counteract the latest flood disaster. He felt that it made better economic sense to spend a nominal sum on a yearly basis when the stream flow was at a low stage to help direct the water at its peak. He called for an annual expenditure of fifty thousand dollars for river channel improvement, which would, in the long run, be cheaper than the current philosophy of flood control. Williams realized that until reservoirs were constructed upstream from the farmlands, the hazards of floods would continue to devastate Indian property as they had in 1941, 1948, and 1952.[21]

The need to channelize the Rio Grande not only affected the village of San Juan but also other land in the Española valley (including the pueblos of Santa Clara and San Ildefonso) and non-Indian-based land as well. The question was what money sources could be found for this restoration. The Army Corps of Engineers could work on localized flood projects, but its expenditures were limited to under one hundred thousand dollars. Having already completed work for the city of Española and submitted a preliminary plan for Santa Clara, the agency's hands were tied for the near future. Recognizing the seriousness of the situation, Williams contacted Commissioner Glenn Emmons. The commissioner's study revealed that the region under construction was outside the middle Rio Grande project so that the Bureau of Reclamation could not offer any relief. Frustrated, Emmons turned to U.S. Senator Dennis Chavez and inquired about the possibility of using state river improvement funds for the San Juan area. So while the problem required immediate action, underfunding held up any possible relief.[22]

By the summer of 1955, irrigation work was being carried out at Taos, Nambe (damaged by flash flood the previous year), San Juan, San Ildefonso, Tesuque, Santo Domingo, and San Felipe. While this work involved flood damage, other pueblos suffered from a lack of water. Drought periods had increased in recent years, and the situation was becoming critical at Laguna, Sandia, Isleta, Zia, San Juan, and Nambe. Laguna had practically gone out of the farming business because its water supply had dried up, and the possibility of developing underground water was minimal. At Zuni the Black Rock dam was not holding much water, and the farmers there were barely holding their own.

Amidst the chaotic vacillating between feast and famine in the weather cycle, the United Pueblos Agency came up with an idea that found hearty agreement with both the engineers' corps and the Reclamation Bureau. The agency assigned one full-time engineer who would coordinate projects of other agencies affecting Indian land. This move allowed the UPA to get into the action at the planning stage, which led them to include Indian desires from the very start of the project.[23]

Although floods on the Rio Grande devastated numerous Pueblo lands, the Jemez River proved to be just as challenging. Flowing into the main artery of the Rio Grande from the northwest, its overflows damaged both Zia and Jemez pueblos. In July 1955, a flash flood ripped through the valley as an unusually intense rain- and hailstorm poured down, wrecking Zia's main canal. Water from arroyos intercepted by the canal and the adjacent hillsides ran into the mother ditch, causing breaches in hundreds of places. In the aftermath, the canal was filled to the brim with silt and other debris, fifty acres of crops were destroyed, and in some areas two feet or more of silt and debris were deposited on the fields. To make matters worse, two more flash storms hit the area during the cleanup, requiring some sections of the canal to be reworked. Repair crews worked two shifts a day, seven days a week, for two weeks before some water made its way to drought-starved crops. Further rehabilitation at the end of the irrigation season was required to return the system to normal, with a projected cost of ten thousand dollars. The damage at Jemez was not nearly as drastic, as the situation was cleaned up in a week with no crop loss.[24]

Flood problems returned to the Jemez valley in the spring of 1958 due to the rapid meltdown of heavy winter snows, and this time Jemez took a bigger hit. Antonio Sando, governor of the pueblo, contacted Senator Chavez about the damage done to roads and individual properties and the possibility of obtaining permanent flood-control works on the reservation. Chavez, who was also concerned about non-Indian residents in the valley, contacted Commissioner Emmons about the possibility of a joint program of bank protection and canal diversion construction involving the Soil Conservation Service, the Army Corps of Engineers, or other government agencies. Two weeks later, Superintendent Williams wrote to the Army Corps of Engineers in Albuquerque about the feasibility of constructing emergency works and possibly a larger-scale flood-control project for the protection of Jemez, Zia, and adjacent non-Indian lands. District Engineer Colonel Albert Reed assigned a member of his staff to accompany a UPA official, Governor Sando, and Zia governor John Paino to survey the damages. Their reconnaissance revealed damage to flood-control works, an irrigation heading, one hundred acres of Jemez pastureland, a farm road that was severed, and the destruction of seven acres of Jemez farmland plus three acres at Zia. While the amount of agricultural acreage was relatively small, the impact on individual Indian farmers was severely felt since the average farm unit on the reservation was only about five acres, and any crop loss dealt a blow to their yearly income. Unfortunately, the Army Corps of Engineers faced monetary limitations and confined its work to limited repairs on damaged flood-control works rather than large-scale improvements. At that time the engineers' major focus of work was on the Rio Grande, while consideration of work on its tributaries, such as the Jemez River, would be taken up at a later time as finances allowed.

Such a hollow promise meant that the UPA was forced into immediate action. The Branch of Land Operations used two bulldozers to erect levees for diversion of water away from the crops as a temporary measure. Additional work was undertaken by the Agricultural Stabilization and Conservation Program (ASC) at the request of Jemez officials. Working in conjunction with the native people, they realigned channels and encouraged the Indians to plant willows along the emergency levees to make them more permanent. Area Director Wade Head felt that since ASC work required native participation, the Jemez would "undoubtedly be more inclined to maintain the improvements when completed."[25]

The obvious answer to these vexations was dam construction. Unfortunately, Pueblo pleas were falling on deaf ears. Given the restrictive cost of construction, large sums of money were simply not being made available. Smaller dams, like those at Zuni, had been constructed, but their impermanent nature tested stability during heavy floods such as the one in 1960.[26]

THE DISTRICT

If any of the pueblos had a distinct edge on the others in the protection from floods and droughts through dam construction, it was those that were situated within the Middle Rio Grande Conservancy District—Cochiti, Santo Domingo, San Felipe, Santa Ana, Sandia, and Isleta. A major part of the original purpose of the district was to provide new and permanent facilities for water storage and flood protection. This would include Pueblo and non-Indian lands between the northern boundary of Sandoval County and the Elephant Butte Reservoir in the south (approximately forty miles north and one hundred south of Albuquerque).

In the early years of the district many promises were made to Pueblo leaders that when construction plans were finished, the various villages would have all of the water they needed. Both Francisco Tenorio of Santo Domingo and Porfirio Montoya of Santa Ana testified in 1946 that they remembered meetings held in 1929 and 1930 in which district leaders promised the Indians sufficient water for their farming needs, but a severe drought that year resulted in numerous complaints from Pueblo leaders and a feeling among the Indians that they had been slighted by district officials. Diego Ramos Chiwiwi of Isleta complained that his pueblo always had problems with water shortages since the district was initiated, but the situation in 1946 was worse than any previous experience as village farmers received hardly any water at all. It was particularly frustrating to him that there was enough water in the canal, but the conservancy ditch rider, who authorized distribution throughout the valley, told various Pueblos that the water was being saved for non-Indians who were taxpayers. This statement came as a shock to Lorenzo Sanchez of San Felipe, who thought he understood the district leader's guarantees to mean that the native farmers would always have priority in water rights. The result of these shortages was a serious decline in crop production for all villages in the district for that year.[27]

After one of the biggest floods ever on the Rio Grande in 1941, the Board of Commissioners of the district requested that the Reclamation Bureau and the Army Corps of Engineers take control of construction of permanent flood protection, silt control, power development, and irrigation improvement throughout the valley. So with this federal assistance, plans were laid out for rehabilitation and development of land and water resources in 1947, and both federal agencies exchanged data concerning costs, benefits, and engineering features. According to their combined report, unless their recommendations were undertaken, the irrigated agricultural economy of the valley would virtually disappear within a half century.

After an inspection by representatives of the two federal agencies and L. G. Boldt of the UPA, they met with Abel Paisano, chairman of the AIPC, and members of the Committee on Irrigation to outline a general plan for improvements in the district's pueblos. A report from R. H. Rupkey, hydraulic engineer for the Bureau of Reclamation, stated that the flood of 1941 had destroyed the pumping plant for the San Felipe canal and left the Indians with little assistance from the district for its reconstruction. In addition, the drainage system at Cochiti was damaged, and the repairs left the pueblo lands with a high water table that needed to be alleviated to expand the cultivated area. Flooding on the Jemez River caused sand to enter the Rio Grande and spill over onto cultivated lands of Santa Ana. Silt entering irrigation and drainage ditches of Isleta caused major water flow problems. In general, the possibility of floods and rising water tables threatened virtually all of the pueblos. Therefore, Rupkey recommended the construction of dams for silt and flood control, which he felt would be of great value for the Indian farmers.[28]

In 1948, the U.S. Congress officially authorized a coordinated plan for the Middle Rio Grande Conservancy District through the passage of the Flood Control Act and expanded coverage through a similar act two years later. Not only were two major federal agencies involved, but they also brought with them financial assistance, the lack of which had thwarted progress in construction projects of the past. The district had originally been formed to reverse the trend of depletion of farmlands in the valley, which had started in the late nineteenth century, but its resources were strained, and Native American progress was slow. Older Indians like Francisco Abeyta of Isleta remembered that the Pueblos farmed so much more when he was younger that they sold their excess crops in Albuquerque's Old Town.

In order to rekindle the Pueblo agricultural program, district officials knew that the problems of water shortages had to be rectified. A problem that had been virtually ignored in the past involved the nonbeneficial consumption of water by bosque areas, which had helped lead to shortages from 1936 to 1945. So one of the first projects in the 1950s was the removal of willow and cottonwood growth from areas along the river that had grown up in brush, trees, and native vegetation. Since the Rio Grande in the middle valley was an area of silt collection, the aggraded area was one of flood danger, increased water losses, and poor drainage of irrigated lands. The problem became particularly acute during the period from 1936 to 1952 with certain areas receiving four times as much silt as others. One solution was the process of channelization, which consisted of material being excavated and used to create a soil bank levee.

Dams were extremely important in the rehabilitation project because they not only helped to prevent flood damage and store water for periods of drought, but they also allowed for sediment control during times of high flow. The Army Corps of Engineers made considerable progress in this area as time went on, but their work was slowed by financial restraints, Indian concerns, and problems with water rights. This latter aspect affected new dams and enlargement of older ones, such as the proposed expansion of the San Gregorio Dam in 1956. Other structures completed in this era included the Jemez Dam, finished in 1953, which provided for flood and sediment control near the mouth of the Jemez River, and the Abiquiu Dam on the Rio Chama, the major tributary of the Rio Grande in New Mexico, completed ten years later.[29]

In 1958, the Army Corps of Engineers conducted a survey for flood control on the Rio Grande and its tributaries in New Mexico above the Elephant Butte Dam and distributed its findings to other federal agencies, including the BIA, in the interest of coordinating activities for the building of the Galisteo and Cochiti flood-control dams and reservoirs on the Galisteo River and the Rio Grande. After Area Director T. B. Hall contacted the UPA, Superintendent Williams submitted his comments, which held that the general plan would be greatly beneficial to the middle valley pueblos—particularly Cochiti, Santo Domingo, and San Felipe, which, lacking sufficient levee, had all lost land from river erosion. Since construction plans involved lands owned by the first two pueblos, the plans had to be discussed with Pueblo representatives so that they could decide whether a land exchange or cash settlement would be preferred for areas lost to rehabilitation development. He also felt that because the other four tribes in the district would also benefit, the corps should present the plan at a meeting of the middle Rio Grande pueblos, and Deputy District Engineer Joseph Albert agreed to this idea.

The meeting was held in March of 1959, and another was conducted at Cochiti the next month to provide the tribal council with a detailed understanding of the situation. With the aid of project maps, corps officials explained the basic features and purposes of the dams and the corps' need to obtain title to the lands affected prior to authorization of construction. This would consist of twenty-six hundred acres of Cochiti lands. Following the presentation, corps representatives escorted the Indians on a field trip to view the general location of the dam construction site. They also explained that the structure would neither impair the delivery of irrigation water nor affect any Indian water rights. Given the typical Pueblo attachment to land, the greatest concern of the Cochitis was the loss of a substantial tract of their property for the construction area. Corps representatives attempted to assuage these fears by suggesting an exchange of land from the public domain in the near vicinity of the pueblo.

It was not just the loss of property that worried the Cochiti people, even though it meant that a large portion of their already limited grazing lands would be inundated. More importantly, tribal leaders lamented that "esoteric areas used by our people for traditional tribal ceremonies will be within the proposed reservoir. The value to us of such an area cannot be expressed in monetary terms." Therefore, they felt that the dam should be located upstream from the present site.

If the dam could not be relocated, Cochiti officials insisted that certain stipulations should be met. First they desired a land exchange of equal or greater value in a location that was satisfactory to the Pueblo authorities. Recognizing that this would require federal legislation, they expected full support from all of the necessary branches of government. In addition Indian leaders insisted that they be allowed to develop recreational facilities in the areas surrounding the proposed reservoir, from which they hoped to collect fees for usage. They also called for personal compensation for individual farmers who lost cultivated land due to the construction. Fully knowledgeable of the necessity to protect their water rights, the Cochitis adamantly stated that they "would violently oppose" any loss of irrigation water which they felt should not be disturbed during construction.

Given the bureaucratic red tape involved in a project of this size and the necessary involvement of state, federal, and native governments, progress was slow. The report on this project was submitted to the agencies involved for review and comments as required by the Flood Control Act. The Army Corps of Engineers felt that the bill authorizing construction would not be submitted to Congress until 1960. It would be a long and arduous process. Three years later, the Cochitis continued to profess their desire "to cooperate with their white brothers in providing the needed protection from flood dangers inherent to the Rio Grande." Their stipulations for agreement remained the same except that they were now calling for the corps and its contractors to give preference in hiring to the Pueblos during construction, operation, and maintenance of the project. Within the next two years, they also made provisions for shoreline developments by the corps and the conservation of fish and wildlife resources. Construction began in February 1965, with the final project, finished eight years later, consisting of an earth-filled dam that was 251 feet in height and 5.1 miles long.[30]

In April of 1959, the Army Corps of Engineers met with the tribal council of Santo Domingo Pueblo to discuss the proposed construction of the Galisteo dam and reservoir about four miles upstream from the river's confluence with the Rio Grande. Two weeks later members of the tribal council met to analyze the impact of a dam within their pueblo. The major concerns were the loss of valuable grazing lands, traditionally sacred areas, and several springs that were known to flow even during periods of severe drought. Tribal leaders also worried about the relocation of the Santa Fe Railroad required by the construction of the proposed dam on the Galisteo River and the resulting land loss for the right-of-way strip. They felt that the pueblo should be compensated for any land loss with areas of equal value. Finally, they were concerned about the lack of flood-control works between the dam and the Rio Grande, which they believed should be included in the original construction, and the unsettled issue of liability in case of flood damage. Like the Cochiti Dam, this one faced numerous delays before it was completed in 1970.[31]

In both of these cases, Pueblo officials demonstrated an increase in political savvy, which allowed them to state what was economically and socially important to them. At the same time they were jockeying for position to achieve maximum benefits with minimal loss. In the past they merely rejected innovations because of cultural tradition; now, in the economic reality of the mid-twentieth century, they sought the benefits of modernization without the loss of physical and religious items that were important to their culture.

SOIL CONSERVATION

Pueblo adoption of modern scientific farming methods was essential to the advancement of their agriculture. One method involved practicing soil conservation. Improving Pueblo farmlands was the first step in preserving and strengthening Pueblo agricultural traditions in the postwar years. The deterioration of reservation land was not an uncommon problem, and as Native American farmers emerged from the war, they faced a crisis that had plagued them for decades: almost two-thirds of their reservation acreage was in the arid regions of the West, and by 1947 roughly only 8 percent had undergone necessary soil conservation regimens. Soil erosion was so rampant that by 1949, native farmers had lost 20 percent of their farmlands to its effects. Commissioner Brophy recognized that if the slow pace of modernization continued, halting the deterioration of Indian land would take well into the next century.[32]

The agricultural problems facing the Pueblo farmers were outlined in the 1951 *Annual Report of the Soil Conservation Operation*. Oliver Hole, soil conservationist for the Northern Pueblos Agency, reported that improper care of agricultural land and irrigation systems had reduced the productivity of Pueblo farms and spawned considerable soil erosion. In fact, soil damage and water shortages had driven farmers to leave fallow a very large section of available farmland. Many farmers had only enough water to sustain perennial crops and raise small gardens. Drought during the previous two years restricted the production of forage, opened fields to wind erosion, and lowered incomes in livestock production. Hole also lamented the Pueblos' limited management of rangeland and their land-distribution practices. Having subdivided their farmland over many years, they grew crops on small tracts scattered all over the reservations and thus limited the income potential of their agriculture. Also coming under Hole's criticism were the tribal governments of the nineteen pueblos. Dominated by conservative elders, Hole felt these political bodies hampered modernization and progress. Pueblo tribal customs, Hole claimed, hindered the development of mainstream education on the reservations, although a large number of Pueblos desired the advantages that it offered.[33]

In 1951, to overcome the so-called backwardness of the Pueblos, the NPA held eight educational meetings attended by 120 farmers. Agency personnel lectured on all phases of soil conservation, crop production, and produce marketing and also developed a farm plan for the Santa Fe Indian School, which had about forty-five acres under irrigation. The school program, which dealt with crop rotation, fertilizer application, soil erosion, and irrigation problems, demonstrated a significant gain in vegetable production. The NPA also administered programs to control large populations of destructive rodents, a particular problem in the Nambe and Pojoaque valleys. County agents supplied poison grain to eliminate kangaroo rats and prairie dogs, especially focusing on areas around erosion-control structures to prevent rodent damage. The problem was serious enough that rodent-control efforts were mounted in conjunction with non-Indian farmers to limit the populations more effectively.[34]

Low rainfall, steep topography, and severe overgrazing generated soil erosion and inedible-plant growth, two severe problems in the pueblos. This, in turn, reduced the value of the rangeland for livestock production, greatly minimizing Pueblo income. Hole felt that proper range management, if followed for a few years, would double carrying capacity. He was pleased, however, that Pueblo dikes, diversion dams, and ponds were protecting irrigation systems and spreading excess water over flat areas. Land leveling, although much needed throughout the pueblos to lessen erosion, stirred little interest because of the expense. Wherever possible, the agency tried to encourage contour farming and irrigation to corral the problem. Bureau of Indian Affairs reimbursement for NPA conservation work came to over twenty-eight thousand dollars for the fiscal year.[35]

In 1951, the U.S. Department of Agriculture sponsored a program to demonstrate the benefits of modern conservation methods to farmers and ranchers throughout New Mexico, choosing the Pojoaque Soil Conservation District as a pilot area. Federal, state, and private agencies tried to apply the program district-wide with the hope of extending it—if it proved to be successful—to other regions. The NPA especially benefited from the project, which included approximately one-third of its land. Having worked with the district for the past two years, the NPA officials embraced the concept of a watershed and initiated the plan and made a number of field trips to familiarize themselves with problems and challenges throughout the district.[36]

Although the Pueblos made some improvements in soil conservation, agents in the early postwar era still faced considerable challenges to modernizing their agriculture. Funding limitations translated into staff shortages at the NPA, which had only one conservationist. The Albuquerque district spared limited engineering help, but such staff loans proved burdensome. A fully staffed agency would certainly advance soil conservation for the northern pueblos, but federal funds were limited. Pueblo tradition and culture—primitive farming methods and religious practices—also remained a roadblock to scientific farming in Pueblo country, in Hole's opinion. Indian schoolchildren were learning soil and moisture conservation, but Hole wanted adult education to complement and aid the scientific-farming program.[37]

By 1954, the NPA enjoyed more cooperation and material assistance from the BIA Branch of Soil and Moisture Conservation in reaching the Indians who desired help. More than in the recent past, the Pueblos demonstrated willingness to accept scientific advice and enthusiasm for soil-conservation practices. Hole attributed the new Pueblo keenness to the attention paid to the Pueblos by other local, state, and federal agencies. The conservation branch program that had the greatest impact on the Pueblos was the demonstration project. Native farmers witnessed how agricultural improvements could materially increase their income even on small farms or plots. Indeed, progressive Pueblos began to consolidate their lands and apply scientific practices that fit their farms and returned profits. Conservationists hoped that native farmers would adopt additional modern techniques in the coming years. Such advances, however, were offset by Pueblo farmers' inadequate standard of living. The lack of sufficient farmland forced many Pueblos to support their families by working away from home. This situation frustrated Hole, who saw many hours of conservation work and agricultural improvements lost to farmer absences lasting days or weeks at a time.[38]

Following the trend of other tribes in the West, individual pueblos initiated local farm-aid associations called "conservation enterprises." In 1956, Isleta and Tesuque formed their organizations, with Nambe and San Juan creating theirs the following year. Tribal councils believed that the new organizations would assist their farmers' use and development of soil and water resources and create greater interest among individual farmers in their villages. Each Pueblo reservation set up five-member governing committees that acted as vendors for all construction, made equipment available, and

Figure 10:
Shucking corn, Jemez Pueblo, ca. 1946–1952. Photo by Ferenz Fedor. Courtesy of the Museum of New Mexico, neg. no. 102004.

furnished improved varieties of seeds and fertilizers to individual farmers. As time went on, the programs included land leveling to expand irrigable areas, lining ditches with concrete, and purchasing modern equipment to ease the burden of local farmers.

Perhaps the most notable success occurred at Isleta, where credit was extended by the conservation enterprise to individuals who repaid their loans through increased income from crop production. Encouraged by the tribal council, Isleta's conservation enterprise began to lease land in 1960 to bring previously rehabilitated but unfarmed acreage into production. Within ten years 482 acres had been leveled, lined with concrete ditches, and leased to farmers. The two people most responsible for Isleta's success were John B. Caldwell, the tribe's soil conservationist from the BIA, and John D. Zuni, who served on the tribe's first conservation services enterprise and remained actively involved in planning and implementing conservation projects. Caldwell had assisted Isleta since the pueblo began its program in 1956 and worked closely with tribal officials, committee members, and individual farmers. He also maintained a close working relationship with local county officials who ran conservation programs. Embracing modern agricultural techniques, Zuni was the first farmer in the pueblo to line ditches with concrete and one of the first to lease land that he eventually reclaimed. His innovative approach directly impacted other farmers at Isleta.[39]

Santa Clara Pueblo also initiated its soil conservation enterprise in 1956, and within three years its farmers were laying irrigation pipe, purchasing seeds, and leveling land. The community's budget for 1960 and 1961 showed a profit from soil-conservation work. However, two years later Santa Clara farmers terminated their original agreement, claiming that the pueblo's conservation enterprise did not provide all desired aid programs. The pueblo wanted services expanded to include rangeland conservation and wildlife and recreation development, programs encouraged nationally by the BIA Branch of Land Operations (BLO).[40]

By 1961, the conservation program had grown more sophisticated. The BLO was now responsible for the management of soil and water resources. The agency consisted of five sections, including Soil and Moisture Conservation, Extension, Irrigation, Range Lands, and Sales. The first group emphasized the introduction of practices new to the Pueblos. Working directly with the farmers, BLO technicians introduced the selection of crop varieties better adapted to the high-desert climate, proper fertilization and cultivation, insect and weed control, and more efficient applications of irrigation water. Conservationists also assisted ranching operations with detention and diversion dams that spread water and controlled erosion, ponds that supplied water to livestock, and reseeding programs that improved rangeland grasses. The United Pueblos Agency, under the BIA, heeded the many calls for additional personnel and supplied six conservationists for eighteen pueblos (Zuni being excluded) with each agent covering two to five reservations in close proximity.[41]

Fencing Pueblo lands was a common issue throughout the region. Unfenced fields and ranges suffered destruction by the sheep and cattle belonging to non-Indian neighbors. Nambe had started but never completed a boundary fence, while San Ildefonso faced the same dilemma with a very different neighbor, Los Alamos National Laboratories, the United States' primary nuclear weapons research facility. For three years (1958–1961), the Atomic Energy Commission (AEC), the agency that oversaw the laboratory, had been negotiating with the Pueblo officials over the clear separation of government land from areas considered sacred on the San Ildefonso reservation. The AEC wanted a cost-sharing venture because neither side could afford the whole survey and fencing project. Both sides felt aggrieved—the commission wanted to eliminate Pueblo trespass and both groups sought to exclude roaming, destructive livestock. Finally, the AEC agreed to pay approximately forty-five hundred dollars for their share and requested that the BIA allocate between twenty-five hundred and three thousand dollars while the San Ildefonso supplied the labor.[42]

Taos Pueblo had a similar fencing problem, but its objective was to expand rangeland. In 1956, the tribal council passed a resolution requesting federal funding for fences and cattle guards on a portion of the reservation purchased for livestock grazing in 1937. Never used by the Taos, the parcel was annually abused by non-Indian trespassers running stock. The UPA determined that it could produce good range forage, and Taos officials wanted to plant cool-season grass, which, when combined with native warm-season species, would lengthen the grazing season and alleviate overgrazing on other parts of the reservation. Initially, BIA officials balked at this proposal. In the past, reseeding projects on unfenced range had failed, and the Taos' proposal made no provision for protecting the reseeded area while they were establishing the grass. The Taos council quickly passed a resolution assuring the BIA that the pueblo would seek measures to protect the range in question, and Superintendent Guy Williams gave his support. The reseeding project, protected by a new fence, was a success. Five years later, the Taos again expanded ranch land through the replacement of sagebrush with wheat grass. Once more they requested federal funds to fence the parcel, having already exhausted tribal money to construct additional fencing along the state highway.[43]

Two Pueblo communities, however, were moving away from executing individual small projects to drafting overall operational plans. Following range soil-site and condition inventories by the BLO in 1960, Taos and Laguna pueblos adopted range-management plans for all or part of their reservations. The United Pueblos Agency was particularly pleased, for the Taos and Laguna efforts were the first plans ever initiated by the agency. Delighted UPA officials attributed the Pueblos' eager cooperation to a transformation of tribal procedures that now ran "counter to century-old traditions deeply engrained in their culture . . . in order for them to compete in [the] modern world."[44]

The initial steps of soil conservation taken during the 1950s—some heretofore deemed untraditional or unaffordable—served as foundations for future agricultural developments on Pueblo reservations. These first efforts effected land consolidation, land leveling, ditch lining, cooperative planning with state and federal agencies, applying modern science, and developing the first overall operational plans for an entire reservation. Some federal agents and tribal officials believed that the Pueblos had turned the corner toward agricultural modernization.

A QUESTION OF WATER RIGHTS

Disputes between Pueblos and outsiders over the sharing of water dated to the Spanish rule and grew worse as the population increased in the late nineteenth century. Officials hoped to satisfy the needs of water users along the Rio Grande with the construction of major dams, and gradually minor water-construction projects were implemented for Indian farmers to make their systems more efficient. But the influx of people would not subside, and water conflicts grew in numbers and intensity. What started as local conflicts for the Pueblos developed far more complexity, both legally and technically, than any situation in the past.

Shortly after the war ended, the arid western pueblos of Acoma and Laguna faced a depletion of the water supply for their irrigable lands due to increased pumping on what was known as the Bluewater Project. This non-Indian construction pumped water from springs that fed the Rio San Jose and raised legal questions for Hydraulic Engineer R. H. Rupkey. Because the pumping had been initiated in early 1946, legal steps to prevent the reduction of water flow could not be taken until actual records were made. During this time lapse, water loss would be well established and difficult to stop. Rupkey felt that Indians would have problems establishing responsibility for water loss since the ownership of pumping plants could change hands many times in just a few years. Recognizing that the state of New Mexico had codes covering this situation, he asked authorities to investigate its specifics to fix responsibility of any water loss on the party causing the problem.[45]

At a meeting of the AIPC's irrigation committee in November of that year, the issue of accurately measuring water was raised again. Geraint Humpherys, district counsel for the Indian service, wrote that it was "thoroughly essential to measure water deliveries into Indians' own ditches, in order to make effective, documented demands for more." He also felt this action would keep the district honest in delivering water to the Pueblos if its officials knew they were measuring. More importantly, the only way to force the district to measure water distribution was for Indians to act first. Rupkey, realizing that the district had installed new recording gauges, agreed with this sentiment but warned that such devices would also show Indian waste, and therefore the Pueblos should be careful not to build up evidence that could be used against them. In the interest of protecting Pueblo water rights, UPA superintendent Eric Hagberg contacted the Soil Conservation Service two years later about complaints from the Geological Service personnel that construction by community ditch groups on canal headings caused the water to bypass gauging stations. Having been established in the Tesuque and Nambe valleys in the late 1930s, these devices were important in measuring water distribution and protecting users' rights. Therefore, Hagberg felt that the SCS should monitor the situation more closely so that an already tense situation would not worsen.[46]

The issue of protecting Pueblo water rights raised its ugly head again when the New Mexico Timber Company began its construction of Gilman Mill pond in 1948. This structure, made for the storage of logs hauled out of the Jemez Mountains, was filled with water diverted from Guadalupe Creek through an old irrigation ditch. In addition to floating the logs, water was also used for a steam boiler that was fed by a pipeline above the reservoir site. Eventually, excess water passed over a spillway and back into the creek. Since this stream was a tributary of the Jemez River, it was part of the system that irrigated the Jemez and Zia pueblos. Of course, any loss to upstream users would be detrimental to Indian farming, especially during times of low flow.

By April of the following year, the state engineers approved the company's application to appropriate water from Guadalupe Creek even though state officials had been alerted about the possible interference with Pueblo water rights. In July a BIA representative visited the state engineer's offices in Santa Fe and informed him that during a trip to the site, the representative had observed that practically all of the water was being diverted to the millpond at a time when the Indians downstream were critically short of water. Dodging any responsibility, the state official recognized that the company's rights were secondary to the Pueblos' based on prior appropriation but claimed that he was in no position to regulate the water in question and hoped that both parties could work out distribution among themselves. Inspectors from the bureau's irrigation division clearly observed that the spillway flow was less than that of the intake diversion and attributed loss to seepage and evaporation. Still, Assistant State Engineer A. F. Brown did not feel that the small amount of water involved required the appointment of a commissioner to regulate the water flow.

A. E. Fife, the bureau's area irrigation engineer, was shocked that the state engineer had even considered the application in the first place because it was a well-known fact that there were no unappropriated surface waters left in the Rio Grande basin. In fact, the conservancy district was gradually building up an indebtedness to downstream users below the Elephant Butte Reservoir, which forced the district to protest any new appropriations. He even suggested that the state engineer refuse the timber company's permit until it purchased sufficient land with water rights to cover its needs at the mill. Further complicating the problem was the fact that the water on the Jemez River and the Guadalupe Creek had not yet been adjudicated, and so nobody knew exactly how much water each user was entitled to. Therefore, Fife called for the state to install continuous recorders on both ends of the pond and a water meter on the boiler's pipeline to determine the exact water consumption by the timber company. He also called for a current survey of land irrigated downstream by Pueblo and non-Indian users. He was convinced that during times of water shortage, white operators upstream from the reservations were diverting more than their share of water even though their rights were secondary to those of Jemez and Zia. Recognizing the seriousness of the situation for the future, Fife felt that the threat to Indian water rights might even demand that court action be taken to establish them on a quantitative basis, and bureau irrigation lawyers were in complete agreement with him.[47]

Undaunted, the timber company continued to abuse its water privilege by going beyond the high-water-only limits imposed by state officials. In the spring of 1951, the governors of both Zia and Jemez complained to UPA authorities, who registered their concerns with State Engineer John Bliss. In addition to the quantity of water being used by the company, Pueblo leaders were also concerned about the quality due to pollution caused by the millpond. State officials latched onto this aspect in order to divert attention from the issue of legal amounts and stated that there was no foundation for quality control. Bureau area director Eric Hagberg cautioned the State Engineer's Office that the real issue was water shortage and resulting crop loss and not pollution. Bliss reassured him that the state recognized the Indians' paramount rights and that the Gilman Mill pond permit was only for surplus waters.[48] While Indian officials were aware of the native rights to water, they were equally concerned that state officials should be not only cognizant of this legality, but also aware that something should be done about it or it would be overlooked in the future.

Still the issue of Pueblo water rights was not going away, and other events in the early 1950s kept it alive. Once again a tributary of the Jemez River, this one known as Jaramillo Creek, was involved in a conflict over diversion by non-Indians. In October of 1951, Frank Bond and Sons of Albuquerque made a formal application with the state engineer to build a dam across the creek to store floodwaters. Given the long history of water shortages at Jemez and Zia pueblos, Fife concluded that Bond would probably contend, like the New Mexico Timber Company, that water would be stored only during periods of surplus. There was, however, no practical way to keep a check on the diversion operations and nothing in the design to bypass the dam during critical months. Pointing to the problem on Guadalupe Creek as "a striking example of the problems of safe-guarding Indian rights when a small appropriator has been allowed to divert water during flood periods by the State Engineer," Fife urged caution. He warned that if the application were approved, Pueblo rights would once again be ignored, and he thought it was "time that we speak very plainly to the State Engineer regarding the realistic implications when he approves an application of this kind."

For a number of days, representatives of both pueblos had been flocking into Fife's office to protest the Bond application and demand a meeting. He asked William Brophy to join him on a trip to one of the pueblos to discuss the problem with the Indians. By January, Fife had worked up a supporting affidavit laying out the Pueblo case for the State Engineer's Office. He contended that since the source of water under application by Bond had already been reserved and appropriated, the approval of his permit would be detrimental to the rights of Zia, Jemez, and all of the middle Rio Grande pueblos.[49]

Yet another threat to Jemez and Zia water sources occurred in 1951 with a report of a transmountain diversion from the Rio Las Vacas, another tributary of the Jemez River, to a stream outside of its drainage known as Nacimiento Creek. In July of that year Pueblos, led by Diego Abeita of Isleta, who served as chairman of the district's Indian irrigation committee, raised objections. About a month later Irrigation Engineer L. G. Boldt accompanied Jemez officials on a trip to the headwaters of the Rio Las Vacas in the Jemez Mountains. They located the diversion ditch on Clear Creek, its westernmost tributary, and observed from a date impressed in the cement that it was constructed in 1945. Two years after this construction, a group known as the Nacimiento Community Ditch Association asked for permission to build a storage reservoir. This situation greatly concerned Fife because water that was destined for the Zia and Jemez farmers was being carried to the Rio Puerco watershed toward Cuba and away from the pueblos.[50] The timber company and Bond cases, along with the situation in Nacimiento, had at least served as a warning to both Indians and bureau officials that threats to their water were real, bound to increase, and in need of immediate reaction.

Of course it was not just the Jemez River that troubled the Pueblos. More important were problems on the Rio Grande. During the 1950 irrigation season, the conservancy district failed to meet its delivery schedules in spite of the fact that there was enough water to do so. This, of course, meant that others were getting their share in spite of paramount Pueblo rights. The loss in crop production was bad enough, but now bureau officials were wondering if the doctrine of prior appropriation had lost its meaning.

A bigger disaster awaited Pueblo farmers the next year when the Rio Grande Compact Commission, an agency formed in 1938 to control water use along the river in Colorado, New Mexico, Texas, and Mexico, voted unanimously to open the gates of the El Vado Reservoir on the Rio Chama in northern New Mexico to deliver water to the Elephant Butte Reservoir in the southern part of the state. This action was taken to satisfy complaints from users below the Middle Rio Grande Conservancy District that they were constantly being shortchanged by upstream users. More shocking to federal officials was the commission's decision not to allow any more storage for the rest of the season regardless of the facts that the national government had contributed about 10 percent of the cost of the dam for the benefit of the Indians and the original compact contained provisions that it would not adversely affect native water rights. Adding to the officials' anger was the fact that neither the bureau nor the Pueblos had been consulted before this drastic action had been taken. The commission held firm in their resolve even after the secretary of the interior made a formal inquiry.

Fife felt that this was just another example of "the Indians being pushed around during [a] period of subnormal supply of water"—a time when he felt primary water rights should be most important. In trying to determine why so many non-Indian interests held so little respect for Indian paramount rights, he concluded that the main reason was "a lack of definite legal determinations" of their water rights, which were "vague and lacked dimensions." Indeed, Pueblo water rights were more or less assumed but lacking a solid legal determination that was needed when a conflict arose. He, Brophy, and the district pueblos urgently encouraged bureau officials to firm up Pueblo rights as soon as possible even if it required the attention of the attorney general.[51]

The dramatic move by Texas officials in 1951 grew out of an agreement between states served by the waters of the Rio Grande known as the Rio Grande Compact of 1938. This came about after Texas initiated litigation in 1935 following the construction of the El Vado Reservoir. The need for such an agreement was the result of a dramatic increase in agricultural activity throughout the Southwest in the early twentieth century. Like the Santa Fe Compact of 1922, which divided the waters of the Colorado River among seven states, this one established shares from Colorado, New Mexico, and Texas based on what was thought to be a normal flow. Unfortunately, problems commenced almost immediately when New Mexico failed to deliver contracted amounts to Elephant Butte the very first year. There was some relief in 1941, a wet year in the Southwest, but after that the state continually fell behind. During the prolonged drought of the 1950s, the situation would only worsen. Texas, which was owed some 263,100 acre-feet of water by 1951, reinstituted its lawsuit that year.[52] The fight was on and the Pueblos stood right in the middle.

Pueblo involvement was based on Article XVI of the compact, which stated that nothing in the agreement should impair the rights of the Indian tribes. In addition to their rights to the Rio Grande flow, the Pueblos also had an interest in the waters stored at El Vado because they were paid for by the federal government on the Indians' behalf. Pueblo rights were considered prior and paramount and could not be lost through nonuse. Even though the state of Texas did not specifically mention the Pueblos in its complaint, their action to release the water in the El Vado Reservoir was detrimental to established rights. The conservancy district urged that the gates of El Vado be closed in 1952 to prevent possible flood damage downstream, but they added that closure would also hold water for Indian use.[53]

In March of 1953, the district's Indian irrigation committee held a meeting in Albuquerque to discuss the *Texas vs. New Mexico* lawsuit and its implications for the Pueblos. Chairman Diego Abeita invited Fred E. Wilson, special assistant attorney general of New Mexico; Martin A. Threet and D. A. McPherson of the conservancy district; BIA area counsel William Brophy; and Area Director C. L. Graves. The Pueblos were trying to decide if they should get involved in the lawsuit and wanted the benefit of everyone's opinion. A hearing was to be held the next month in Santa Fe to determine whether or not the United States was an indispensable party, and Wilson felt that the Indians should wait for this decision before bringing pressure on the federal government to intervene. He also felt that "responsible men" from the three states could revise the compact to include Indian rights. He was especially concerned that if the state of Texas got an injunction against anybody using El Vado water in the district until the Texas debt was made good, then nobody above Elephant Butte could divert any water. If the Pueblos actually got into adjudication concerning their specific water rights, they would have to determine exactly how many acres they were farming. The engineering and legal expenses would be huge. Wilson did not feel that the Pueblos had been overlooked years ago when the compact had been made but that the organizers were not "looking to the future and measuring consequences of the situation."[54]

In the fall of 1953, Indian Commissioner Glenn Emmons met with representatives of individual pueblos to discuss their concerns. Most of the villages complained about a lack of water for agriculture. Abel Padilla of Tesuque claimed that farming activity had declined for this reason, and Abel Sanchez of San Ildefonso stated that irrigation development was needed to expand the farmland base, thereby providing jobs for young men who had just returned from service during the Korean War. Diego Abeita of Isleta revealed that several years of drought had increased everybody's awareness of water storage and conservation and recommended that the conservancy district store water at El Vado for the needs of the district's pueblos. He felt that the biggest service the commissioner's office could provide would be to keep a close eye on the water litigation because "everybody is suing everybody else."[55]

In December of 1954, Special Master John R. Green, appointed by the Supreme Court in the dispute, conducted a special hearing in St. Louis, Missouri. Green had previously recommended a dismissal of the Texas lawsuit because the federal government was a necessary party due to the fact that the Pueblos were involved in the water rights claim and they were wards of the United States. Eugene T. Edwards, who represented the state of Texas, claimed that the court could make a ruling without damaging Pueblo rights, which his state would safeguard according to the compact, and this would remove the federal government from the case. Water lost to the Pueblos, he claimed, could be made up to non-Indians in the district through proper repair and maintenance. Edwards stated that he could find no evidence that the Pueblos had contributed to El Vado's construction and that their water should properly be taken from the district's share. However, Fred Wilson felt that the federal government was still indispensable in the case, but he was concerned about non-Indian losses if the Pueblos got unlimited water from the district's allowance.[56] So while both states agreed that the Pueblos were legally entitled to Rio Grande water, neither wanted it to come from their share.

A compromise that would eliminate litigation involved who would control the floodgates at El Vado. As early as 1952, the district had suggested that the State Engineer's Office should handle the job, subject to approval of the Indian bureau, but the BIA claimed no authority in the matter. Instead, the Reclamation Bureau took over this control of the district's operations, and it handled its job so well that the Supreme Court decided to delay its decision on the federal government's involvement until late 1956. The previous year's water surplus of twenty thousand acre-feet had encouraged this action, and Fife hoped that the controversy could be settled without going to trial. In February 1957, the Supreme Court finally dismissed the suit, claiming that the federal government, because it administered irrigated Indian lands and owned structures in the district, was an indispensable party and thus should not have been disregarded.[57]

So for the time being, water shares on the Rio Grande seemed to be apportioned. Ironically, what started out as a dispute between states resulted in a situation that involved the federal government and all because of the interests of the Pueblo Indians who had all but been forgotten. Earlier postwar disputes, which seemed minor at the time, were harbingers of things to come. These events had alerted all parties involved that

Indian water rights were well established and could not be dismissed. Unfortunately the precise amounts due were still vague and with population increases sure to follow, future litigation was a certain consequence. It would, however, be more contentious, expensive, and drawn out.

For almost a generation following World War II, Pueblo agriculture faced problems that were neither new nor easy to overcome. The chronic underfunding that had always plagued Indian programs continued to limit overall progress. The Pueblos benefited from small gains in range management, district irrigation, and local conservation, but a shortage of technical assistance kept Pueblo-wide farming and pastoral development to a minimum. Floods that had devastated the region were eased somewhat by dam construction, but this relief brought with it the question of water rights, which would only get more complex and expensive in the future. Restrictions based on Pueblo religious and governmental traditions still hampered BIA personnel and programs. Moreover, the farm programs of the Pueblos declined, and the termination policy was an assurance that they would continue to move in that direction. Ironically, as Pueblo farming went down, the water still being used on their reservations became more important, and the rights to it would drag the Pueblos and their regional neighbors into a contest that would not only determine these legalities for native New Mexicans but for other Indians as well.

CHAPTER NINE

Pueblo Agriculture and Economic Development (1962–1980)

SELF-DETERMINATION FOR PUEBLO FARMERS

Complaints about termination and its devastating consequences echoed across Indian country to the nation's capitol, but with the election of John F. Kennedy, a new day dawned for Native Americans. The president, dismayed by conditions of poverty throughout the country, including Indian reservations, sought to improve economic development in native homelands and extend self-determination to native peoples. Following the theme of the United Nations in pronouncing political independence and economic growth for third world nations, he sought to extend the goals of economic development and self-determination to native peoples. In 1961, Secretary of the Interior Stewart Udall assembled a task-force report on Indian affairs that emphasized the development of tribal resources, both human and natural. More than a decade would pass before termination ended as a federal policy, but the next three administrations would continue to push for the economic advancement and self-determination of Native Americans.[1]

What began as antipoverty programs under Kennedy expanded into the War on Poverty during Lyndon B. Johnson's administration. The Office of Economic Opportunity (OEO)—created by the Economic Opportunity Act of 1964—oversaw the programs and encouraged local communities to take on decision-making power. However, intense lobbying by various Indian organizations was needed before Congress extended the OEO benefits to Native Americans with the creation of a special "Indian desk."[2]

As a result of Johnson's OEO program, community action projects encouraged tribal councils to develop their own economic programs. Commissioner of Indian Affairs Philleo Nash requested that, with the assistance of local BIA superintendents, all reservations nationwide initiate a ten-year development program to determine the needs of individual tribes. In June 1964, the BIA held a conference in Santa Fe to disseminate information on the program. BIA superintendents in New Mexico were encouraged to inform tribes about available services. Designed to be "people oriented," the program emphasized community action and cooperation between the BIA and local Indians. Accordingly, the superintendents determined that overall Pueblo goals should include the development of irrigation, range, and arable lands. They next surveyed each pueblo to learn its specific priorities for new programs and funds. The following chart lists demographic, unemployment, and labor statistics for each pueblo.

According to the statistics provided by the report, about 1 out of every 4 Pueblo tribal members was now living off the reservation. Almost 15,000 of the nearly 20,000 Pueblos (excluding Zunis) remained in their homelands trying to eke out an existence. Santo Domingo, Isleta, and Santa Ana had the highest percentages of home residents while Cochiti, San Juan, Sandia, and Nambe had the lowest. Unemployment for all Pueblos was nearly 28 percent with conditions best at Picuris and Taos and worst at Pojoaque. The jobless rate for most villages ran between 24 and 31 percent. Out of a working force of 4,752, the number of Pueblos working full time was 3,432. Some of these were involved in employment in the tribal government, arts and crafts, tourism, and recreation, but the vast majority were probably in the

TABLE TEN

PUEBLO LABOR STATISTICS—1964

PUEBLO	MEMBERSHIP	ON RESERVATION	%	LABOR FORCE	WORKING	UNEMPLOYED %
Acoma	2,355	1,674	71%	526	410	22%
Cochiti	614	346	57%	132	90	32%
Isleta	2,193	1,974	90%	673	559	17%
Jemez	1,528	1,076	70%	356	224	37%
Laguna	4,432	2,956	67%	1,030	721	30%
Nambe	230	135	59%	44	30	28%
Picuris	164	100	61%	35	30	15%
Pojoaque	64	41	64%	10	4	60%
Sandia	215	124	58%	33	25	24%
San Felipe	1,350	1,000	74%	430	301	30%
San Ildefonso	272	224	82%	62	43	30%
San Juan	1,150	670	58%	227	159	30%
Santa Ana	413	366	89%	124	86	31%
Santa Clara	723	535	74%	152	120	21%
Santo Domingo	1,977	1,938	98%	476	309	35%
Taos	1,390	896	64%	272	196	18%
Tesuque	207	142	69%	39	27	31%
Zia	458	377	82%	131	97	26%
TOTAL	*19,735*	*14,574*	*74%*	*4,752*	*3,432*	*28%*

time-honored profession of farmer and/or rancher. Assuming that three out of every four were still in agriculture, this would mean that there were approximately 2,500 farmers in 1964. These same villages in a survey of 1890 listed over 1,350 ranchers and farmers out of a population of 5,666 (excluding Zuni) representing 74 percent of the workers. So while the population had more than tripled during the previous seventy-five years, the number of people in agriculture had probably not even doubled.

To combat high unemployment on the reservations, the pueblos listed a number of agricultural goals and priorities. Many cited the need to improve and increase range and agricultural land with better irrigation systems and flood-control programs. Others submitted specific wish lists. Isleta wanted to prepare its remaining 4,100 acres for irrigation and to drill auxiliary wells for specialty crops. Acoma sought irrigation facilities to utilize water from the Rio San Jose. Jemez hoped to rehabilitate its entire 2,500 acres of irrigable land, and Picuris wanted to do the same with 220 acres in addition to developing over 14,000 acres of rangeland. San Juan's priority was the improvement of 2,000 acres of irrigable land, and the pueblo also hoped to acquire land adjacent to its reservation. Santo Domingo set a target date of 1975 for developing over 3,000 acres of farmland and 65,000 acres of rangeland for nine hundred animals. Taos hoped to add 5,000 more acres of irrigable land and Tesuques 800 more acres to their respective reservations.[3] Of course, none of these wishes would become reality without funding. Indeed, lack of money would be a major obstacle to Johnson's antipoverty programs, including those in Pueblo country. In spite of the fact that the number of farmers was dwindling proportionally to population growth, the Pueblos, still clinging to their traditional economy, saw expanding agriculture as a way to combat chronic unemployment and poverty in their homelands.

They were not alone in this thinking. A study done in 1967 by the Kirschner Management and Economic Consultants of Albuquerque focusing on water-use projections for Nambe, Pojoaque, San Ildefonso, San Juan, Santa Clara, and Tesuque for the rest of the century noted that the most important part of their economy was agriculture. In spite of the fact that the firm found that farming was relatively undeveloped, the demand for irrigation water exceeded all other water needs combined. These six pueblos had more than seventy-six hundred acres of irrigable land but irrigated only a little more than thirteen hundred acres in 1965. Still generally tied to subsistence agriculture as they had been for generations, they were certainly not utilizing available farmland

and water resources and, more importantly, these activities had decreased. In 1940, the villages irrigated over two thousand acres, and a quarter of a century later, this figure had fallen to seventy-six hundred.

The investigators queried how this could happen in an age when the demand for agricultural output was growing and pressure was increasing on available farmland both regionally and nationally. They projected that the demand for agricultural land, as the population grew, would lead to a deficit of cropland acreage by the year 2000. This, in turn, would mean that all areas of arable land would become a potentially valuable resource. They cited the "occupational revolution" in the postwar years that had drawn off Pueblo talent to Santa Fe, Los Alamos, and Española as a partial reason for the decline in farming. However, since labor and capital (in terms of federal financing programs available to Indians) were both accessible, they felt that the underutilization of agricultural resources could be attributed to improper economic planning. While the investigators did not expect a sudden agricultural revitalization in the six pueblos, they urged that maximum utilization of farmland become a desirable policy goal.[4]

During the 1960s, the federal government began to increase assistance for the development of agricultural resources on Indian reservations. By 1968, to that end, almost half of the BIA's budget was being pumped into other federal agencies, including the Department of Agriculture. Although that department did not administer any specific reservation programs, it did oversee others that benefited Native Americans. For example, many New Mexican tribes participated in soil and water conservation projects that aided the Rio Grande valley.[5]

The spirit of cooperation between federal and state agencies was highlighted in 1963 when Secretary Udall signed an agreement with the Pojoaque–Santa Cruz Soil and Water Conservation District in New Mexico. For the first time the Interior Department indicated a willingness to work with a local district on an area development project. Previously, there had been some cooperation between districts and the Agriculture Department, but the Interior Department had disregarded those conservation efforts to concentrate on its own regional initiatives. The National Association of Soil and Water Conservation Districts hailed the agreement as a big step toward future cooperation in developing western land. For the Pueblos, who came under the control of the Interior Department and also fell under the jurisdiction of state conservancy districts, this action meant cooperation between typically conflicting agencies in establishing uniform goals and procedures for conservation projects without any conflicts of interest.[6]

Recognizing that federal assistance alone could not solve the region's economic depression, the Pueblos joined their non-Indian neighbors in various Resource Conservation and Development (RC&D) projects that were authorized under the Food and Agricultural Act of 1962. Under the leadership of the Soil Conservation Service, the northern New Mexico RC&D work was one of ten pilot projects approved by the secretary of agriculture in 1964. Realizing their common problems, the people of the region, in conjunction with the Soil Conservation Service, the New Mexico state engineer, the Bureau of Indian Affairs, and other local agencies and civic organizations, pooled their resources to combat poverty in their locale. The effort was a perfect example of how local people with a voice in determining their economic future could work with government agencies to accomplish their goals.

Given the untapped resources of the region, RC&D's purpose was the conservation and development of those resources. This included the establishment of facilities for marketing local fruits and other specialty crops, improvement of community irrigation systems and rangelands, and flood prevention. The project specifically focused on developing the overall resources of the eight pueblos in the region. In addition to increasing employment and per capita income, the RC&D hoped to stabilize the agricultural economy through effective land use and conservation practices as well as to increase the value of crops. As a result of the RC&D's work, the BIA reported some impressive accomplishments by the northern pueblos, including range and irrigation improvements and watershed protection projects.[7]

As early as 1965, Walter W. Olsen, general superintendent of the UPA, reported that all eight northern pueblos had submitted project proposals under the RC&D work plan and that accelerating regular BIA programs designed to create new jobs and opportunities in the region was spurring forward Pueblo development initiatives. Specifically, San Ildefonso was working with the Pajarito community in lining nine thousand feet of irrigation ditches—a project that was noteworthy in its interracial cooperation. Santa Clara had approved a similar project of thirty-five hundred feet and, with the help of the BIA, installed a mile-and-a-half pipeline that opened a new grazing area on the reservation. Nambe also completed a watershed project in which twelve hundred acres were cleared, seeded, fenced, and outfitted with three livestock water tanks and twelve erosion-control dams.[8]

The work of the Soil and Moisture Conservation Service (SMC) went so well that it offered examples for others to follow. In 1963, the SMC allocated over $268,000 for UPA programs, breaking down the sum into land-use planning (28 percent), soil improvement (20 percent), water management (44 percent), and operation and maintenance (8 percent). The pueblos recognized that they could use funds only in areas where proper management practices were followed and that the details of the plan had to be approved by their people. Santa Clara was allocated $27,000 for a showcase project, and two years later, in 1963, the conservation service of the pueblo entertained the idea of paying for equipment to level eight acres of land for an experimental farm. In 1966, the BIA set up a conservation-training program for a group of African students who toured the Southwest that summer. They visited Taos, Jemez, Zia, and other pueblos with a special emphasis on conservation.[9]

Thus, the 1960s represented a major shift in emphasis for the Pueblo farming programs. Previously thwarted by the termination program, the pueblos found a new spirit of cooperation with the federal government. Johnson's War on Poverty initiatives helped rejuvenate Pueblo agricultural activity, emphasizing the maximum utilization of resources and the cultivation of specialty crops. The new cooperative efforts not only brought together state and federal agencies, but also the Pueblos and their Hispanic neighbors. This new combination of groups saw that they had more commonalities than differences and assisted one another in natural resource development. More importantly, the voice of local people was now being heard. Progress would be slow, but a new age of self-determination was beginning.

With detailed plans and mutual cooperation, it appeared that authorities needed only to find the human and financial resources to bring the dreams to reality. Still, there were major gaps to bridge. Traditional Pueblo subsistence farming continued to present a major barrier to a full-blown modern agricultural program. More problematic was that maximization of land and water resources required huge capital outlays from federal authorities to a small group of Native Americans who heretofore had been largely ignored. In addition, as the decade of the sixties wore on, tax dollars would shift from the War on Poverty to the war in Vietnam. The policy of "guns and butter" would produce far more of the former than the latter.

PUEBLO RANCHERS

As an adjunct to farming, livestock operations had never received equal attention. Development of rangelands, fencing, stock tanks, and well drilling initiated in the early twentieth century were all cost-prohibitive ventures and did not have the cultural and religious traditions of crop growing. However, the two related ventures meshed and often involved programs that were interrelated. For example, the need for fencing at Tesuque in 1966 was requested not only for range management but also for the protection of gardens and croplands.[10] The emphasis on livestock programs varied from one pueblo to another, but the need to develop in this area was a growing concern.

Federal government fencing programs dated back to the New Deal period. Construction under the Civilian Conservation Corps provided some relief in this area, but the pueblos of Jemez, Santa Ana, and Zia felt that boundary fences on their grants were unnecessary at that time because of the insufficient development of range-water facilities. As time passed two changes brought about a reconsideration of this situation—the development of new water facilities and the creation of new boundaries. The first simply meant that new areas for grazing had been opened up and required the protection of range resources. The second had to do with the addition of a portion of the Espíritu Santo Grant, held in trust for Jemez and Zia, and the increased traffic along State Highway 44, which bordered the Santa Ana and Zia pueblos.

In terms of the grant, the UPA authorities informed the Pueblos that necessary water developments on this land could not be constructed until the area was fenced for exclusive livestock usage. The problem, of course, was financing. Over thirty-four miles of boundary fences were required to divide the Indian sections of the grant land and deny uncontrolled use by non-Indian livestock. The Jemez contributed sixty-two hundred dollars for the project but were unable to supply additional funding; this meant that a large section of their grant lands was unusable. Assistance from the Indian bureau proved to be futile since fencing projects were not allowed under the terms of the Area Redevelopment Act.

The highway problem also resulted in frustration. At the time of its construction, Route 44 presented few problems because it was lightly traveled, especially at night. But with the development of oil, gas, and uranium resources in the San Juan basin all that had changed; traffic had not only increased dramatically but also involved heavy trucks that often traveled after dark. The loss of Indian livestock had amounted to almost fifty head per year for the past nine years, and more serious was the danger to the traveling public. The Pueblos required twenty-seven miles of fencing along their borders, but the state highway officials had no plans for fencing a road after its construction.

Thus thwarted in their efforts, the governors of the three pueblos petitioned Deputy Commissioner of Indian Affairs John O. Crow, who contacted Senator Clinton P. Anderson and Chief Highway Engineer T. B. White; the Pueblos were in touch with the representatives Joseph Montoya and Thomas Morris. The congressmen were unable to garner any financial assistance from the Interior Department or the legislature and Crow's efforts turned up empty. The commissioner finally contacted the Pueblo governors, and while he recognized the need for fencing, he lamented that the CCC days had passed. The current policy was for land users to provide for their own needs, and he felt that livestock owners should be willing to contribute to the fence project to prevent the loss of their animals. He did suggest that the Pueblos take full advantage of the Agricultural Conservation Program to receive cost sharing for the project. Clearly, these events took place in an age of transition between the forces of termination and self-determination, and the Pueblos were caught in the middle. In the future, the Indians and the State Highway Commission would rely on cooperative fencing agreements to solve their common problems.[11]

While fencing was first on the list of rangeland development, water sources had to be found to feed livestock and irrigate seeded land. In 1964, the state of New Mexico initiated a water resource study and part of the information required by the State Engineer's Office included the total number of active stockwater ponds and their average size by surface area. For the Pueblos there amounted to 223 ponds that averaged three-fourths of an acre.[12] Maintenance of these was minimal compared to other water sources, and the cost after construction was not a particular burden for the tribes.

A more cost-prohibitive activity for Pueblo ranchers involved rangewater wells, windmills, and pipelines. As part of the termination movement, these types of activities were to be taken over by the tribes, even though many could not afford it. In 1964, the Indians of the Southern Pueblos Agency agreed to assume maintenance for all wells, springs, and pipelines constructed after the fiscal year 1958. The BIA continued to be responsible for the older operations, which included 146 wells, forty springs, and five pipelines. The Indians considered the maintenance cost of the remaining 58 range-water wells to be prohibitive, and in 1966 the All Indian Pueblo Council, deeply concerned about the problem for livestock owners, passed a resolution urging that the bureau absorb this burden. Chairman Domingo Montoya cited the financial position of subsistence ranchers and their lack of training and know-how necessary to do the job. Rejecting this proposal, acting commissioner Theodore Taylor emphasized that the assumption of these responsibilities would be gradual and that each pueblo would be given individual consideration.[13]

In order for the Pueblos to take control of their range improvement programs, they first had to learn the steps needed to be taken according to government planners. In the summer of 1969, the Tesuque council invited officials of the Northern Pueblos Agency to discuss a range improvement program. Agency officials explained the need for a complete soil and range survey and a well-prepared range management plan prior to range improvements. Acknowledging these necessities, the council requested that work commence as soon as possible. By the following year, the soil survey was finished and range plans were being prepared. Similar work was done at Taos in 1970, and five years later the Branch of Land Operations expended $12,500 for fencing, grass seed, soil survey, and range well maintenance.[14]

The problem of developing water sources was particularly acute for the vast western pueblos whose arid regions lacked a major source of water. Acoma Pueblo was particularly upset by development cost and passed their own resolution in 1969. Deputy Commissioner T. W. Taylor cautioned Governor Sam Victorino that President Nixon's recent order for a 75-percent reduction in federal construction expenditures had a major impact on funds available. Undaunted, the Acoma Tribal Council pointed out that developing irrigation facilities from the Rio San Jose and a range management program had been part of the ten-year planning report submitted to Secretary Udall in 1964 and that they had submitted a Long Range Plan to the UPA in 1958 requesting funds for irrigation. The need for outside funding was crucial since tribal members' income was less than $750 per year, and in 1964, Acoma was listed as one of sixteen tribes in the nation considered poverty-stricken. Acoma officials listed forty-three different needs, cited by village irrigation bosses, involving concrete ditches, pipelines, windmills, storage tanks, spring repair, and water trough construction.

Unfortunately, the UPA Natural Resource Office was equally limited in funding for range improvement programs. Six positions in the Range Water Program had been cut in 1967 and 82 percent of the bureau's 1970 budget was allocated for maintenance. The natural resource manager took little pity on the tribe and pointed out that Acoma livestock men had received $215,000 in gross income in 1969 and that the $250 average maintenance for the eight range wells for which they were responsible was well within their budget. He also pointed out that cooperation between the bureau, who furnished a foreman and material, and the tribe's workers resulted in the repair of storage tanks each year. Many of the forty-two projects cited by the Acoma council, he noted, were already completed within budgetary limitations, including wells and springs for which the tribe assumed responsibility. He urged Acoma officials to update their progress report and realize that portions of the ten-year economic development program were out of date. In the spirit of lingering termination attitudes, he noted that "the people using and benefiting from the range should gradually assume greater management responsibility" because whenever Native American range users become financially involved in improvements, they took more interest in the program and range management programs became "'theirs', not the 'Government's.'"[15]

Perhaps aware that a more adroit approach was necessary, Acoma governor Merle Garcia contacted New Mexican congressmen in 1976 to seek support for Acoma's soil and moisture conservation program. He tactfully indicated that the pueblo lacked the wherewithal to bring the resources back to their best condition and to heal bare areas that were once good meadows and fields. He stressed the need for technical assistance from the federal government to improve livestock operations that would allow ranchers to achieve some kind of "reasonable success." Tribal officials reduced their ranching wish list down to nine items, which included dams, ponds, reseeding, brush control, fencing, and a range condition survey. Working through Washington bureaucrats proved effective, as their correspondence with bureau officials led to closer cooperation with Acoma tribal leaders in establishing priorities for natural resource development.[16]

Zuni, perhaps the most arid of all the pueblos, had unique problems in the area of range improvements. In 1970, their acting superintendent Ira Dutton requested information on rangeland terraces. Being in a low-rain zone and plagued with high-intensity storms, he felt that terracing would decrease the heavy runoff that frequently occurred on the reservation. However, Forest Service and Bureau of Land Management reports revealed that costs for contour furrowing were too high to benefit low-value rangeland and should only be used on valuable irrigated farmland. Rejected in those efforts, Zuni officials pushed forward. Two months later the tribal council authorized Governor Robert Lewis to enter into a contract with the BIA to construct erosion control features, create stockwater and diversion dams, and make range and irrigation improvements.

Zuni officials felt that the real problem was the result of outsiders not being aware of progress they had made in range management. Having taken over the supervision of the government's programs on their reservation, they sought methods to insure that future stockmen in their growing population would learn modern methods of land improvement. They pointed out that where these types of programs had been carried out, the resulting range conditions improved to good or excellent. Unfortunately, few if any scientific records were kept, and the constant change in personnel meant that there was little evidence for improvements that had taken place. But modernization had come about, resulting in livestock increases that amounted to 18,000 sheep, 150 goats, 808 cattle, and 400 horses.

To advertise these gains, the tribe planned a demonstration project, which would expose for the first time the basic elements of proven conservation to other tribes, neighboring ranchers, and Zuni youths with the hope that everyone could increase their knowledge of soil and water conservation. This project area, located on both sides of the main highway through the reservation, was easily accessible and the tribal council planned to put signs along the road detailing treatment methods with updated results. Zuni officials planned to make film, slides, photographs, and field trips available to their schools—all at their own expense. They hoped that their efforts would produce individual enthusiasm, participation, and support critical to the success of range programs.[17] Previously rejected by bureau officials, the Zunis hoped to prove that their cause was just and that through their own independence, local ranchers could adopt their ways to modern science.

In some ways the problems for Pueblo ranchers were more difficult than those of strict agriculturists. The vastness of the areas they controlled meant higher cost in applying soil and moisture conservation methods. Plants and animals need water, but stock required feed that was expensive to produce.

Also, most of the Pueblos had dealt with crop raising for a longer period, and it was more intimately intertwined with their religious customs. Still, the development of both of these aspects of their economy in the late twentieth century were necessary to employ those Pueblos who wanted to remain tied to the land.

EXTENSION SERVICE FOR PUEBLO FARMERS

The Pueblos' success in rapidly changing modern scientific farming depended on acquiring an awareness of the latest methods available. As federal policy moved away from termination and toward self-determination, many federal agents believed that the best way to capitalize on the new independence was education at all levels. In the late nineteenth century, agents had attempted to initiate various innovations designed to modernize Pueblo agriculture, but this action went against the grain of local tradition. It was not until the New Deal era that modernization began to take hold, and in the 1950s soil conservation spread through the Pueblo region with some villages initiating their own services. Still, the more tradition-bound farmers held to their ways, and for them the goal of reaching their highest potential remained distant.

The federal government recognized the need for agricultural education for America's farmers in the early twentieth century, and in 1914, Congress passed the Smith-Lever Act, which created the Cooperative Extension Service designed to take knowledge directly to the country's rural people. The idea was to translate the products of research into laymen's terms and make them available for farmers to make higher-quality decisions. The extension service survived and expanded because it improved the quality of life for farmers by being responsive to their needs and using objective research to design educational programs for them. The three basic areas of service included assistance in farming and ranching, home economics, and 4-H clubs.

While extension service personnel flirted with modernization efforts on some Pueblo reservations in the 1930s, they met with limited success. Initially, the Indian bureau was responsible for providing Native Americans with technical assistance, but in the 1960s state universities were contracted to provide the services. Under state statutes, New Mexico State University (NMSU) was charged with the responsibility of disseminating information to local tribes, and in 1956, the first contract was made with the Jicarilla Apaches.

Figure 11: Delfin Lovato, All Indian Pueblo Council Chairman (San Juan—Santo Domingo). *Pueblo News* photo, PN 0121. Courtesy of the Indian Pueblo Cultural Center.

Gradually, the university's services expanded to twenty-one different tribes, and within fifteen years, its budget was over two hundred thousand dollars. By 1974, the BIA office in Albuquerque was receiving almost 10 percent of the bureau's total funding for extension programs.

A. E. Trivis, who administered the program until 1971, ran an aggressive program that was administered by local extension agents working directly with the tribal members. They met with tribal officials and bureau workers to establish annual programs. Special emphasis was given to intensive farm activities, including advice on obtaining loans for production improvements and encouraging Native American participation in both county and statewide extension activities. Similar government programs before World War II were staffed by agents who were typically incompetent, underpaid, and ill concerned with the plight of Pueblo farmers. After the war, however, the agents were qualified, full-time employees who provided services aimed at the needs of the Pueblos.

A yearly evaluation of the program helped to determine its effectiveness, changes in tribal priorities, and the allocation of funds for the upcoming year. NMSU also provided specialized training courses, specialists and research people (thirty-five by 1976) available to travel to reservations, and publications of special interest and value to Pueblo farmers. The university provided these services, including supervisory costs, with no charge to the bureau, but as they assumed an increasing role in extension services, their program was restricted by limited budgets. Thus, falling behind in the race to achieve Indian agricultural modernization, in 1975 the extension service requested funding from the Department of the Interior through the bureau as a means of funding the education program. The service suggested that in this way land grant universities could help Indian farmers achieve the national goals of independence and self-determination.[18]

The budgets of the NMSU extension program reflected that, early on, Zuni was more actively involved in the program than all of the other pueblos combined. The Zunis received 42 percent of the 1969 budget, compared to 27 percent for the whole UPA. The money helped finance two agricultural agents while other pueblos had none. Concerned about the details of the program, newly elected AIPC chairman Benny Atencio contracted Director Trivis in 1971 about a possible meeting to discuss the program. A week later Vice Chairman Delfin Lovato suggested that BIA representatives and other state extension service personnel be included in the gathering slated for March 4.

The meeting was probably more contentious than any of the non-Indians expected. After Trivis outlined the broad aspects of the program, William C. Gutzman, area range conservationist and extension supervisor, explained the operations in his area. He pointed out that at that time, the federal government was contracting with eighteen states to provide extension services to Native Americans. With special emphasis on New Mexico, Gutzman noted that Indians in the area were currently receiving more than five million dollars in income from their products.

Before the entire program was explained, Pueblo leaders launched into an open criticism of the extension services. Lovato felt that the program's resources were not producing much income for individual Indians and that native leaders were doing all of the work for which six agents were getting credit. Atencio wanted to know why the state could not provide services for Indians without contractual money, and he complained that bureau officials were not cooperating in water resource inventory. Both officials also asked why no Indians had been hired in technical extension positions like "agricultural agent" and what had happened to two positions that had previously been housed within the AIPC.

Non-Indian officials countered these assertions. Gutzman pointed out that great potential existed for increased income for Pueblo farmers and ranchers, and James Cornett, area natural resource manager, emphasized the necessity of making beneficial use of water rights through increased production lest the Indians lose them sometime in the future. Unaware that there was any lack of cooperation on the bureau's part, he understood that Washington officials were in charge of the water program. Mark Stevens quizzed Lovato on the livestock situation, and when Lovato confessed a lack of knowledge on the subject, the SPA natural resource manager listed an inventory of eight thousand cattle and more than eighteen thousand sheep for which Pueblo livestockmen were constantly requesting assistance to manage herds and ranges. Gutzman addressed the new contract program by noting that Indians would get the same services as other citizens but that Congress and the bureau felt they should focus on problems unique to the reservations.

Gradually, Pueblo employment became the central focus of the discussion. First Gutzman noted that Commissioner Robert Bennett, an Oneida, requested that the bureau get out of the extension business and contract positions so that Indians could get services equal to other Americans; thus temporary positions were to be filled by technicians. Non-Indian officials told the Pueblo leaders that no qualified Native Americans could be found to fill the job of extension agent. Atencio flashed back that his extensive agricultural experience should make him a candidate, and Lovato stated that he did not accept the argument and mentioned the fact that the OEO program was successful in finding Indian employees. Basically, these Pueblo officials gave the others the impression that they were more interested in Indian employment than hiring technically trained college men.[19]

This confrontation was happening on the heels of the beginnings of affirmative action and the rise of Native American nationalism throughout the country. Even the Pueblos, generally known for their cooperation and passivity, attached themselves to the movement. Non-Indian officials may have been surprised and somewhat disappointed at the reaction, and understandably so, but self-determination ideally embraced independence and self-worth, and in retrospect they should have expected this reaction in its initial stages. Years of following federal dictates had worn Pueblo patience thin, and centuries of self-reliance were beginning to reassert themselves.

The early complaints of Pueblo officials that the extension program lacked Indian employees, and the counterclaims by non-Indian officials that technically trained agents were a necessity, were resolved by an innovative NMSU program initiated in 1975. The Indian Resources Development and Internship Program, funded by the Kellogg Foundation and the Four Corners Regional Commission, was designed to develop Indian professionals in agriculture and technology. Director Russell Kilgore called it "a massive effort" to prepare Native American youth for professional studies, involve them in career-related jobs while in college, and match them with jobs in government, tribal affairs, and industry after graduation. By 1979, twenty-five Indian students were on cooperative work assignments with the BIA, on commercial farms, and in agribusinesses in several western states. The university also conducted a feasibility study for initiating an agricultural business enterprise for Acoma Pueblo the next year. Officials concluded that potential existed to significantly expand production from irrigated agriculture, but that both the BIA and the tribe should take an active part in putting the program together.[20] Certainly a new spirit of cooperation had developed since the early years of the program.

The extension service program moved ahead but with mixed acceptance. By 1975, the Zuni budget increased only slightly while the Southern Pueblos Agency's budget grew to almost five times what it had been for the entire UPA during the six previous years. Participation in the program was strictly voluntary, and the eight northern pueblos decided not to receive services. Tribal leaders of the SPA, however, were impressed with the information they received and surprised with the development in local Pueblo leadership. They also wanted future assistance with setting up new farming operations, local crop demonstration plots, and improved water distribution systems. With special emphasis on agricultural development for the young, Zunis requested a youth agent for their reservation. Unfortunately, the national economic recession of the late 1970s undermined the extension program. Ironically, federal budget cuts came at a time when the Pueblo interest in the extension service was peaking. Underfunding, therefore, threatened recent gains made by the tribes.[21]

Extension service assistance embraced a wide spectrum of activities. One of the most cost-effective ideas was the use of new types of fertilizers. Agent Fred Alber reported some spectacular results in 1976 that resulted in a fantastic increase in demand the next year. Nearly fifty farmers in the program requested almost 350 tons of fertilizer; their interest was in response to a report showing that in some cases a dollar spent returned more than five times as much in increased harvest. While results had exceeded expectations, there were problems involving procedures. Because the fertilizer experiment was done in cooperation with the Tennessee Valley Authority (TVA), all experimenters were required to follow systematic test demonstration plans in order to qualify for the low-cost educational program. Many Pueblo farmers ignored the rules concerning test demonstration plots and simply applied the fertilizer without following instructions. In the third year of the program, a meeting was held to emphasize the need to comply with regulations, and the results of the next year made a strong impression on the farmers. With the accumulation of raw data, the TVA made plans to publish a feature story in their national magazine complete with pictures and comments from Pueblo farmers. Other publications such as the *19 Pueblo News* and statewide extension service reports contained major features on Indian farm activity.

Another area of interest for the extension program involved a gardening program. The first annual gardening clinic was held in March 1977, when horticulturist Dr. Ricardo Gomez held a workshop for the southern pueblos. The program was so well received that Agent Alber tapped into an outside budget to purchase a tractor for Zia, along with implements, seed, fertilizer, and insect spray. He taught the farmers how to operate and maintain the new equipment and methods for controlling insects. The latter was such a major concern that Extension Entomologist John Durkin conducted three insect-control workshops, and more demonstrations followed the next year. These efforts paid off, as the Pueblos were now able to identify and anticipate certain insect damage and take action to prevent it. It was especially gratifying to extension personnel that Pueblos schooled in this subject were teaching others what they had learned. Again, this instruction proved cost-effective, as the Pueblos had lost several thousand acres of alfalfa to weevil infestation. Grasshoppers, traditionally a serious pest for Indian farmers, also did extensive damage especially in 1979—a time agents dubbed "the year of the grasshopper." While native farmers were now taking action to minimize insect damage, agents were still concerned about lingering dependency—the expectation for extension employees to "do it for them." They planned to persist in their efforts to turn this situation around.

A rather unique aspect of the extension efforts focused on fruit trees. While the Spaniards had introduced them centuries before, they never seemed to become an important part of agricultural production and in some cases simply died out. All of this changed in 1973 when 2000 fruit trees were ordered for the southern pueblos, and additional ones were ordered in subsequent years. In 1977, the service acquired 527 more trees for 100 Indian families. Agents set up twenty-one planting demonstrations at six pueblos, and many families planted them immediately. The agents also instructed them in pruning and watering and handed out an NMSU circular on the subject. In addition to fruit trees, Ernest Lovato of Santo Domingo requested help with landscaping with the expressed intention of making his area a showplace in the village to inspire others to follow his example. The fruit tree program continued to expand; two years later 215 families planted 800 trees, and agents made plans to continue the effort in 1980.

Extension programs expanded to include youth in livestock activities. Work in the 4-H program involved livestock programs such as a slide series on animal health, a horse clinic, and pig raising—all of which helped to increase youth confidence and interest. The emphasis on livestock encouraged the extension staff to organize cattlemen in a unified effort to market their animals. This was especially true of excess horses that were taking up rangeland that could be used more profitably by cattle and sheep. Pueblo ranchers, at the urging of agency personnel, attended various livestock schools and seminars where marketing was a major emphasis. The main thrust of these meetings was to encourage the formation of co-ops in order to maximize profits.[22]

The work of the extension service, however, made significant progress in a relatively short time. The program worked so well at Zuni that tribal officials decided to sever ties with the service. By 1981, the Zunis had terminated their agreement with NMSU and closed the extension office. Zuni had always been the most independent of the pueblos (a condition fostered by both geographical isolation and distinctive linguistic traits), and its leaders believed that after fifteen years of extension assistance, they could competently handle their own business. However, Area Director Sidney Mills found their decision distressing. He was aware that Extension Agent Elmer Allen had helped the Zunis make vast improvements in their livestock program and develop their local leadership, but Mills lamented that the Zunis would sorely miss the extension program in the future.[23]

In addition to the university's program, the Pueblos also received technical assistance directly through the federal government. In 1975, Area Director Patrick L. Wehling contacted all New Mexican pueblos about their participation in the Agricultural Conservation Program. Working in conjunction with the Soil Conservation Service, the BIA urged interested Pueblo farmers to prepare a conservation plan as a prerequisite. New Mexico officials distributed handbooks that explained the program to interested parties. Along with federal and local authorities, the pueblos established guidelines to facilitate sound resource management systems through conservation and erosion control.[24]

Additionally, Commissioner Morris Thompson of the BIA made natural resource management a national objective. He believed that establishment of firm policies would improve management of resources that were important to both the federal government and Native Americans. He stressed soil and range inventories, technical education programs, and comprehensive management plans for Indian ranchers. He also wanted to provide a technically trained agricultural specialist to inform Indian landowners about farmland resources, federal cost-sharing programs, and management options. In a short time policies for improving natural resource management were being implemented at all pueblos. Each agency's highest priority was hiring an agricultural specialist. By 1977, a number of the pueblos had made plans to hire a natural resource manager with command of the native language if possible.[25]

Technical assistance programs elevated the sophistication of Pueblo agriculture to a new level. Tradition-bound farmers held sway for generations after the vanguard of American occupation arrived. Not until the New Deal period did the farmers begin to ease their grasp. Basically ignored in the postwar termination era, much Pueblo farming activity remained stagnant, and large bodies of resources lay undeveloped. With the advent of the self-determination policy, federal officials hoped to increase the economic independence of Pueblo farmers. Nothing was more beneficial in this effort than the work of the extension service, whose officials and agents determined early on that the key to unlocking the door of opportunity for native farmers was education. Educational opportunities, however, could only be cultivated through federal funding, which would remain a challenge.

THE FLOOD PROBLEM

For as long as native people had farmed the river valleys of the American Southwest, they had faced the threat of overflow during times of high spring runoff and summer monsoon storms. Damage could range from troublesome to severe, but no matter what the extent, the unpredictability of mother nature caused considerable erosion on Indian property. Moreover, attempts to alter the situation often proved futile, and proposed solutions were considered exorbitant. So common was the occurrence that many people simply learned how to live with the problem. However, when damage was significant and could not be ignored any longer, many Pueblos sought funding to cure these vexations on a permanent basis with the hope that crop loss and ruined fields could become a thing of the past.

Although not located on a major stream, Zuni continually fell victim to the ravages of nature. Numerous small dams had been constructed over the years, but they often proved ineffective or subject to structural damage. Having suffered considerable damage in 1957 and 1961, the pueblo's dams were inundated in August of 1963 after ten days of heavy rains. Bolton Dam suffered a thirty-foot-wide and fifteen-foot-deep breach, and some damage was done to Eustace Dam as a result of the heavy rains in the Pescado Creek watershed. Numerous irrigation structures were damaged in another heavy storm the next year.

Repetitious flood problems convinced the tribal council in 1969 to apply for federal assistance through the Soil Conservation Service under the Watershed Protection and Flood Prevention Act of 1954. Also known as Public Law 566 (PL-566), it provided for planning and construction of needed flood-control features on small watersheds. The law called for cooperation between federal, state, and local agencies and emphasized the need to clear up all water rights questions before construction.

Two years later, Zuni governor Robert E. Lewis petitioned the Four Corners Regional Commission to fund the necessary planning operations for the Oak Wash watershed. Having already contacted the necessary federal and state agencies and completed a feasibility study, Zuni officials had worked out a cooperative funding proposal for thirty thousand dollars, of which they were asking the commission to finance about one-third. They hoped to protect some 845 acres of productive farmland within the Zuni Irrigation Project and add a similar amount as well. Within two months the funding was approved, but archaeological clearance delayed construction; more than twenty different ruins were involved, including Yellow House, the largest and most historic one. As the project became more complicated, expenses mounted, and the master plan eventually reached a projected cost of seven million dollars by 1975, of which the U.S. Senate approved 4.8 million. Governor Edison Laselute felt that this would "allow farming to return again to its proper importance in the overall economy of Zuni."[26]

The major thrust of the dam construction program in the Rio Grande valley involved structures on its tributaries. In 1964, the Army Corps of Engineers announced the proposed Rio Puerco–Rio Salado Flood Control Project, which included portions of the Laguna and Isleta reservations. The first of the these dams, known as the Hidden Mountain site, was to be located on the Rio Puerco about forty-two miles above its confluence with the Rio Grande, while the other, the Loma Blanca site, was supposed to be five miles upstream on the Rio Salado. The key purpose of both dams was flood control and sediment storage. Both reservoirs would be operated in such a manner as to hold floodwater for two days, during which sediment would settle, allowing clear water to be released downstream. Since neither of the streams was permanent, 90 percent of both reservoirs was dedicated to holding sediment; designers hoped to prevent its deposition in the Rio Grande's Elephant Butte Reservoir.

Like many projects of this nature, these proposals had positive and negative features. On the one hand, the Hidden Mountain Reservoir held the potential for the development of a valuable permanent pool for fishing for both reservations. This public fishing and recreation area would be particularly appealing because of its proximity to Albuquerque. Although the Rio Grande valley had long been an important migration route and wintering area for waterfowl, much of this habitat had been destroyed by channelization, marsh drainage, and irrigation diversion. The proposed permanent pool would provide a badly needed waterfowl area. The downside of the proposal had to do with the process of containing the water. Because the Rio Grande Compact prohibited floodwater storage until New Mexico paid its water debt to Texas, no water rights were available at that time to form a pool. Therefore, the corps would only provide for a pool if the Indians could furnish a source of water for filling and maintaining it. Without this reservoir at the Hidden Mountain site, however, an area of over ten thousand acres would be covered by floodwater and silt that was sterile and not productive for forage in the near future. Without the necessary water supply, the bureau rejected the idea.

Following an environmental impact statement, the corps made another offer in 1972 that was discussed by representatives of the Southern Pueblos Agency and the councils of Isleta and Laguna. They felt that other public lands should be acquired for the two pueblos in exchange for an easement (rather than fee title, which would cause them to lose title to the land) and that the Indians should be compensated for land loss due to construction. Having dealt with similar matters in the past, the Pueblos clearly understood the worth of their property. They also wanted to retain use of the reservoir for livestock, recreation, and other economic development opportunities. They desired an entry road to the pool for these purposes and also asked that an adequate water supply be retained. Recognizing their legal rights in this matter, both Indian groups asked to retain their water rights to irrigate their land and that archeological investigations be completed. The pueblos were quick to realize that the corps' plan would take their land for the benefit of downstream users without providing them any permanent features. Finding this situation both offensive and unfeasible, they both rejected the proposal.[27]

Obviously, the Pueblos were much more concerned about controlling flood problems for their own lands. A rather unique situation involved the San Ildefonso reservation and the Pajarito village (one of the so-called Spanish exceptions) within its confines. For years floods had occurred in the area, in spite of several Pueblo earth dams built for containment, and caused water to flow across State Highway 30. Particularly heavy rains in 1965 brought complaints from the village and eventually resulted in a proposed channel from the highway through the reservation and into the Rio Grande. The tribal council granted the necessary easement, but when the conveyance channel failed to control floods in following years, San Ildefonso, in frustration, called on outside assistance, including Governor Bruce King. Since a substantial portion of the water contributing to the problem came from Forest Service land, the solution required combined assistance from the BIA, U.S. Forest Service, State Highway Department, Soil Conservation Service, non-Indians in the village, and the tribe. Unfortunately the resulting work protected the highway but failed to stop flooding on adjoining lands.[28]

Pojoaque watershed was another region that was continually being devastated by flooding in the early 1960s. Floods ravaged croplands and irrigation systems, but control measures proposed for the area did not meet the criteria for PL-566 consideration, so authorities sought assistance from the Northern Rio Grande Resource Conservation and Development Project. Finally in 1970, the Army Corps of Engineers met with the NPA and representatives from Nambe, Tesuque, San Ildefonso, and Pojoaque. The watershed drainage embraced an area of about two hundred square miles with the worst flooding taking place from the Pojoaque Creek's intersection with the Tesuque River down to the Rio Grande. The estimated average annual damage was $125,000 in terms of irrigation works, farmlands, sediment, and crop damage. In order to halt the continual destruction, the Pojoaque Creek required the construction of levees, dikes, and jetties. Land Operation Officer Bob Measles suggested that he meet with the corps and SCS to determine what construction was needed and how it would affect Indian lands. All of this would take time and money, and the area was still subject to flood damage by the late 1970s.[29]

Numerous pueblos passed resolutions in the late 1960s and early 1970s to ask for flood-control projects. In 1968, Sandia and San Juan requested assistance from the Soil Conservation Service under PL-566, with the latter asking for a footbridge across the floodwater diversion to be included for certain religious ceremonies. Santa Clara and Tesuque added their names to the list the next year. Other pueblos, including Santo Domingo, Sandia, and Cochiti, called for field surveys or aerial photography as the initial step for flood damage prevention. Later, Santo Domingo asked for emergency bank protection, and Taos requested assistance repairing earth dams hit by floods. Extremely heavy snowpacks in the winter of 1972–1973 led bureau officials to sound the call of alarm throughout the region. Later that summer State Engineer Jack Koogler contacted Area Director Walter Olsen about the National Flood Insurance Program and the eligibility of the Pueblos.[30] The problem was vast, and immediate assistance was necessary to curb a situation that held back progress for Indian farmers.

Traditionally, the Jemez-Zia area suffered some of the worst flooding, and such was the case in June 1969, when heavy storms damaged both reservations. Jemez's main canals suffered approximately thirty major breaks, two miles of canals were silted full, and there was additional structural wreckage. Jemez farmers lost 50 percent of their corn, 75 percent of wheat and gardens, and 80 percent of their alfalfa. In total over seven hundred acres of crops were extensively damaged. While Zia's canals remained intact, their crop loss was significant, including almost half of the corn, alfalfa, and garden products, which amounted to more than one hundred acres. As with other pueblos, part of the Jemez problem had to do with drainage systems needed along the major highway through the village, and bureau officials suggested construction along State Highway 4. District Engineer Tom Cord said that water passing over roads was a normal pattern that had existed for many years. Despite contacts made with Commissioner Louis Bruce and U.S. Senator Joseph Montoya, money for this expensive project could not be found, and construction was delayed indefinitely.[31]

Typically, Pueblos registered complaints about flooding on their lands, but in at least one case, non-Indian landowners adjacent to the reservation blamed their neighbor for the problem. Both Max Lucero and Gene Weinheimer of Albuquerque cited floodwaters from Sandia that were descending onto their land, damaging the surrounding property. Since the water source originated on Indian land, they believed it was impossible to proceed with any flood-control work. Because the Indian bureau had no funded program to solve this problem, acting superintendent Mark Stevens suggested existing congressional authority PL-566, which was set up for situations like this.

A local SCS Planning Unit had already initiated studies on the Sandia watershed in 1970, and the pueblo's council had approved the study. Within a month, a meeting was held with Sandia and government officials along with the local people. Given the unique proximity of the problem, the planning unit also discussed possible assistance through the county commission and the Albuquerque Metropolitan Arroyo Flood Control authority, but it was later found out that the problem area was out of its region, and to include it would require an amendment by the state legislature. Worse yet, Area Director Loyd E. Nickelson determined that it would take at least five years to get the flood-control project going.[32]

Chronic underfunding, which had always stymied progress in the past, continued to be a problem. Previously, the pueblos and Indian officials sought money from a single area, but now they branched out to a variety of sources. While construction costs had risen and continued to block progress, they now found the added expense of archaeological clearance, feasibility studies, and watershed investigations. In other words, the whole process had become more bureaucratized. This situation not only meant cost overruns but also a much greater time span before construction commenced. But the Pueblos and government officials had also become more sophisticated in dealing with flood control, rejecting proposals that offered them few benefits, alerting tribes when spring runoffs endangered the region, and seeking out flood insurance programs. Dams seemed to be the answer to their almost yearly flood threats, but their construction raised another paramount issue concerning water rights that would pose a major legal threat to the expansion of the Pueblo farm program.

THE WATER RIGHTS PROBLEM

In the post-World-War-II-era Pueblo economy, nothing became more important than the right to water. Without it, economic development would be greatly hindered. In the first half of the twentieth century, when the population of the American Southwest was small, questions concerning this issue were relatively few. But during the 1960s and 1970s, the rate of population increase in the urban areas of the Rio Grande valley was among the highest in the nation. By 1980, more than half of the state's population lived in the upper Rio Grande basin, and by the end of the century, it was expected to almost double. Accompanying this unprecedented growth in the cities was a depressed economic condition in the rural sections; with the limited amount of water and the lack of arable land being threatened by the population boom, the number of people who could depend on agriculture for a living was in decline. In addition to insufficient quantity of water, there were also problems with salinity, sediment, nonbeneficial consumption, and floods creating property damage.

As population and economic growth increased, so did the struggle over water control. The complexity of the problem resulted in a mixture of federal, state, and local agencies involved in the planning and utilization of the region's most precious resource. The point of contention involved which of these had the highest authority. Of course the Rio Grande Compact had adjudicated the amount of water to be used by the states (including Mexico) through which the river flowed, but because of reservoirs like El Vado and Elephant Butte constructed with federal money, it was also governed by federal reclamation law. In addition, federal flood-control law applied because reservoirs like Cochiti were built for that purpose. The situation became even more complicated when water was transferred from the San Juan River into the Rio Grande watershed following congressional approval in 1962.

But where did the Pueblo piece fit into this complex puzzle? The answer lay in the ambiguous and contentious area of Indian water law. The famous *Winters* case of 1908 had given Indians prior and paramount rights to all of the water necessary to develop their reservations. These rights were reserved when a reservation was created, could expand in the future as the tribe's need increased, and could not be lost even if Indians failed to use the water. Thus state laws of prior appropriation were abrogated, and Native Americans had a totally unique set of water rights that fell outside of the doctrine that governed most of the western United States. The ambiguities of this decision were modified in another Supreme Court decision of 1963, called *Arizona v. California*, which tempered *Winters* by stating that the amount of water reserved was for "practicably irrigable acreage" on reservations—the vagueness of which required some interpretation by future courts. Still, the future needs of reservations, as well as their present ones, were reaffirmed.[33]

Even when the Supreme Court created the so-called Winters Doctrine, Pueblo officials were confused as to its implications for native New Mexicans, and this situation would only become more complex later on. The reason for this lay in the unique history of the Pueblo Indians. Without question the Pueblos were aboriginal users of New Mexico's water, having developed the oldest irrigation systems in America, which they used for centuries before the Spanish arrival. Under Spain, the Pueblos were treated as wards of the Crown, and when Mexico took over the region, they became citizens according to the constitution of 1824. By the Treaty of Guadalupe Hidalgo in 1848, through which the United States acquired New Mexico, the federal government agreed to preserve all rights granted under the two previous administrations; indeed, most of the Pueblo grants were confirmed in 1858. Unfortunately for the Pueblos, they fell into a state of limbo between the *Joseph* case of 1876 and its reversal in the *Sandoval* decision of 1913, during which time they lost a great deal of land to non-Indians. To compensate them for their loss, Congress created the Pueblo Lands Board in 1924, and when these new land claims proved to be dissatisfactory, another act was passed in 1933 to provide for additional compensation. Added to this mix, of course, was the extension of the Middle Rio Grande Conservancy District to the Pueblos in 1928, which established paramount water rights for the six pueblos therein. The question of how the Winters Doctrine applied to the Pueblos, if it did at all, was confused by their undulating legal status; the fact that their lands had not been set aside by federal treaty, statute, or executive order; and the confusion over the recognized legal starting date for their land claims.

The Interior Department, realizing that the water question was at a crisis stage, created the Indian Water Rights Office in 1971. When Secretary Rogers Morton made the announcement, he added that the new office would probably utilize the services of controversial bureau water rights attorney William Veeder. Considered a renegade by many Anglo and conservative Indians, Veeder was viewed with caution because of his mixed record in water rights cases, and some felt that he had actually lost ground for Indians. His appointment, coming at a time of friction between younger and often more rebellious Native Americans and traditional tribal leaders, cause a startled and apprehensive attitude among the Pueblos. The six district pueblos, being the main Indian water users in the state, called an emergency meeting to discuss the initiative. Their focus was Veeder's appointment, which drew strong criticism from Cochiti and Isleta—the pueblos using by far the most water. Long recognized for their sense of dignity, the Pueblos were outraged when Veeder partisans stormed the BIA headquarters in a protest against John O. Crow, assistant to the commissioner, which led to the arrest of a number of militant Indians. More importantly, they did not believe Veeder's claim of representing Indian water interests. Tesuque was particularly affected by Veeder's approval of a water contract that allowed the Colonias de Santa Fe land development to usurp some of their water rights.

The Pueblo water issue came to a head when Congress approved the San Juan–Chama Project in June of 1962. In the past, downstream users, including those in New Mexico, failed to get enough water to satisfy their needs, and the new project was designed to deliver to the state its share of water from the upper Colorado River. It would come from the San Juan River, a tributary of the main stream in southwest Colorado, and would be funneled through a series of tunnels through the continental divide and thence to the Rio Chama, a tributary of the Rio Grande. The water would not start to flow until 1971, and most of the early operation was used to fill the Heron Reservoir near the Chama's headwaters and add to the storage of other downstream bodies of water, but by 1980 it was at full capacity.[34]

In 1961, the State Engineer's Office commenced proceedings to adjudicate water rights for the expressed purpose of administering water that would eventually come into the Rio Grande from the San Juan–Chama Project. Previously, locally elected officials of various community ditches determined amounts of water for their users, which meant that there was no formal system of distributing water rights. With new water coming in and population demands putting a strain on the system, a more sophisticated legal system needed to be put in place. So the state initiated a lawsuit for this purpose in which most non-Indian water users on the Chama were named as defendants.

Figure 12: William Veeder, Indian water rights expert, July 27, 1978. Pueblo News photo, PN 0243, section 14. Courtesy of the Indian Pueblo Cultural Center.

The first pueblo affected by this action was San Juan, which was irrigating approximately two hundred acres from the Rio Chama via a canal serving Pueblos and non-Indians. Almost two thousand more acres on the grant were being watered from the Rio Grande. The UPA called an emergency meeting on this matter and set up priority schedules for completion of irrigation reports for various pueblos, and San Juan was given top priority. Similar procedures were set up for the Nambe-Pojoaque-Tesuque drainage basin in 1962, and two years later the state engineer filed for adjudication of the water rights for Santa Clara and San Ildefonso. The Pueblos conducted studies, surveys, and investigations to gather information for the possible defense of their water rights.

The adjudication proceeding led to an immediate clash between the State Engineer's Office and the Pueblos in which the former claimed that the Indians' rights should be determined by the state law of prior appropriation while the latter felt they had a right to as much water as necessary to irrigate all of their farmlands based on the Winters Doctrine. The state based its case on the idea that the Pueblos had been under the Spanish and Mexican legal systems, both of which had traditionally recognized the doctrine of prior appropriation. The state also claimed that the Indians lost any reserved rights to water through payments made to them under the two Pueblo Lands Acts—something with which the Pueblos would not agree. State lawyers realized that if the Pueblos were granted Winters Doctrine rights, a large number of non-Indian water users could lose their rights to water.

After a thorough investigation of the area, which included maps showing agricultural usage and a survey on the extent of water rights, the State Engineer's Office filed suit in the U.S. district court of New Mexico. This action, filed in 1966, became the famous *New Mexico v. Lee Aamodt* (his name being the first of the non-Indians listed alphabetically) or simply the *Aamodt* case. In addition to the pueblos of San Ildefonso, Pojoaque, Nambe, and Tesuque, the litigation also involved more than 2,250 non-Indians. Based on its role as trustee for the Pueblos, the United States filed a motion to dismiss the action for lack of jurisdiction of the court and then entered a motion to intervene in the suit. As a result the Pueblos and the federal government joined together as plaintiffs, which split the case into two proceedings designed to determine non-Indian and Pueblo water rights.

The initial *Aamodt* ruling of the court in 1973 was in favor of the state, claiming that the New Mexican doctrine of prior appropriation governed Pueblo water. The federal government, acting on the Pueblos' behalf, appealed the case to the U.S. Court of Appeals for the Tenth Circuit, which reversed the lower court's ruling in favor of the Pueblos in 1976. In this decision, the court of appeals claimed that the Pueblos, not being adequately represented by government counsel, were entitled to private legal aid—something that the district court had denied them. Then the court ruled that the Pueblo Lands Act did not cause the Pueblos to lose their water rights and, finally, that the Pueblo rights to water, although not "technically applicable" under the Winters Doctrine, were not subject to the jurisdiction of the state. The court of appeals asked the federal district court to determine water allocation.

In order to quantify these rights, the court recognized three classes of non-Indian appropriators. The first consisted of claimants who used water prior to the 1858 Pueblo land title confirmation. The laws of Spain and Mexico were to be used to settle conflicts between the Pueblos and non-Indians in this group, and the final determination was referred back to the district court. The second class consisted of those non-Indians who held lands after 1858, but the court, lacking information on this group, referred their determination back to the district court for a solution. Finally, the court established the third group as those who obtained land through the Pueblo Lands Act, in which case the court felt the Pueblo water rights were superior.

Failing to get a rehearing by the circuit court, the State Engineer's Office appealed to the U.S. Supreme Court, but after review the Court decided not to hear the case.[35] Thus effectively blocked, the Pueblos, the state, and non-Indian users returned to the U.S. district court for final settlement, but many questions remained. In the next decade more rulings would come, and the *Aamodt* case continues in adjudication limbo today.

Since the situation in the middle Rio Grande valley was governed by a unique set of circumstances, the six pueblos in the area, in conjunction with the Indian bureau and the Bureau of Reclamation, worked out an agreement on resolving water rights disputes in 1981. In an attempt to avoid adjudication, they established procedures for determining how much water the conservancy district would have to store in El Vado Reservoir before the irrigation season began in order to meet Indian crop needs. This represented the first written agreement on Pueblo water storage since the district contract was signed in 1928. It stemmed from a request by the Interior Department's engineer Everett Polanco for the district to store forty thousand acre-feet, which the officials refused to do, claiming that he had gone beyond his authority, but Pueblo Irrigation Committee chairman Paul Shattuck of Isleta felt that Indian farmers had been slighted in the past and wanted to be assured that released water would be designated specifically for the six pueblos. In fact the district had closed down canals leading to their land. Pueblo officials wanted more authority over water release, and the new agreement would set up their own gauging stations to help determine their entitlements. Both groups appealed their claims to the Interior Department, and the agreement was seen as a means of settling the dispute. Unfortunately, the district was not in on the negotiations, and no one knew if the new contract would bring about a resolution.[36]

The Jemez River basin, water source for Jemez, Zia, and Santa Ana, became the next major battleground in the Pueblo water rights war. The first pueblo was by far the biggest user with historic rights to irrigate some 3,000 acres, while the others had 300 and 180 acres, respectively. Since their water rights had never been legally settled, there were constant problems of allocation and distribution, and in 1980 the Pueblos took action to get a complete stream adjudication. The case focused on the Nacimiento Creek, San Ysidro Community Ditches, the state, and timber and geothermal interests, but the loss of water to outsiders was at least a century old. The Nacimiento community began diverting water in the 1880s, and by 1896 they had reduced Jemez to 2,100 irrigated acres. The community constructed a storage reservoir in 1948 that was expanded by the New Mexico Game and Fish Department seven years later, all of which ignored Indian rights. Although the state had at least approved of these changes, numerous other diversions were never part of the official record.

While the amount of water used for agricultural purposes in the basin remained relatively the same, that portion used by the Pueblos gradually declined. Zia's irrigated acreage had actually increased through the bureau's efforts in 1922 to about 392 acres, and they also built a small reservoir in 1977, but they were only irrigating 99 acres at that time. Jemez, which had only received adequate water once in thirty years, had been drastically reduced to 424 acres. Santa Ana's original irrigated land on the Jemez had dried up during the Spanish era, but it still had a water right in this area. The Pueblos brought suit against energy firms that had begun development of geothermal deposits in 1980, but the loss to the Indians was so threatening that government officials felt that the large number of people using domestic wells might have to be included in the suit.

The Indian bureau prepared a water development plan that they hoped might turn the Pueblo agricultural decline around. Because the water supply was depleted during the summer months, due mostly to haphazard individual operations and indiscriminate upstream use, the bureau proposed dam construction to store winter and spring runoffs, called the Jemez River Project. Studies conducted previously emphasized its cost-effectiveness. The additional water would greatly increase the amount of arable acreage by a factor of four, thus helping to alleviate the high unemployment rate among the Pueblos; recreational use would also help to diversify their income. The major problems with this idea were site selection and a strong negative reaction from environmental groups. However, government officials pointed out that without the construction of the San Ysidro dam, a Pueblo victory in the courts would impose an extreme hardship on non-Indian users.

Whereas Zia initially supported the dam construction, Santa Ana was strongly opposed because they felt that the Indians were not being given any opportunity to comment on the proposal. They particularly objected to the idea that their claims were being compromised before the stream had been adjudicated. Area Director Sidney Mills agreed with this assessment and called for immediate court action since the state offered the Pueblos little protection and under the present system the Indians could not enforce their water rights during a shortage. Moreover, threats from the Union Geothermal Project were mounting, and the Pueblos needed to force some action to halt their development. Gradually frustrated by the situation, the three pueblos filed suit against some six hundred individual and corporate water users in the Jemez watershed in 1983 seeking damages for alleged unlawful diversions.[37]

Concomitant with the Jemez Basin case was the third major water crisis, involving disputes between non-Indians and the pueblos of Acoma and Laguna concerning the Rio San Jose. The case was very similar to the Jemez one in that water loss extended back to the nineteenth century, the Indians were constantly decreasing their arable acreage, and the decline of their agricultural activity paralleled water use caused by an increasing growth of modern, diversified non-Indian industry. When Field Solicitor Lotario D. Ortega initially filed a litigation report in 1980, the two pueblos sought only damages against upstream users, but they soon called for a general adjudication of the Rio San Jose watershed, and the area office agreed with this change. Since both watersheds were being investigated at the same time, money needed for historical research designed to prove past irrigation by the Indians put a strain on financial resources. More importantly, the matter required expeditious action since the statute of limitations imposed a deadline by the end of 1982.

The source of contention was the Bluewater Creek, a major tributary of the Rio San Jose. Around 1884, American settlers built a small dam about ten miles northwest of Grants that had to be reconstructed in 1928. This seventy-foot-high structure provided irrigation water for some nine thousand non-Indians and was operated by the Bluewater-Toltec Irrigation District. A second diversion was added in 1890 at Ojo de Gallo Spring about five miles south of Grants. The Indian bureau protested the Bluewater Dam and although their irrigation engineer Paul Hodge wrote a report that clearly established that the reservoir was depriving both pueblos of a major source of water, no action was taken. Conditions worsened with the drought of the 1940s, and as surface waters decreased, non-Indian farmers began to develop groundwater wells during the next two decades. As pumping increased, the water table lowered and local springs, which formerly provided almost half of the river's flow, dried up. Recent mining operations, power development, and a uranium mill added to the water shortage, and sewage discharges from Grants decreased the water quality.

Given the long history of progressive encroachment in the Rio San Jose basin and the need of both pueblos to expand their agricultural program, a water suit was necessary to protect their rights to the only perennial source of irrigation water. The SPA, recognizing the threat posed by recent developments, offered a water rights seminar to make tribes aware of their rights and prepare them for claims against trespass. As with the Jemez case, government officials knew that the state would offer little protection on unadjudicated streams, and they claimed that lack of legal definition would hinder non-Indian development in the energy-rich basin. In the past the Pueblos had shared shortages with non-Indian users even though they had senior rights, but with the *Aamodt* case as their inspiration and the crop loss in the postwar era estimated at $3.6 million, they were becoming more assertive lest they lose a major part of their water entitlements. Therefore in 1982, the federal government, acting for both Acoma and Laguna, filed suit against the irrigation district and approximately fourteen hundred other defendants seeking damage for water trespass. However, this fell short of a general adjudication, and the waters remained muddy on this issue.[38]

Pueblo agriculture faced a plethora of challenges during the 1960s and 1970s. Beginning with self-determination, the Pueblo voice, idle for so long, was reawakened. Federal bureaucratic policies and dictates remained a constant in Pueblo life, but they were increasingly designed to prepare the Pueblos for the complex future that awaited them. Governmental officials at federal, state, and local levels put a new emphasis on cooperation, although state officials were often less helpful than their federal counterparts. Both ranchers and farmers faced the prospect of mastering a new level of sophistication in their attempts to modernize Pueblo agriculture. The combination of economic planning, resource management, range surveys, and archaeological and environmental laws replaced the rather haphazard modernization programs of the past. Some Pueblos were ready to engage the challenges of the modern world and to gain economic benefits for their people, but as a whole, the Pueblos still rejected ideas and methods that were incommensurate with Pueblo tradition and life.

All of this agricultural transformation was being played out against a tumultuous background of dissent that reverberated throughout the country at this time. The voices of youthful criticism could be heard in water rights cases along with the need for greater Pueblo involvement in their farm program. Ironically, some of the most strident complaints came against the extension program, which probably did more to advance Indian interest and participation than any other organization. But adjustments were accomplished through internship programs and better legal representation, and Pueblo nationalism was somewhat tempered by traditional leadership that sought to achieve similar goals with a more dignified approach. Pueblo officials, however, still faced overwhelming obstacles: flood control, population growth, economic development, and the resulting struggle over the region's most precious resource—water. Surprisingly, the keenest interest in these threats to Pueblo agriculture occurred at a time when farming was in decline. Farming had long been the hallmark of Pueblo self-reliance, but the forces set in motion after the war were threatening Pueblos' most traditional occupation.

CHAPTER 10

Modern Pueblo Agriculture (1980–Present)

THE NEVER ENDING WATER STORY

At the very core of modern Pueblo agriculture is the question of water—not only the right to it but also its legitimate delivery. No Pueblos were more concerned with these aspects than those who fell under the auspices of the Middle Rio Grande Conservancy District. Their rather random inclusion within this agency bore mixed results for the six tribes. On the one hand they wound up being the biggest water users of all the New Mexican pueblos and thus benefited economically. However, the grandiose promises concerning water delivery to the Indians were never realized, leaving them feeling frustrated and cheated. In reality these tribal groups were in the right place at the right time when the district was initially created. As a state agency, however, it demonstrated little commitment to native farmers and apparently felt them more of a nuisance than a partner.

Nothing demonstrated this better than Pueblo claims against the district for negligent operations of their irrigation and drainage systems which surfaced in 1980–81. This resulted in detailed investigations by John R. Baker for the All Indian Pueblo Council in which he laid out a history of irrigation in the region since the turn of the century and the possibility of claims against those who had failed to meet expectations. He gave special attention to Sandia and Santa Ana but the other four pueblos were considered as well.

Concerning Sandia, Baker cited problems with the canals and drains that were not functioning as releases for subsurface water in spite of past promises to the contrary. Complaints of high water tables began to surface in 1967 due to significant annual crop damages and interruption of land-leasing income. Even those lands in production were experiencing reduced crop yields, especially deep-rooted alfalfa plants.

Conditions were similar at Santa Ana, where initial construction by the MRGCD in 1930 established new high line canals as an adjunct to the original Indian-built ones, and acres farmed increased by 9 percent. In addition the district was obligated to maintain the new systems and even compensated Santa Ana when seepage problems occurred in 1937. But the problem worsened in the next decade and crop losses mounted in the 1950s and 1960s. During drought years, like 1946, promised water was not delivered in spite of superior rights, adding another blow to Indian farmers. Dam construction was designed to cure this problem, but ironically Cochiti Dam, completed in 1975, added to the high water table because of the clear water released, which tended to seep from earthen canals, and the higher rate of flow over a longer period of time. This condition impacted all of the district's pueblos, and many were citing the district for negligence.

Since the MRGCD seldom if ever responded to requests in writing, Baker suggested an annual meeting between their officials and those of the affected pueblos to reach an understanding of operation and maintenance responsibilities that were owed the Indians. This might also include annual tours of inspection so that district officials could observe problem areas and thus be witnesses to relief due to the Pueblos, and thereby be forced into fulfilling their obligations. Aware that they could no longer rely on anyone but themselves to protect their interests, the Pueblos, after seeing the Baker reports, decided to take matters into their own hands.[1]

In June of 1981, Lotario Ortega, always the watchdog on legal matters, sent a four-volume study to federal authorities to help resolve damage claims on behalf of the district's six pueblos. These demands were based on inadequate maintenance and operation of drainage and irrigation systems on Indian lands, which resulted in an estimated loss of fifty million dollars based on crop loss and land restoration costs. Rather than pursuing the litigation route, which would involve expensive legal bills and could lead to rewards that the district could ill afford, Ortega suggested that the secretary of the interior request legislation that would fund a district rehabilitation project for about half the cost of the proposed damage claims. He felt that this physical solution was more equitable in that it would provide significant benefits for Pueblo and non-Indian residents in the form of better water delivery and lower water tables. Unfortunately, these claims were rejected for failing to meet requirements of the statute of limitations, but Pueblo officials were encouraged to pursue more current claims which required further documentation.[2]

Rebuffed in their efforts to improve the delivery system, the Pueblos focused on fulfillment of their paramount water rights, which were agreed on by the federal government and the district in 1980. At issue was the exact amount of water to be set aside for the Pueblos. The AIPC claimed that forty thousand acre-feet should be stored in El Vado Reservoir for Pueblo farmers, but the district, citing studies made in 1953 and 1962, claimed that even for extreme drought conditions, less than half that amount was necessary for Indian lands having storage rights. The Bureau of Reclamation agreed with this assessment and set aside half the requested amount for the Indians. The Interior Department felt that the only conceivable cause of water shortage would be extreme negligence on the part of the Pueblos and claimed that excess storage would adversely affect non-Indians in New Mexico and Texas.

Richard Hughes of the Albuquerque law firm representing the Pueblos protested the lower water amount, but his complaints were ignored by Interior secretary James Watt. Hughes was especially upset that the Indian bureau was not given any role in the matter; rather, the controversial figures, which were the key issue in the district's appeal, were supplied by the Reclamation Bureau—an agency notoriously hostile to Indian water rights.

He suggested to Watt that an administrative law judge be designated to conduct a hearing involving all parties to clear up the matter. As a way of bringing resolution to this significant question, Hughes offered a compromising solution, but he was up against state and federal agencies that had stood in the path of Pueblo water rights in the past as well as a cabinet member who had been openly unfriendly toward Native Americans.[3]

In addition to the concerns of the district's pueblos, those in the Pojoaque-Nambe-Tesuque stream systems continued to seek redress in the *Aamodt* case. Nineteen years after the case was filed in 1966, U.S. District Judge Edwin Mecham issued a complicated ruling that seemed to favor both sides. The judge concluded that the pueblos of Nambe, Tesuque, Pojoaque, and San Ildefonso had aboriginal or first stream rights but that this only applied to lands irrigated between 1846 (the end of the Mexican war, by which America acquired the Southwest) and 1924 (when the Pueblo Lands Board was created). Mecham also ruled that non-Indian priorities began when non-Indians first irrigated their lands.

In rendering this decision, Mecham generally rejected the Winters Doctrine advanced by the Pueblos, which gave Native Americans rights to irrigate all their lands even if they were not currently using them. He stated that this only applied to Nambe's "grant" lands but not to the "reserve" lands or to the other Pueblos. The former involved Pueblo property recognized during Spanish rule while the latter referred to areas added later on by the U.S. government. This rather stunning conclusion was a huge blow to Indian water rights and would have major repercussions concerning water allocation throughout the West.[4]

So in the areas of utmost importance to the advancement of their agricultural program, the Pueblos were being thwarted by state and federal agencies and one of the most significant court decisions on water rights in the nation's history. Already the victims of crop losses in the past, amounting to high financial deficits that seemed impossible to reclaim, they faced a grim future that offered little hope of expanding their most traditional occupation and relieving a burdensome unemployment problem. More importantly, all of this came about because their rights to land and water, which had been guaranteed federal protection from the beginning of American involvement, failed to receive necessary defense. Without the watchful eye of the U.S. government, a farm program already in decline faced perilous times ahead.

The adjudication of the waters of northern New Mexico also signaled the end of a centuries-old distribution system—the acequia. Long the hallmark of Spanish tradition, owners of community ditches ignored the legalities of prior appropriation, and opted for a system that emphasized community need in times of drought. The ideal gave way to the real. The implementation of conservancy districts in 1927 to provide flood control, storage, and irrigation-distribution systems significantly empowered these public corporations. As political subdivisions of the state, the districts gradually replaced the age-old acequia method and with this transformation, individual rights overtook those of the community.[5] Exact quantification of these amounts continues to defy the scientific exactitudes often demanded by the courts, but the multiple needs for modern water use demand legalities over custom.

THE DECLINE OF PUEBLO AGRICULTURE

The general downward trend of Pueblo agriculture in the postwar years is best measured by observing the change in the number of irrigated acres on individual reservations. A breakdown of these statistics during the New Deal and post–World War II period is provided in the following table.[6]

For most pueblos the decline in irrigated acreage is shocking. The majority show a loss of more than 60 percent. Only Isleta, Sandia, Santa Clara, and Zuni demonstrate increases from the first measurement in 1938 to the last, with Isleta, Sandia, and Santa Clara showing a remarkable turnaround after decades of decline. The presence of Isleta and Sandia in the Middle Rio Grande Conservancy District certainly contributed to their acreage increases. Zuni's upward trend stemmed more from its expansive ranching than from its agricultural efforts. Still, irrigated acreage for two tribes went down by over 80 percent, and three others showed losses of over 70 percent. While the Pueblo-wide percentage of decline from 1938 to 1964 was 32 percent, if the gainers are subtracted from the total, the loss for the remaining pueblos at the last time they were measured is 6,174 acres or 60 percent.

The overall decline of agriculture was not unique to the Pueblos in the Southwest. In 1973, the Four Corners Regional Commission's study on agriculture revealed a similar decline for Anglo, Hispanic, and Indian farmers in the area. Agricultural employment went down from 40,600 in 1970

TABLE ELEVEN
PUEBLO IRRIGATED ACRES

PUEBLO	1938	1949	1952	1964	1981	CHANGE 1938–1964*	PERCENT
Acoma	1,015	1,054	789	723	293	-722	-71%
Cochiti*	776	727	586	505	188	-588	-76%
Isleta*	3,340	3,212	2,289	2,788	3,503	+163	+5%
Jemez	1,391	1,486	1,362	742	275	-1,116	-80%
Laguna	1,456	943	812	222	—	-1,234	-85%
Nambe	291	182	122	78	—	-213	-73%
Picuris	188	180	177	101	—	-87	-46%
Pojoaque	39	24	18	15	—	-24	-62%
Sandia*	833	741	351	1,320	1,335	+502	+60%
San Felipe*	1,499	1,249	1,340	651	469	-1,030	-69%
San Ildefonso	265	208	170	130	—	-135	-51%
San Juan	932	716	621	636	—	-296	-32%
Santa Ana*	559	583	488	376	199	-360	-64%
Santa Clara	430	415	312	807	—	+377	+87%
Santo Domingo*	1,799	1,397	1,108	793	873	-926	-51%
Taos	2,171	1,445	1,561	1,790	—	-381	-18%
Tesuque	177	191	185	68	—	-109	-62%
Zia	309	328	346	181	118	-191	-62%
TOTAL	***17,470***	***15,081***	***12,637***	***11,926***	—	***-5,544***	***-32%***
Zuni	2,511	2,757	—	—	—	+246	+10%

* For pueblos with no 1981 data, 1964 numbers are used. Zuni is excluded for lack of data.

to 32,300 in 1980, and the agricultural sector was predicted to show the slowest growth when compared to other economic areas. The shrinking agricultural sector would force many local people (like the Pueblos) to leave their native villages and towns to seek employment elsewhere. Only the Navajo Irrigation Project offered hope for agricultural expansion in the Southwest. In the New Mexico sector of the Four Corners, agriculture's share of the total earnings dropped drastically from 14.9 percent in 1950 to 7 percent in 1970. Adding to the problem was the quality of surface and subsurface water, which contained significant quantities of salt.[7]

Numerous explanations for the slide in Pueblo agriculture have been offered over the years. In 1976, historian Joe Sando of Jemez Pueblo, author of two books on the Pueblos, cited water shortages, soil erosion, and population increases to explain the decline, but he also added the disruptions and upheavals of World War II to his explanatory mix. Pueblo soldiers learned new skills, returned to take advantage of the GI Bill, and became skilled workers and professionals. In addition he blamed the soil-bank program that emerged in the 1950s. Under the program, the Pueblos, like other American farmers, were compensated for leaving their lands fallow. Sando observed that by this time, subsistence farming was a thing of the past. Finally, he regretted the introduction of welfare, which he claimed "created a new kind of Pueblo person, one who does not work."[8] Certainly this last remark reflects the drop in centuries-long self-reliance that had been the backbone of the Pueblo character.

Two years later civil engineers William J. Balch and John W. Clark of New Mexico State University analyzed the forces of agricultural decline in Pueblo country. Their explanations echoed reasons cited immediately following World War II. In their opinion, the transfer of land from one Pueblo generation to the next was the principle cause for declining Pueblo agriculture. Although lands belong to the individual pueblo, their rights of use were inherited, and this practice over the generations had left modern farmers with small, widely scattered land holdings. The resulting subsistence farming was on the decline because of the large effort required for a small economic return. Balch and Clark simply reinforced what small farmers throughout the country already knew: they were going out of business because they could not afford the technology employed by larger corporate farms. The civil engineers also cited a technology employed by larger corporate farms. Finally, they cited a shortage of irrigation water and general decline of interest in farming. However, they did point out that while agriculture was not the dominant way of life it once was among the Pueblos, the indigenous religious ceremonialism connected with agriculture was still an important part of their lives.[9]

A report on Santa Ana by John Baker in 1981 studied trends in its agricultural lands from 1936 to 1980. Farm acreage ebbed and flowed from the late 1930s until 1953 when it dropped substantially from 644 to 460 acres. Over the next twenty-seven years, acreage continued to decline—292 acres in 1962, 215 acres in 1976, and finally 179 acres in 1980. It was hard to pinpoint the exact cause of this decline; certainly, off-reservation work and the high cost of farm equipment attributed to the loss, but water-flow problems also added to the situation, especially siltation, droughts, floods, and usurpation. In his 1983 study of Acoma, Robert R. Lansford of the Southwest Research and Development Company noted that the largest acreage decreases occurred after 1954 but finally bottomed out in 1975. Farm acreage slightly increased in 1980.[10]

Although these scholars help to explain trends for the pueblos in general, they do not clear up why some Pueblo reservations expanded their agriculture while others contracted their programs. In a number of pueblos, there was a shift from a theocratic to a more democratic form of government. The secular governing bodies often stressed economic modernization over traditional ceremonial practices that slowed agricultural change and growth. This pattern was especially apparent at Santa Clara Pueblo. Other pueblos—such as Sandia and Isleta, whose farm acreage expanded—could rely on longer growing seasons because of lower elevation and more naturally level land on their reservations. Zuni's fundamental change from traditional agriculture to livestock, typical for western pueblos, resulted from a new importance placed on market economies and the loss of water resources to siltation and salinization caused by clear cutting at higher elevations.

Many of the agricultural losers turned to other occupations. People from Acoma and Laguna were drawn to the lucrative but dangerous uranium mining industry. Jemez, already troubled by an irregular topography, turned to the burgeoning arts-and-crafts industry. Its people also sought, along with other northern pueblos, employment at Los Alamos National Laboratories. Lying near a reservoir, Cochiti moved into outdoor recreation and invested in a housing development plan that failed.

A more proximate cause of its declining agricultural program was a high water table created by water leaking under Cochiti Dam. On the other hand, Santo Domingo, one of the few pueblos to maintain its traditional ways, stressed livestock expansion near the Jemez Mountains.[11]

Certainly the changes wrought by World War II, and the movement of people into the Southwest, put a strain on water use and agricultural lands throughout the Four Corners region. Educational opportunities opened up to Pueblo youth and, combined with relocation programs, took many of the best and the brightest away from the reservation farms. A new day had dawned for Pueblo farmers, dominated by court fights, modern equipment, and fewer people to produce crops. Receding into the past were the days of subsistence agriculture. The new focus would be on cash-producing crops—the biggest change being the replacement of traditional corn with alfalfa as the pueblos' major crop. If things continued on this course, the oldest continuous irrigation tradition in America would face possible extinction, and the religious ceremonialism that accompanied Pueblo agriculture, the very core of Pueblo existence, would play out against a hollow background.

NEW-AGE PUEBLOS

The idea of focusing in on high market-value crops began in 1967 when Domingo Montoya, chairman of the All Indian Pueblo Council, contacted Superintendent Kenneth L. Payton of the UPA about the possibility of setting up vegetable demonstration plots in all of the Pueblo villages. The demonstrations were designed to include all phases of gardening—selecting seeds, preparing seed beds, planting, fertilizing, cultivating, irrigating, harvesting, and marketing. Individual pueblos would choose an agricultural leader to guide younger people through the process. In the spirit of economic development, an important adjunct to this plan was to form a Pueblo cooperative that would obtain mechanized farming equipment and assistance for marketing chile and other crops. With large portions of their reservations going fallow, the goal was to interest Pueblo youth in a modernized method of agriculture that would open up new areas of employment.[12]

By the 1980s, the Pueblos initiated a movement to return to their agricultural roots under Southern Pueblos Agency guidance. Sandia started a pick-and-grow vegetable operation, and San Felipe conducted a trial between traditional and modern farming methods. After years of encouragement from tribal leaders, Picuris began to clear new areas and plant alfalfa, wheat, and garden vegetables with the hope of eventually marketing outside their village. Governor Bernard Duran and Lieutenant Governor Gerald Nailor, however, worked together in persuading members of Picuris, one of the smallest New Mexican pueblos, to turn the clock back and return to their traditional ways. They hoped to make Picuris's people more self-sufficient and ensure tribal water rights by continuing to use their ancient irrigation systems. Against the tide of shrinking, federally funded Indian programs during the Reagan administration, the eight northern pueblos were encouraged by a grant from Health and Human Services to create thirty-four jobs designed to put sixty acres of tribal land into crop production.[13]

The 1990s witnessed an invigorated return to the land by a number of Pueblos. The San Juan Agricultural Cooperative, launched in 1992 by tribal members, was responsible for a revival of farming in the San Juan River valley. Pueblos now farmed land that had been barren for decades, and the cooperative, which also operated a food processing plant, marketed their crops. Funded by the New Mexico Community Foundation, the cooperative hoped to provide a sustainable economic program that would embrace the Pueblo traditions. As manager Jeff Atencio declared, "If we lose our farming, we're going to lose a big part of our religion." The co-op members' ties to the past were revealed in the name of their product line: "Pueblo Harvest Foods." Their line included dried green chile and stew, smoked tomatoes, chicos, pozole stew; another of squash, beans and corn; and dried apples, cantaloupe, and honeydew melons—all very marketable. They were sold in almost fifty stores in a dozen states including a cooperative in Albuquerque and markets in Santa Fe and Taos.[14]

Revitalization of agricultural traditions at Zuni was an important part of the Zuni Sustainable Agriculture Project (ZSAP). ZSAP was funded by a grant from the Ford Foundation to bring back agriculture on a big scale under the direction of Donald Eriacho. A component of the Zuni conservation project was to restore land and water for future generations and promote family farming and gardening. Another project managed by Zunis was the Zuni Folk Varieties Project, designed to identify seeds their ancestors carefully developed for the pueblo's unique climatic conditions. The village also initiated the Zuni Irrigation Association, education programs, and cooperative research with outside scientists to advance agriculture on the reservation.

Other Pueblos joined the back-to-the-land movement as well. At Tesuque, Clayton and Margaret Brascoupe planted two big fields of corn, beans, and squash—crops historically associated with Pueblo agriculture—as part of a farming project to raise half of their family's food. Their seeds and methods were more traditional than those advocated in scientific farming. The Brascoupes considered gardening and nature as great teachers that generate "respect and the desire to help others." Leonard and Elsie Viao of Laguna complemented the return to native tradition by raising corn to cover the shrine during the pueblo's annual festival.[15]

For many people the trek back to the past arose from dissatisfaction with social and political conditions that encouraged and allowed them to stray away from their Pueblo heritage. As Cochiti tribal councilman Marcello Suina stated, "We lost the way we lived." For Cochiti, the return to Pueblo culture began in 1969 when the pueblo leased its land to a California developer, who then subleased the property for residential housing construction. Cochiti's people were lulled into thinking that the housing development would put them on easy street. In 1984, the investor declared bankruptcy and the pueblo bought back the lease, but by then their last alfalfa crop had rotted from water released by the Cochiti Dam seepage. The pueblo became disillusioned and sought to turn things around. Middle-aged Cochiti tribal leaders, armed with college degrees and business experience, attracted the attention of Congress, which authorized a twelve-million-dollar settlement between the tribe and the Army Corps of Engineers to fix the damage caused by the groundwater leaking under Cochiti Dam. By 1995, the farmland was dry enough to sow. Unfortunately, many people, especially young ones, had lost interest in farming, and the pueblo had to hire a non-Indian, educated in agriculture, to steer its residents back to their agricultural way of life. Still, tribal councilman Andy Quintana believed that many would find their way back to the soil, for there was "always something to learn from the land... always some kind of strength to be drawn from it."[16]

Perhaps no single crop represented the turn toward marketing in the new-age agriculture better than blue corn, a product that had been grown by Indians of the Southwest for centuries. Its color connoting harmony, longevity, and good luck, blue corn was considered a sacred plant by many Pueblos. In the modern era, however, health-conscious consumers, seeking an organically grown crop, encouraged New Mexican farmers to grow almost one thousand acres of blue corn, which produced about half of the nation's supply. Since it contained 20 percent more protein, 50 percent more iron, and twice as much manganese and potassium as yellow corn, consumers were willing to pay a premium price for the blue variety.

The Pueblos became attracted to blue corn and even held a seminar at Santo Domingo for farmers who wanted to learn more about its production. Leading the way was Santa Ana, which received a twenty-thousand-dollar grant from the Ford Foundation in 1992 for a project that would combine the revival of traditional farming practices with new economic opportunities. After decades of allowing land to go fallow, the farmers at Santa Ana dedicated one hundred acres to the cultivation of blue corn, alfalfa, and vegetables. They also had a grain mill that produced blue cornmeal, atole, and a salted parched-corn snack food. In the process, Santa Ana revived two traditions—farming and self-reliance. The movement away from federal aid and toward economic independence was aided by Santa Ana's effort to attract the attention of a British business known as the Body Shop that sold skin and hair products at 860 stores in forty-two countries. The English enterprise worked for some six years to help developing communities turn traditional crops into profitable ingredients for its cosmetics. The Body Shop's engagement with Santa Ana led to seven blue corn items, including moisturizer, soap, and body oil, that were sold in 130 stores throughout the United States.[17]

THE LEGACY

The mere mention of the Pueblo Indians of New Mexico conjures up an image of irrigation farmers living in multistoried adobe houses and practicing ancient religious ceremonies. In some ways this picture is both prehistoric and modern. One side displays the link to the past while the other represents the connection with the present. It is, of course, their agriculture that bonds the two and provides a common thread between traditions and modernization. Like other Native Americans, the Pueblos have been challenged to walk life's road with one foot in each world. For some the task proved daunting and required that they lean in one direction or the other. Agriculture, so tied to the Pueblo persona, provides a perfect reflection of this cultural dilemma.

From their very first contact with European people, the Pueblos provided a unique contrast with stereotypical Indian images. These native inhabitants were non-nomadic farmers more concerned with crop production than warfare. More importantly, the arid climate of the Southwest required unique watering techniques not found in eastern America or most of Europe. The efficiency of irrigation canals also necessitated a total communal effort that forged a bond among Pueblo farmers. All of this was greatly admired by Spanish immigrants and explorers who had faced similar problems in the Old World. However the religious practices that accompanied these ventures were not so well received by Christians who viewed them as relics of a superstitious past. Only a major rebellion and a fear of outside invasions forced a Spanish attitude change that allowed for accommodation. So it was the outsiders who were forced to alter their views while the Pueblos clung firmly to the past.

Early American observers were equally impressed with the Pueblos' culture—their work ethic, nobility, home construction, and agricultural pursuits. Some found the latter lacking in sophistication and unequal to the production standards of citizen farmers back East. If only, they reasoned, the native New Mexicans could adopt the innovations of modern science, then surely productive results would follow. Of course, this change required a complete transformation from traditional customs, and it was often met with stiff resistance. Only dramatic circumstances could bring this about, and it came in the form of major worldwide events that forced the Pueblos into the modern world.

The Pueblos had always been willing to make certain adjustments to outside influences as long as their impact on cultural traditions was minimal. Thus they adopted certain foods from the Spaniards and were willing to join the fight against nomadic invaders who threatened the welfare of both groups. They also demonstrated a willingness to fight in the courts of foreigners who gained control of the area when their water and land were threatened. But the impact of the Depression and the Second World War wrought pressures that proved to be too great for Pueblo resistance. Gradually, the Pueblo farm program moved closer to acceptance of capitalism. Like the Spaniards before them, the American officials realized that accommodating Pueblo traditions made for an easier road toward their idea of progress.

If the war and economic hardship opened the door to modernization, the period that followed led to a major decline in Pueblo agriculture that threatened its very existence. It was, however, one of the great ironies of their history that while their farm program was falling apart, the Pueblos' rights to water became of paramount importance. Court cases concerning this very grave matter, as yet unresolved, have and will change the adjudication of water in the American West—the result of which affirm some Indian rights and destroy others. It was the very tenacity of the Pueblos that brought this question to the fore—a tendency that represents the quintessential element of their character.

Of course the Pueblos also benefited from the assistance of caring people outside of their culture. These people represented a wide array of concerned observers who made a total commitment to the betterment of Pueblo life and included H. F. Robinson, Sophie Aberle, John Collier, and Lotario Ortega. Many employees of the extension service and agency personnel also contributed their skills and concerns in the battle against forces that sought to undermine native agricultural programs. Fortunately, these people came along at a time when the Indians needed them most, and their actions encouraged Pueblo leaders to raise their voices against those who sought to overrun this small but spiritually strong group in the ancient villages of northern New Mexico. And raise them they did. One of the major themes in the postwar era was the renewal of Pueblo leadership to a more vocal position. The more complex and threatening the situation got, the more the leaders took hold of the controls and sought to steer clear of harm's way.

What will be the future of Pueblo agriculture? Will the Pueblos continue to embrace two worlds at the same time or become overwhelmed by modernization? As younger generations continue to lose their languages, surf the Net, intermarry, watch television, and move into the fast-paced economy of modern America, will they lose interest in their roots so deeply embedded in their native soil? Will gambling casinos, first introduced in 1994 at Sandia[18] and now becoming common on Pueblo reservations, replace traditional occupations? Who will labor in the hot and dusty fields while air-conditioned gaming facilities beckon? More importantly, what will become of traditional religious ceremonialism that has accompanied Pueblo farming activities since ancient times?

Answers to all of these questions are purely conjectural, of course, but one should never underestimate the power of Pueblo cultural traditions.

A half-century ago some experts predicted the demise of Pueblo religion, but these projections have proven to be false. While the split between modern and traditional pueblos remains a reality, there has been a revival of ceremonial life even though some of it focuses on entertainment rather than religion.[19] In all probability, farming will never achieve the zenith of the past, but it is difficult to imagine the extinction of Pueblo agriculture. As modern communication links obliterate regional and cultural distinctions, however, so too will the Pueblos become more American and less Native American. But the powerful ties to the past, though stretched thin in the distant future, will remain. Without them, Pueblo religion would lose its meaning, and so too would the descendants of prehistoric southwestern farmers who once carved out earthen canals to provide for their very existence.

NOTES

INTRODUCTION

1. Theodore E. Downing and McGuire Gibson, eds., *Irrigation's Impact on Society* (Tucson: University of Arizona Press, 1974). This book covers irrigation systems throughout the world. Special attention is given to the Anasazi Southwest in R. Gwinn Vivian's article "Conservation and Diversion: Water Control Systems in the Anasazi Southwest" (95–112).
2. Joseph R. Long, *A Treatise on the Law of Irrigation: Covering All States and Territories* (Denver: W. H. Courtright, 1916), 1–2.
3. Clesson S. Kinney, *A Treatise on the Law of Irrigation and Water Rights and the Arid Region Doctrine of Appropriation of Waters*, 4 vols. (San Francisco: Bender-Moss, 1912), 1: 101–17. In addition to laws concerning irrigation and water rights in different countries throughout the world, the author gives a concise synopsis of ancient irrigation systems in the countries mentioned.
4. Michael S. Berry, *Time, Space, and Transition in Anasazi Prehistory* (Salt Lake City: University of Utah Press, 1982), 31. Previous writers theorized that maize was introduced in the Southwest as early as 3500 B.C., but Berry states that it did not occur until a few centuries before the birth of Christ. For a discussion of the domestication of plants in New Mexico see Richard B. Woodbury and Ezra B. W. Zubrow, "Agricultural Beginnings, 2000 B.C.–A.D. 500," in *Handbook of North American Indians*, 20 vols., ed. Alfonso Ortiz (Washington, D.C.: Smithsonian Institution, 1979), 46–50.
5. David E. Stuart, *Anasazi America: Seventeen Centuries on the Road from Center Place* (Albuquerque: University of New Mexico Press, 2000), 35–39.

For a mention of the Mogollon ancestry of the Zuni, see E. Richard Hurt, "Historic Zuni Land Use," in *Zuni and the Courts: A Struggle for Sovereign Land Rights*, ed. E. Richard Hurt (Lawrence: University Press of Kansas, 1995), 9; see also Barry M. Pritzker, *A Native American Encyclopedia: History, Culture, and Peoples* (New York: Oxford University Press, 2000), 6.

6. Edgar L. Hewett, *Handbooks of Archaeological History* (Albuquerque: University of New Mexico Press, 1945), 80.
7. Charles Avery Amsden, *Prehistoric Southwesterners from Basketmaker to Pueblo* (Los Angeles: Southwest Museum, 1949), 103–4; Andrew L. Knaut, *The Pueblo Revolt of 1680: Conquest and Resistance in Seventeenth-Century New Mexico* (Norman: University of Oklahoma Press, 1995), 59–62; Stuart, *Anasazi America*, 42.
8. Arthur H. Rohn, "Prehistoric Soil and Water Conservation on Chapin Mesa, Southwestern Colorado," *American Antiquity* 28 (April 1963): 441–46.
9. Paul R. Franke and Don Watson, "An Experimental Corn Field in Mesa Verde National Park," *The University of New Mexico Bulletin*, no. 296 (1936): 35–41.
10. Stuart, *Anasazi America*, 43–44, 47.
11. Ibid., 51–54, 56–57, 61, 63.
12. Ibid., 65–67, 85–87, 123–25, 127, 136, 140.
13. Thomas E. Mails, *The Pueblo Children of the Earth Mother*, 2 vols. (Garden City, N.Y.: Doubleday, 1983), 1: 285; Linda S. Cordell, "Prehistory: Eastern Anasazi," in *Handbook of North American Indians*, 20 vols., ed. Alfonso Ortiz (Washington, D.C.: Smithsonian Institution, 1979), 9: 151; Florence Hawley Ellis, "Archaeological History of Nambe Pueblo, 14th Century to the Present," *American Antiquity* 30 (July 1964): 34.
14. Stuart, *Anasazi America*, 150.
15. Stephen C. Jett, "Pueblo Indian Migrations: An Evaluation of the Possible Physical and Cultural Determinants," *American Antiquity* 29 (January 1964): 290–93. Jett feels that the evacuation was too orderly to suggest a starving group. He also feels that without the horse, raiders could not have overwhelmed the village structures of the Anasazi. Indeed, proof that these raiders were in this area at the time is lacking.
16. Kirk Bryan, "Pre-Columbian Agriculture in the Southwest, as Conditioned by Periods of Alluviation," *Association of American Geographers* 31 (December 1941): 224–25.
17. Ira G. Clark, *Water in New Mexico: A History of Its Management and Use* (Albuquerque: University of New Mexico Press, 1987), 5.
18. Nancy Bonvillain, *Native Nations: Cultures and Histories of Native North America* (Upper Saddle River, N.J.: Prentice-Hall, 2001), 313–14; Stuart, *Anasazi America*, 162–63. Typically, the eastern group is referred to as the Rio Grande pueblos while those in the west include Zuni (sometimes along with the Hopi in Arizona) and often Acoma and Laguna.
19. Stuart, *Anasazi America*, 152–55, 162–63.
20. Ibid., 180.

CHAPTER ONE

1. Herbert E. Bolton, *Coronado on the Turquoise Trail: Knight of Pueblos and Plains* (Albuquerque: University of New Mexico Press, 1949), 13–14.
2. Edgar Lee Hewett, *The Physiography of the Rio Grande Valley, New Mexico, in*

Relation to Pueblo Culture (Washington, D.C.: U.S. Government Printing Office, 1913), 18–19. The author gives an excellent account of meteorological conditions of the region.

3. Bolton, *Coronado on the Turquoise Trail*, 68.
4. Viceroy Antonio de Mendoza to Francisco Vasquez de Coronado, 1540, in *Narratives of the Coronado Expedition, 1540–1542*, eds. George P. Hammond and Agapito Rey (Albuquerque: University of New Mexico Press, 1940), 84–85.
5. Juan Jaramillo, "Narrative Given by Captain Juan Jaramillo of His Journey to the New Land in New Spain and to the Discovery of Cibola Under General Francisco Vasquez Coronado," in *Narratives of the Coronado Expedition*, ed. Hammond and Rey, 297.
6. Pedro de Castañeda, "Narrative of the Expedition to Cibola, Undertaken in 1540, in which are Described All Those Settlements, Ceremonies, and Customs," in *Narratives of the Coronado Expedition*, ed. Hammond and Rey, 52.
7. Ibid., 218.
8. Bolton, *Coronado on the Turquoise Trail*, 184–85.
9. Jaramillo, "Narrative Given," 300.
10. Castañeda, "Narrative of the Expedition," 255–59.
11. Lynn I. Perrigo, *The American Southwest: Its Peoples and Cultures* (New York: Holt, Rinehart and Winston, 1971), 26; Herbert Eugene Bolton, *Coronado: Knight of Pueblos and Plains* (Albuquerque: University of New Mexico Press, 1971), 192–93, 201–15.
12. Herman Gallegos, "Gallegos' Relation of the Chamuscado-Rodríguez Expedition," in *The Rediscovery of New Mexico, 1580–1594: The Explorations of Chamuscado, Espejo, Castaño de Sosa, Morlete, and Leyva de Bonilla and Humaña*, ed. George P. Hammond (Albuquerque: University of New Mexico Press, 1966), 82–108. See also J. Lloyd Mecham, "The Second Spanish Expedition into New Mexico," *New Mexico Historical Review* 1 (July 1926): 272–87.
13. Antonio de Espejo, "Brief and True Account of the Exploration of New Mexico, 1583," in *Spanish Exploration in the Southwest, 1542–1706*, ed. Herbert Eugene Bolton (New York: Charles Scribner's Sons, 1916), 178. In addition to these foods, Espejo mentions wild onions near Acoma (183).
14. J. Lloyd Mecham, "Antonio de Espejo and His Journey to New Mexico," *Southwestern Historical Quarterly* 30 (July 1926): 127–29.
15. Diego Perez de Luxan, "Account of the Antonio de Espejo Expedition into New Mexico, 1582," in *Rediscovery of New Mexico*, ed. Hammond, 182–86.
16. Ibid., 201–4.
17. Gaspar Castaño de Sosa, *A Colony on the Move: Gaspar Castaño de Sosa's Journal 1590–1591*, trans. Dan S. Matson (Salt Lake City: Alphabet Printing, 1965), 100. While Sosa was not specific about the Indian farming implements, Matson feels that they included digging sticks, stone or wooden hoes, and maybe stone axes. In addition, Pueblo pottery was used to store seeds for the next year's crops.
18. Gaspar Castaño de Sosa, "Memoria," in *Rediscovery of New Mexico*, ed. Hammond, 100.
19. Castaño de Sosa, *A Colony on the Move*, 5–6.
20. Don Juan de Oñate, "Contract of Don Juan de Oñate for the Discovery and Conquest of New Mexico," in *Don Juan de Oñate: Colonizer of New Mexico*,

1595–1628, 2 vols., eds. George P. Hammond and Agapito Rey (Albuquerque: University of New Mexico Press, 1953), 1: 44–54. Although iron implements came into New Mexico at this time, their numbers were limited and their use was not widespread.

21. Martin Lopez de Gauna, "Instructions to Don Juan de Oñate," in *Don Juan de Oñate*, ed. Hammond and Rey, 1: 65–68.
22. Don Juan de Oñate, "Act of Taking Possession of New Mexico, April 30, 1598," in *Don Juan de Oñate*, ed. Hammond and Rey, 1: 334–39.
23. Don Luis de Velasco to the Viceroy, 22 March 1601, in *Don Juan de Oñate*, ed. Hammond and Rey, 2: 609–10.
24. Don Francisco de Valverde, "Investigation of Conditions in New Mexico, 1601," in *Don Juan de Oñate*, ed. Hammond and Rey, 2: 633–34.
25. Ibid., 645–54.
26. Don Luis de Velasco, "Breaking Camp at San Gabriel," in *Don Juan de Oñate*, ed. Hammond and Rey, 2: 672–75.
27. Fray Juan de Escalona to the Viceroy, 1 October 1601, in *Don Juan de Oñate*, ed. Hammond and Rey, 2: 692–96. Concerning the hurried enforcement of tribute, Edward Spicer concluded that the Pueblos did not receive the benefit of advance missionary parties as tribes farther to the south had. This meant that the Pueblos did not have a gradual adjustment to the offensive practices of the tribute system. See Edward H. Spicer, *Cycles of Conquest: The Impact of Spain, Mexico, and the United States on the Indians of the Southwest, 1533–1960* (Tucson: University of Arizona Press, 1962), 167.
28. Geronimo Marquez, "Report of the People Who Remained Behind in New Mexico, 1601," in *Don Juan de Oñate*, ed. Hammond and Rey, 2: 714–22.
29. Francisco de Valverde, "Inquiry Concerning the New Provinces in the North, 1602," in *Don Juan de Oñate*, ed. Hammond and Rey, 2: 851.
30. Viceroy Martín Lopez de Gauna, "Governor Peralta's Instruction," in *Don Juan de Oñate*, ed. Hammond and Rey, 2: 1087.
31. France V. Scholes, "Church and State in New Mexico 1610–1650," *New Mexico Historical Review* 11 (January 1936): 32–33.
32. Ibid., 147–55.
33. France V. Scholes, "Civil Government and Society in New Mexico in the Seventeenth Century," *New Mexico Historical Review* 10 (April 1935): 105–6.
34. Diego del Rio de Loza, "Report to the Viceroy by the Cabildo of Santa Fe, New Mexico, February 21, 1639," in *Historical Documents Relating to New Mexico, Nueva Vizcaya, and Approaches Thereto, to 1773*, 3 vols., ed. Charles Wilson Hackett (Washington, D.C.: Carnegie Institution of Washington, 1937), 3: 71. Hereafter cited as *Historical Documents.*
35. Francisco de la Mora, "Petition to Mexico, February 4, 1639," in *Historical Documents*, ed. Hackett, 3: 117–18.
36. Fray Estevan de Perea, "Case Brought against Juan Lopez, October 30, 1633," in *Historical Documents*, ed. Hackett, 3: 131.
37. George McCutchen McBride, *The Land Systems of Mexico* (New York: American Geographical Society, 1923), 45; H. Allen Anderson, "The Encomienda in New Mexico, 1598–1680," *New Mexico Historical Review* 60 (October 1985): 353–73.
38. Alonso de Billanvebay Gonzalo Lopez, "Petition of the Procuadores of New

Spain and the Great City of Mexico," in *Historical Documents*, ed. Hackett, 1: 135–39.

39. McBride, *Land Systems of Mexico*, 106–10. Realizing the limitations of an arid environment, the Spaniards employed a long-lot system of land grants in New Mexico. This assured the farmers equal distribution of water. See Alvar W. Carlson, "Long-Lots in the Rio Arriba," *Annals of the Association of American Geographers* 65 (March 1975): 48–55; for encroachment as a source of the Pueblo Revolt of 1680 see Michael C. Meyer, *Water in the Hispanic Southwest: A Social and Legal History, 1550–1850* (Tucson: University of Arizona Press, 1984), 51–52.
40. Richard E. Greenleaf, "Land and Water in Mexico and New Mexico 1700–1821," *New Mexico Historical Review* 47 (April 1972): 88–89.
41. Herbert O. Brayer, *Pueblo Indian Land Grants of the "Rio Abajo," New Mexico* (Albuquerque: University of New Mexico Press, 1938), 9–10. See also Marc Simmons, *Spanish Government in New Mexico* (Albuquerque: University of New Mexico Press, 1968), 189. These offices were not always differentiated in frontier provinces.
42. Ramón A. Gutiérrez, *When Jesus Came, the Corn Mothers Went Away: Marriage, Sexuality, and Power in New Mexico, 1500–1846* (Stanford: Stanford University Press, 1991), 304; G. Emlen Hall, *Four Leagues of Pecos. A Legal History of the Pecos Grant, 1800–1933* (Albuquerque: University of New Mexico Press, 1984), 13–14.
43. Gaspar Pérez de Villagrá, *History of New Mexico* (Los Angeles: Quivira Society, 1933), 144. See also Charles H. Lange, *Cochiti: A New Mexico Pueblo, Past and Present* (Austin: University of Texas Press, 1959), 98–100; Pritzker, *A Native American Encyclopedia*, 6.
44. Alvin Sunseri, "Agricultural Techniques in New Mexico at the Time of the Anglo-American Conquest," *Agricultural History* 47 (October 1973): 334; Spicer, *Cycles of Conquest*, 541. See also Hackett, *Historical Documents*, 3: 73; for Spanish food storage see Van Hastings Garner, "Seventeenth Century New Mexico, the Pueblo Revolt and Its Interpreters," in *What Caused the Pueblo Revolt of 1680*, ed. David J. Weber (Boston: Bedford/St. Martins, 1999), 67.
45. Sunseri, "Agricultural Techniques," 330–31.
46. Marc Simmons, "Spanish Irrigation Practices in New Mexico," *New Mexico Historical Review* 47 (April 1972): 141–43.
47. Charles H. Lange, "An Evaluation of Economic Factors in Cochiti Pueblo Culture Change" (Ph.D. dissertation, University of New Mexico, 1951), 85. See also Edward P. Dozier, "Rio Grande Pueblos," in *Perspectives in American Indian Cultural Change*, ed. Edward H. Spicer (Chicago: University of Chicago Press, 1961), 101–2.
48. Clark, *Water in New Mexico*, 7.
49. Ibid., 8; for the division of labor see Bonvillain, *Native Nations*, 314.
50. Scholes, "Church and State," 145–46.
51. France V. Scholes, "Troublous Times in New Mexico, 1659–1670," *New Mexico Historical Review* 12 (April 1937): 158–59.
52. France V. Scholes, "Troublous Times in New Mexico, 1659–1670," *New Mexico Historical Review* 12 (October 1937): 407–9.
53. Donald Nelson Brown, "Picuris Pueblo in 1890: A Reconstruction of Picuris

Social Structure and Subsistence Activities," in *Picuris Pueblo Through Time: Eight Centuries of Change in a Northern Rio Grande Pueblo*, eds. Michael A. Alder and Herbert W. Dick (Dallas: Southern Methodist University Clements Center for Southwest Studies, 1999), 30.

54. Andres Hurtado, "Declaration of Captain Andres Hurtado, Santa Fe, September 1661," in *Historical Documents*, ed. Hackett, 3: 272.
55. Juan Bernal, "Letter of Fray Juan Bernal to the Tribunal, April 1, 1669," in *Historical Documents*, ed. Hackett, 3: 272.
56. Francisco de Ayeta, "Petition of Father Fray Francisco de Ayeta, Mexico, May 10, 1679," in *Historical Documents*, ed. Hackett, 3: 302.
57. Francisco Gómez, "Auto and Declarations of Maestre de Campo Francisco Gómez, Santa Fe, August 13–20, 1680," in *Revolt of the Pueblo Indians of New Mexico and Otermín's Attempted Reconquest, 1680–1682*, 2 vols., trans. Charmion Clair Shelby, ed. Charles Wilson Hackett (Albuquerque: University of New Mexico Press, 1942), 1: 17. Hereafter cited as *Revolt of the Pueblo Indians*; for a thorough look at the reasons for the rebellion see Weber, *What Caused the Pueblo Revolt of 1680.*
58. Alonso Garcia, "Auto of Alonso Garcia, La Isleta, August 14, 1680," in *Revolt of the Pueblo Indians*, ed. Hackett, 1: 69.
59. Don Antonio de Otermín, "Auto of Don Antonio de Otermín," in *The Spanish Archives of New Mexico*, 2 vols., ed. Ralph Emerson Twitchell (Cedar Rapids, Iowa: Torch Press, 1914), 2: 16–20. Hereafter cited as *The Spanish Archives.*
60. Don Antonio de Otermín, "Interrogatories and Depositions of Three Indians of the Tehua Nations," in *The Spanish Archives*, ed. Twitchell, 2: 52–63.
61. Francisco Zavier, "Auto, La Isleta, December 9, 1681," in *Revolt of the Pueblo Indians*, ed. Hackett, 2: 218.
62. Francisco Zavier, "Declaration of the Indian, Juan, Place on the Rio del Norte, December 18, 1681," in *Revolt of the Pueblo Indians*, ed. Hackett, 2: 236.
63. Francisco Xavier, "Continuation of the March, December 13–18, 1681," in *Revolt of the Pueblo Indians*, ed. Hackett, 2: 236.
64. Alonso Rael de Aguilar, "Entry into the Pueblos of Cochiti and San Felipe," in *First Expedition of Vargas into New Mexico, 1692, Diego de Vargas*, trans. and ed. J. Manuel Espinosa (Albuquerque: University of New Mexico Press, 1940), 74–77. Hereafter cited as *First Expedition of Vargas.*
65. Alonso Rael de Aguilar, "Entry into the Pueblo Pecos," in *First Expedition of Vargas*, trans. and ed. Espinosa, 121–23, 198–99.
66. Don Diego de Vargas to the Conde de Galve, El Paso, 12 January 1693, in *First Expedition of Vargas*, trans. and ed. Espinosa, 281–87.

CHAPTER TWO

1. Pritzker, *A Native American Encyclopedia*, 7.
2. E. N. van Kleffens, *Hispanic Law Until the End of the Middle Ages* (Edinburgh: Edinburgh University Press, 1968), 122–26. The term "Spanish law" is used here for descriptive purposes. Actually, the laws were of Castilian origin.
3. Samuel C. Wiel, *Water Rights in the Western States* (San Francisco: Bancroft-Whitney, 1911), 748. In addition to a thorough background on prior appropriation, riparian rights, and western water laws, the author uses a state-by-state approach to describe water statutes in the western United States.

4. Long, *Law of Irrigation*, 26. Long's work is an excellent account of irrigation law with a special emphasis on the doctrine of appropriation and the adjudication of water rights in the western states.
5. Kinney, *Law of Irrigation*, 1: 256. Although this series was written over ninety years ago, it is still considered the classic work on irrigation throughout the world. The author includes a history of irrigation systems, modern irrigation practices, and the laws that affected irrigation in different parts of the world.
6. Kleffens, *Hispanic Law*, 37–38.
7. Ibid., 70–81.
8. Betty Dobkins, *The Spanish Element in Texas Water Law* (Austin: University of Texas Press, 1959), 64–68. Dobkins's book is the most detailed analysis in English on the subject of Iberian and Spanish colonial water law and its application in the Southwest.
9. Gaines Post, "Roman Law and Early Representation in Spain and Italy, 1150–1250," *Speculum* 44 (April 1943): 216–17.
10. Robert Ignatius Burns, "Irrigation Taxes in Early Mudejar Valencia: The Problem of the Alfarda," *Speculum* 44 (October 1969): 562–63.
11. Dobkins, *Spanish Element*, 71–72.
12. Kinney, *Law of Irrigation*, 1: 982–83; Dobkins, *Spanish Element*, 76–77. See also Helen L. Clagett, "Las Siete Partidas," *The Quarterly Journal of the Library of Congress* 22 (October 1965): 342–44. The author supplies a background to the compilation of the Partidas and its impact on law throughout the world.
13. Thomas F. Glick, *Irrigation and Society in Medieval Valencia* (Cambridge, Mass.: Belknap Press of Harvard University Press, 1970), 150–51.
14. Dobkins, *Spanish Element*, 89.
15. *Recopilación de Leyes de los Reynos de las Indias*, 4 vols. (Madrid, 1681), Book 2, Title 1, Law 2.
16. Ibid., Law 1.
17. *Colección de Documentos Ineditos, Relativos al Descubrimiento, Conquista y Organización de las Antiguas Posesiones Españolas de América y Oceania, Sacados de los Archivos del Reino, y Muy Especialmente del de Indias*, 42 vols. (Madrid, 1864–1884), 16: 142–87.
18. *Recopilación*, Book 4, Title 1, Law 10; Title 4, Laws 5 and 8; Title 7, Law 26.
19. Ibid., Title 12, Laws 4, 5, and 9.
20. Ibid., Laws 12 and 13; Meyer, *Water in the Hispanic Southwest*, 50.
21. *Recopilación*, Book 4, Title 12, Laws 16, 17, and 18.
22. Ibid., Book 2, Title 1, Law 4.
23. Ibid., Book 6, Title 1, Laws 21 and 23.
24. Ibid., Title 3, Law 8; Book 4, Title 12, Law 18.
25. Ibid., Book 3, Title 2, Law 63.
26. Ibid.
27. Ibid., Book 4, Title 17, Laws 5, 7, and 9.
28. Ibid., Title 12, Laws 17 and 19.
29. Greenleaf, "Land and Water in Mexico," 88–97. For the ordinance of 1781 see R. Douglas Hurt, *Indian Agriculture in America: Prehistory to the Present* (Lawrence: University Press of Kansas, 1988), 78.
30. Kinney, *Law of Irrigation*, 1: 123.
31. Simmons, "Spanish Irrigation," 139–40.

32. William B. Taylor, "Land and Water Rights in the Viceroyalty of New Spain," *New Mexico Historical Review* 50 (July 1975): 195–99.
33. Michael C. Meyer, "The Legal Relationship of Land to Water in Northern Mexico and the Hispanic Southwest," *New Mexico Historical Review* 60 (January 1985): 61–79.
34. Taylor, "Land and Water Rights," 200–205; for an excellent account of the criteria upon which legal disputes were resolved see Meyer, *Water in the Hispanic Southwest*, 145–64.
35. Dobkins, *Spanish Element*, 98–99.
36. Greenleaf, "Land and Water," 98–99.
37. Ibid., 99–100.
38. Myra Ellen Jenkins, "Spanish Land Grants in the Tewa Area," *New Mexico Historical Review* 47 (April 1972): 114.
39. Marc Simmons, "Settlement Patterns and Village Plans in Colonial New Mexico," *Journal of the West* 8 (January 1969): 11–13. See also Carlson, "Long-Lots," 48.
40. Diego de Vargas to Viceroy Galvez, October 1692, in *The Spanish Archives*, ed. Twitchell, 78–84.
41. Francisco Cuervo y Valdés, order to Spanish citizen, August 1705, in *The Spanish Archives*, ed. Twitchell, 1: 21–25.
42. Ralph E. Twitchell, *Spanish Colonization in New Mexico in the Oñate and De Vargas Periods* (Santa Fe: Historical Society of New Mexico, 1919), 20.
43. Captain Alonso Rael de Aguilar to Acting Governor Juan Páez Hurtado, in *The Spanish Archives*, ed. Twitchell, 1: 394–96.
44. Vincente Durán de Armijo to Governor Gaspar Domingo de Mendoza, 1743, in *The Spanish Archives*, ed. Twitchell, 1: 21–25.
45. Rafael Sena to Governor Albino Perez, 1835, in *The Spanish Archives*, ed. Twitchell, 1: 271–72.
46. Order of the Viceroy, October 22, 1704, in *The Spanish Archives*, ed. Twitchell, 1: 397. See also "Fray Juan Sanz de Lezaun's account of lamentable happenings in New Mexico," in *Historical Documents*, ed. Hackett, 3: 468–72.
47. Father Juan Miguel Menchero to Governor Codallas y Rabál, 1748, in *The Spanish Archives*, ed. Twitchell, 1: 156.
48. "Proceedings in the Establishment of Nuestra Señora de los Dolores de Sandia," 1748, in *The Spanish Archives*, ed. Twitchell, 1: 235–36. See also Salvador Martínez, "Relative to a claim against the Fray Juan Miguel Menchero," in *The Spanish Archives*, ed. Twitchell, 2: 220–22.
49. Luis Romero, "Permission given him to sell a piece of land," in *The Spanish Archives*, ed. Twitchell, 1: 311.
50. El Capulin, "Proceedings as to Occupation of Miguel Romero," in *The Spanish Archives*, ed. Twitchell, 1: 403–7.
51. "San Ildefonso Lands," 1763, in *The Spanish Archives*, ed. Twitchell, 1: 403–7.
52. Juan Griego and Juliana Sais and Francisco Sais to Diego Arias de Quiros, Santa Fe, 1718, in *The Spanish Archives*, ed. Twitchell, 1: 197–99.
53. "Grant, Canada de Santa Clara, 1724," in *The Spanish Archives*, ed. Twitchell, 1: 280.
54. "The Use of the Water in the Santa Clara River, 1734," in *The Spanish Archives*, ed. Twitchell, 1: 282.

55. Daniel Tyler, *The Mythical Pueblo Rights Doctrine: Water Administration in Hispanic New Mexico* (El Paso: Texas Western Press, 1990), 29, 36, 40.
56. Malcolm Ebright, "Water Litigation and Regulation in Hispanic New Mexico, 1600–1850," *New Mexico Historical Review* 76 (January 2001): 3–10, 32–33.

CHAPTER THREE

1. Mark Simmons, "History of Pueblo-Spanish Relations to 1821," in *Handbook of North American Indians*, 20 vols., ed. Alfonso Ortiz (Washington, D.C.: Smithsonian Institute, 1979), 9: 186–87.
2. Fray Joaquín de Jesús Ruiz, "The Form of Government used at the mission of San Diego de los Jemez and San Agustín de la Isleta," in *Historical Documents*, ed. Hackett, 3: 505. See also Dozier, "Rio Grande Pueblos," 146. For increased production and Spanish occupation of abandoned Pueblos see Simmons, "Pueblo-Spanish Relations," 9: 182, 190.
3. Oakah L. Jones, *Pueblo Warriors and Spanish Conquest* (Norman: University of Oklahoma Press, 1966), 134–35. See also Alfred B. Thomas, *The Plains Indians of New Mexico, 1751–1778: A Collection of Documents Illustrative of the History of the Eastern Frontier of New Mexico* (Albuquerque: University of New Mexico Press, 1940), 17–181. The author includes documentation of Comanche attacks on Pecos and Nambe in the 1770s while the native farmers were tending their fields.
4. For prehistoric trade see Charles H. Lange, "Relations of the Southwest with the Plains and Great Basin," in *Handbook of North American Indians*, 20 vols., ed. Alfonso Ortiz (Washington, D.C.: Smithsonian Institute, 1979), 9: 204; and Fred Eggan, "Pueblos: Introduction," in *Handbook of North American Indians*, 20 vols., ed. Alfonso Ortiz (Washington, D.C.: Smithsonian Institute, 1979), 9: 224. For the trade fairs see Simmons, "Pueblo-Spanish Relations," 9: 189–90.
5. Eleanor B. Adams and Fray Angelico Chavez, trans., *The Missions of New Mexico, 1776: A Description by Fray Francisco Atanasio Domínguez With Other Contemporary Documents* (Albuquerque: University of New Mexico Press, 1956), 50–217. The authors supply a pueblo-by-pueblo account of the local church, convent, pueblo, and population. For remarks on Taos, Santa Ana, and Laguna by Fray Morfi see Alfred B. Thomas, ed. and trans., *Forgotten Frontiers: A Study of the Spanish Indian Policy of Don Juan Bautista de Anza, Governor of New Mexico, 1777–1787* (Norman: University of Oklahoma Press, 1932), 96, 98, 103. The study contains some original documents from the period, including a geographical description of New Mexico in 1782 by Fray Morfi. For remarks on Isleta by Bishop Tamarón see Pedro Tamarón Romeral, *Bishop Tamarón's Visitation of New Mexico, 1760*, ed. Eleanor B. Adams (Albuquerque: Historical Society of New Mexico, 1954), 71. For the abandonment of Galisteo see Edward P. Dozier, *The Pueblo Indians of North America* (New York: Holt, Rinehart and Winston, 1970), 64.
6. Oakah L. Jones, *Los Paisanos: Spanish Settlers on the Northern Frontier of New Spain* (Norman: University of Oklahoma Press, 1979), 133, 141–43; Lange, *Cochiti*, 90; Adams and Chavez, *The Missions of New Mexico*, 95, 109, 217, 252, 312, 313, 323.
7. For integration of the races see Jones, *Los Paisanos*, 127; for population statistics see Knaut, *The Pueblo Revolt*, 186.

8. D. W. Meinig, *Southwest: Three Peoples in Geographical Change, 1600–1700* (New York: Oxford University Press, 1971), 27–32. The author supplies a graph of population distribution in the region from 1600 to 1960. See also Dozier, *The Pueblo Indians*, 130.
9. Nemesio Salcedo to the acting governor of New Mexico, 11 August 1809, in *The Spanish Archives*, ed. Twitchell, 1: 30; Myra Ellen Jenkins, "The Baltasar Baca 'Grant': History of an Encroachment," *El Palacio* 68 (spring 1961): 57–59.
10. Juan de Aguilar to Governor Facundo Melgares, 17 August 1818, in *The Spanish Archives*, ed. Twitchell, 1: 30–31.
11. "Proceedings in a dispute between the Indians of Santa Ana and those of San Felipe," May 1813, in *The Spanish Archives*, ed. Twitchell, 1: 422–28, 1: 437–39.
12. "Petition in regard to land near the pueblo of Santa Clara," 1817, in *The Spanish Archives*, ed. Twitchell, 1: 77–78.
13. José Gutiérrez to Governor Alberto Máynez, 3 March 1816, in *The Spanish Archives*, ed. Twitchell, 1: 433–34.
14. "Proceedings in a dispute between the Indians of Taos and some Spanish citizens," 1815, in *The Spanish Archives*, ed. Twitchell, 1: 428–32.
15. Felix S. Cohen, *Handbook of Federal Indian Law* (Albuquerque: University of New Mexico Press, 1942), 383.
16. Spicer, *Cycles of Conquest*, 335–36.
17. Brayer, *Pueblo Indian Land Grants*, 17–19. In the 1850s, W. W. H. Davis, U.S. attorney to the Territory of New Mexico, wrote that if the Pueblos were considered citizens, so were all other Indians, including wild tribes. He felt that Mexico always considered the Pueblos as wards of the government.
18. Daniel Tyler, "Mexican Indian Policy in New Mexico," *New Mexico Historical Review* 55 (April 1980): 106, 109, 115.
19. G. Emlen Hall and David J. Weber, "Mexican Liberals and the Pueblo Indians, 1821–1829," *New Mexico Historical Review* 59 (January 1984): 7–8; Tyler, "Mexican Indian Policy," 104, 106.
20. David J. Weber, *The Mexican Frontier, 1821–1846: The American Southwest under Mexico* (Albuquerque: University of New Mexico Press, 1982), xvii.
21. W. W. H. Davis, *El Gringo, Or, New Mexico and Her People* (Santa Fe, N.Mex.: Rydal Press, 1938), 328.
22. Myra Ellen Jenkins, "Land Grant Problems," in *New Mexico Past and Present*, ed. Richard N. Ellis (Albuquerque: University of New Mexico Press, 1971), 100.
23. Mark Simmons, "History of the Pueblos Since 1821," in *Handbook of North American Indians*, 20 vols., ed. Alfonso Ortiz (Washington, D.C.: Smithsonian Institute, 1979), 9: 207.
24. Tyler, "Mexican Indian Policy," 105; Hall and Weber, "Mexican Liberals," 8–11, 16–22.
25. Wiel, *Water Rights*, 68–69. See also Long, *Law of Irrigation*, 50.
26. Kinney, *Law of Irrigation*, 1: 996.
27. Lynn I. Perrigo, trans., "Revised Statutes of 1826," *New Mexico Historical Review* 27 (January 1952): 70.
28. "Provincial Deputation, 1822–1825," in *The Spanish Archives*, ed. Twitchell, 1: 377–80.
29. "Petition for lands between the pueblos of Santo Domingo and San Felipe, 1831," in *The Spanish Archives*, ed. Twitchell, 1: 87–88.

30. Jenkins, "The Baltasar Baca 'Grant,'" 53, 59–63.
31. "Indians of the Pueblo of Taos and the People of Arroyo Seco, 1823, Water Rights," in *The Spanish Archives*, ed. Twitchell, 1: 380.
32. "Indians of Laguna vs. Joaquín Pino," in *The Spanish Archives*, ed. Twitchell, 1: 182–83; "Petition of Three Lagunas, 15 June 1827," in *The Spanish Archives*, ed. Twitchell, 1: 441–42.
33. "Protest by the Indians of the Pueblo of Isleta, 27 March 1845," in *The Spanish Archives*, ed. Twitchell, 1: 448–49.
34. Elliott Couves, ed., *The Expeditions of Zebulon Montgomery Pike*, 3 vols. (Minneapolis: Ross and Haines, 1965), 2: 606, 740.
35. Daniel Tyler, "New Mexico in the 1820's: The First Administration of Manuel Armijo" (Ph.D. dissertation, The University of New Mexico, 1970), 82–83.
36. Lansing B. Bloom, "New Mexico Under the Mexican Administration," *Old Santa Fe* 1 (July 1913): 4.
37. Hubert Howe Bancroft, *History of Arizona and New Mexico, 1530–1888* (San Francisco: History, 1889), 302.
38. H. Bailey Carroll and J. Villasana Haggard, trans., *Three New Mexico Chronicles: The Exposición of Pedro Bautista Pino, 1812; The Ojeada of Lic. Antonio Barreiro, 1832; And the Additions by José Agustín de Escudero, 1849* (Albuquerque: Quivira Society, 1942), 27–35. One of the crops mentioned by Pino was strawberries (104).
39. Ibid., 38.
40. Ibid., 88. Tesuque's population statistics were added with those of Santa Fe and therefore are not included in this chart.
41. Ibid., 30. Barriero also remarked on the good reasoning power and judgment of the Pueblos. He also spoke of their extreme honesty and truthfulness. In spite of this high praise, Barriero felt that the Indian race was a vanishing one and predicted that the Pueblos would soon disappear. Another observer who praised the character of the natives at this time was Thomas James, who felt that in terms of virtue, courage, generosity, reliability, and ingenuity, the Pueblos were far superior to their Mexican counterparts. See Thomas James, *Three Years Among the Indians and Mexicans* (Philadelphia: J. B. Lippincott, 1962), 90. Exploring New Mexico in 1845, James W. Abert noted that while they were mostly agricultural people, the Pueblos still supplemented their diet by hunting. See James William Abert, *Expedition to the Southwest: An 1845 Reconnaissance of Colorado, New Mexico, Texas, and Oklahoma* (Lincoln: University of Nebraska Press, 1999), 35.
42. Josiah Gregg, *The Commerce of the Prairies*, ed. Milo Milton Quaife (Lincoln: University of Nebraska Press, 1967), 142–43.
43. Ibid., 77–81.
44. Ralph Emerson Twitchell, *The Leading Facts of New Mexican History*, 2 vols.(Cedar Rapids, Iowa: Torch Press, 1912), 2: 56–60. See also Perrigo, *The American Southwest*, 148. The most detailed account of this rebellion can be found in Phillip Reno, "Rebellion in New Mexico—1837," *New Mexico Historical Review* 40 (July 1965): 197–213. See also Weber, *The Mexican Frontier*, 261–65.
45. Perrigo, *The Southwest*, 154.
46. W. H. Emory, "Notes of a Military Reconnoissance (sic) from Fort Leavenworth in Missouri to San Diego in California," in *Senate Executive*

Document no. 7, 30th Cong., 1st sess., 1848 (Washington, D.C.: Wendell and Van Benthuysen, 1848), 38–40.

47. J. W. Abert, "Report of the Secretary of War, February 4, 1848," in *Senate Executive Document* no. 23, 30th Cong., 1st sess., 1848 (Washington, D.C.: Wendell and Van Benthuysen, 1848), 41–55.
48. J. H. Simpson, "Report of Lieutenant J. H. Simpson of an Expedition into the Navajo Country," in *Reports of the Secretary of War, Senate Executive Document* no. 64, 31st Cong., 1st sess., 1850 (Washington, D.C.: The Union Office, 1850), 63–64, 129.
49. Alvin R. Sunseri, "New Mexico in the Aftermath of the Anglo-American Conquest, 1846–1861" (Ph.D. dissertation, Louisiana State University, 1973), 93.
50. Alexander W. Doniphan, Willard P. Hall, and Stephen Kearny, comps., *Laws of the Territory of New Mexico*, trans. David Waldo (Santa Fe, N.Mex.: Oliver Perry Hovey, 1846), 83.
51. Ibid., 114. Ayuntamiento was the same municipal council sometimes referred to as a cabildo.
52. James S. Calhoun to Orlando Brown, 28 January 1850, in *The Official Correspondence of James S. Calhoun While Indian Agent at Santa Fe and Superintendent of Indian Affairs in New Mexico*, ed. Annie Heloise Abel (Washington, D.C.: U.S. Government Printing Office, 1915), 119.

CHAPTER FOUR

1. David Rich Lewis, *Neither Wolf Nor Dog: American Indians, Environment, and Agrarian Change* (New York: Oxford University Press, 1994), 4, 6, 7–10.
2. Hurt, "Historic Zuni Land Use," 11.
3. Ed Newville, "Pueblo Indian Water Rights: Overview and Update on the Aamodt Litigation," *Natural Resources Journal* 29 (winter 1989): 254.
4. Dozier, *The Pueblo Indians*, 130. The number of Pueblos increased from about nine to ten thousand.
5. Jenkins, "The Baltasar Baca 'Grant,'" 88; Perrigo, *The American Southwest*, 176–77.
6. James S. Calhoun to William Medill, 1 October 1849, in *The Official Correspondence*, ed. Abel, 33.
7. James S. Calhoun to Orlando Brown, 29 March 1850, in *The Official Correspondence*, ed. Abel, 173. In 1851 Laguna had encroachment problems with Hispanics, Navajos, and another pueblo (Acoma) (339–41).
8. Sunseri, "New Mexico in the Aftermath," 56–58.
9. Annie Heloise Abel, ed., "Indian Affairs in New Mexico Under the Administration of William Carr Lane," *New Mexico Historical Review* 16 (July 1941): 335–50.
10. U.S. Department of the Interior, Office of Indian Affairs, *Annual Report of the Commissioner of Indian Affairs for 1853*, 440; U.S. Department of the Interior, Office of Indian Affairs, *Annual Report of the Commissioner of Indian Affairs for 1854*, 173; for Anglos ignoring Indian agriculture to acquire the land, see Thomas R. Wessel, "Agriculture, Indians, and American History," in *The American Indian Past and Present*, ed. Roger L. Nichols (New York: Alfred A. Knopf, 1986), 1–9; for early observations on Pueblos see Henry R. Schoolcraft, *Historical and Statistical Information Respecting the*

History, Condition and Prospects of the Indian Tribes of the United States, 6 vols. (Philadelphia: Lippincott, Grambo & Company, 1854), 4: 216.

11. Robert W. Delaney, *Study Guide for Indians of the Southwest* (Durango, Colo.: Center of Southwest Studies, Fort Lewis College, 1971), 17–18. Santa Ana's grant was confirmed in 1869, Laguna's in 1898, and Zuni's in 1913. These villages received patents in 1883, 1909, and 1933, respectively. See also Brayer, *Pueblo Indian Land Grants*, 21

12. Ben M. Thomas to J. L. Smith, 19 June 1877, Miscellaneous Letters Sent by the Pueblo Indian Agency 1875–1891, Microfilm copy, Federal Records Center, Denver, Colorado, Roll number 941, Reel number 8. Hereafter cited as MLS, FRC, Denver. Since all roll numbers for this microfilm are 941, only the reel number (RN) will be given; Ben M. Thomas, 18 August 1882, MLS, FRC, Denver, RN 6.

13. Ben M. Thomas to H. M. Atkinson, 13 December 1882, MLS, FRC, Denver, RN 6. The Spanish documents concerning this claim dated from 1763 to 1788; Pedro Sánchez to General C. Howard, 8 June 1883, MLS, FRC, Denver, RN 6.

14. José Segura to the Commissioner of Indian Affairs, 6 December 1890, MLS, FRC, Denver, RN 10.

15. Ben M. Thomas to H. Price, 12 April 1883, MLS, FRC, Denver, RN 6; M. C. Williams to Commissioner of Indian Affairs, 10 May 1888, MLS, FRC, Denver, RN 8.

16. Davis, *El Gringo*, 66–69. Bandelier says that the corn was usually irrigated only once or twice and wheat eight or ten times. See Charles H. Lange and Carroll L. Riley, eds., *The Southwestern Journals of Adolph F. Bandelier* (Albuquerque: University of New Mexico Press, 1966), 151. Bandelier generally agrees with Davis's measurements. Measuring the acequia madre that went through Santa Ana, Bandelier found it to be one and one-half meters wide, one-half to one meter deep and over seven miles long. The field plots were four to ten meters wide and twenty to twenty-five meters long (111).

17. James F. Meline, *Two Thousand Miles on Horseback, Santa Fe and Back: A Summer Tour Through Kansas, Nebraska, Colorado, and New Mexico in the Year 1866* (New York: Hurd and Houghton, 1867), 158–65, 222, 229–30.

18. Lange and Riley, *Journals of Adolph F. Bandelier*, 117, 126, 185–86.

19. Jesse Green, ed., *Zuni: Selected Writings of Frank Hamilton Cushing* (Lincoln: University of Nebraska Press, 1979), 251.

20. Ibid., 251–56, 261–67.

21. Ibid., 292–95, 270–71.

22. Ibid., 182; Schoolcraft, *Indian Tribes of the United States*, 221.

23. Hurt, *Indian Agriculture*, 96–99.

24. Dolores Romero to Secretary of the Board of Indian Commissioners, 5 December 1885, MLS, FRC, Denver, RN 8; Frank D. Lewis to Laguna Indian School, 29 January 1890, MLS, FRC, Denver, RN 9.

25. Ben M. Thomas to the Governor of Laguna, 3 October 1878, MLS, FRC, Denver, RN 2; Ben M. Thomas to José Santos Tortolita, 4 April 1879, MLS, FRC, Denver, RN 3.

26. Dolores Romero to J. D. C. Atkins, 30 March 1886, MLS, FRC, Denver, RN 8.

27. James S. Calhoun to L. Lea, 14 August 1851, in *The Official Correspondence*, ed. Abel, 339.

28. Ben M. Thomas to José Candelano Ortiz, 21 August 1882, MLS, FRC, Denver, RN 6.
29. Ben M. Thomas to José Garcia, 1 April 1879, MLS, FRC, Denver, RN 3; Ben M. Thomas to E. A. Hayt, 20 March 1878, MLS, FRC, Denver, RN 2.
30. Dolores Romero to J. D. C. Atkins, 12 April 1886, MLS, FRC, Denver, RN 8; M. C. Williams to the Commissioner of Indian Affairs, 26 January 1889, MLS, FRC, Denver, RN 9.
31. Pedro Sánchez to H. Price, 19 December 1883, MLS, FRC, Denver, RN 7.
32. Ben M. Thomas to Joseph B. Holt, 19 July 1879, MLS, FRC, Denver, RN 3.
33. Ben M. Thomas to F. F. Ealy, 2 July 1879, MLS, FRC, Denver, RN 3.
34. Pedro Sánchez to Herbert Welsh, 22 August 1884, MLS, FRC, Denver, RN 7; José Segura to L. Bradford Prince, 21 August 1890, MLS, FRC, Denver, RN 10; U.S. Department of the Interior, Office of Indian Affairs, *Annual Report to the Commissioner of Indian Affairs for 1875*, 130; U.S. Department of the Interior, Census Office, *Report on Indians Taxed and Indians Not Taxed in the United States (Except Alaska) at the Eleventh Census: 1890*, 408. The alleged accuracy of this report is based on the large amount of statistical data included in this lengthy account of the Pueblos. Voluminous numbers do not necessitate factual information, but it is apparent that Agent Poore sought to be as precise as possible in his report.
35. Hurt, *Indian Agriculture*, 100–101; James Calhoun to William Medill, 29 July 1849, in *The Official Correspondence*, ed. Abel, 19.
36. Hurt, *Indian Agriculture*, 126–27; Ben M. Thomas to the Commissioner of Indian Affairs, 15 September 1880, MLS, FRC, Denver, RN 4; M. C. Williams to the Commissioner of Indian Affairs, 21 April 1887, MLS, FRC, Denver, RN 9.
37. Ben M. Thomas to Commissioner of Indian Affairs, 19 December 1877, MLS, FRC, Denver, RN 2; Ben M. Thomas to E. A. Hayt, 15 February 1878, MLS, FRC, Denver, RN 2. Thomas also stated that progress at Laguna could be measured by the fact that they were "giving up devil worship" and paying "attention to things rational."
38. Ben M. Thomas to John Menaul, 26 November 1878, MLS, FRC, Denver, RN 2; Ben M. Thomas to Pedro Pino, 4 September 1878, MLS, FRC, Denver, RN 2; Ben M. Thomas to E. A. Hayt, 28 May 1878, MLS, FRC, Denver, RN 2.
39. Ben M. Thomas to H. Price, 25 May 1882, MLS, FRC, Denver, RN 5; Ben M. Thomas to James H. Willson, 16 August 1882, MLS, FRC, Denver, RN 6; Ben M. Thomas to J. M. Shildes, 23 March 1883, MLS, FRC, Denver, RN 6.
40. Pedro Sánchez to R. W. D. Bryan, 28 January 1884, MLS, FRC, Denver, RN 9.
41. M. C. Williams to S. W. Allro, 5 December 1888, MLS, FRC, Denver, RN 9.
42. W. P. McClure to the Commissioner of Indian Affairs, 11 September 1889, MLS, FRC, Denver, RN 9; W. P. McClure to the Commissioner of Indian Affairs, 5 December 1889, MLS, FRC, Denver, RN 9.
43. Ben M. Thomas to J. S. Shraver, 13 October 1881, MLS, FRC, Denver, RN 5.
44. M. C. Williams to the Commissioner of Indian Affairs, 4 December 1886, MLS, FRC, Denver, RN 8; Ben M. Thomas to the Commissioner of Indian Affairs, 24 April 1882, MLS, FRC, Denver, RN 5; Ben M. Thomas to the Committee on Irrigation, 26 July 1880, MLS, FRC, Denver, RN 4; Ben M. Thomas to H. Price, 28 July 1882, MLS, FRC, Denver, RN 6; Ben M. Thomas to C. H. Howard, 16 May 1882, MLS, FRC, Denver, RN 5; Ben M. Thomas to

H. Price, 8 February 1883, MLS, FRC, Denver, RN 8; Ben M. Thomas to R. W. D. Bryan, 23 October 1882, MLS, FRC, Denver, RN 8; Ben M. Thomas to R. W. D. Bryan, 13 March 1883, MLS, FRC, Denver, RN 6; Dolores Romero to the Commission on Indian Affairs, 29 January 1886, MLS, FRC, Denver, RN 8.

45. Pedro Sánchez to Herbert Welsh, 22 August 1884, MLS, FRC, Denver, RN 7; Pedro Sánchez to Senator Dawes, 30 January 1885, MLS, FRC, Denver, RN 8; Dolores Romero to J. D. C. Atkins, 1 December 1885, MLS, FRC, Denver, RN 8; Dolores Romero to Francisco Martin, 19 September 1885, MLS, FRC, Denver, RN 8.

46. U.S. Department of the Interior, Office of Indian Affairs, *Annual Report to the Commissioner of Indian Affairs for 1890*, 173.

47. Pete A. Chavez, *District Agent Interviews with Pueblo Leaders*, March and April 1952, Letters Received, Office of Rights Protection, Records of the Bureau of Indian Affairs, Record Group 75, Albuquerque Area Office, Albuquerque, New Mexico.

48. Bancroft, *History of Arizona and New Mexico*, 645.

49. U.S. Department of the Interior, Office of Indian Affairs, *The Commissioner of Indian Affairs for 1854*, 173; U.S. Department of the Interior, Office of Indian Affairs, *Annual Report of the Commissioner of Indian Affairs for 1851*, 281; A. W. Whipple, "Report of Explorations for a Railway Route, Near the Thirty-fifth Parallel of North Latitude, from the Mississippi River to the Pacific Ocean," in *Reports of Explorations and Surveys, House of Representatives Executive Document* no. 91, 33rd Cong., 2nd sess., 1856 (Washington, D.C.: A. O. P. Nicholson, 1856), 49; Edward Everett Dale, *The Indians of the Southwest: A Century of Development under the United States* (Norman: The University of Oklahoma Press, 1949), 57–59; Lansing B. Bloom, "Bourke on the Southwest," *New Mexico Historical Review* 11 (January 1936): 121.

50. Dolores Romero to J. D. C. Atkins, 15 April 1886, MLS, FRC, Denver, RN 8; Ben M. Thomas to Edward P. Smith, 15 January 1875, MLS, FRC, Denver, RN 1.

51. Ben M. Thomas to the Governor of Laguna, 28 January 1879, MLS, FRC, Denver, RN 3; Ben M. Thomas to the Governor of Jemez, 24 January 1879, MLS, FRC, Denver, RN 3; Ben M. Thomas to the Governor of Zuni, 28 January 1879, MLS, FRC, Denver, RN 3.

52. Ben M. Thomas to José L. Perea, 28 January 1879, MLS, FRC, Denver, RN 3; Ben M. Thomas to E. A. Hayt, 9 August 1879, MLS, FRC, Denver, RN 3.

53. Ben M. Thomas to H. Price, 17 November 1882, MLS, FRC, Denver, RN 6.

54. Ben M. Thomas, "Received of William Wall," 6 January 1879, MLS, FRC, Denver, RN 3; Dolores Romero to J. D. C. Atkins, 9 February 1886, MLS, FRC, Denver, RN 8; Pedro Sánchez to A. B. Havens, 11 December 1884, MLS, FRC, Denver, RN 8; Dolores Romero to J. D. C. Atkins, 29 August 1885, MLS, FRC, Denver, RN 8; M. C. Williams to Soloman Bibo, 12 March 1887, MLS, FRC, Denver, RN 8.

55. Ben M. Thomas to José Garcia, 24 April 1879, MLS, FRC, Denver, RN 3; Ben M. Thomas to José Dolores Sandoval, 12 May 1882, MLS, FRC, Denver, RN 5; José Segura to the Governor of Nambe, 19 July 1890, MLS, FRC, Denver, RN 10.

56. Ben M. Thomas to Joaquín Mestos, 7 July 1875, MLS, FRC, Denver, RN 1.

57. Ben M. Thomas to Candelanio Estrada, 22 August 1877, MLS, FRC, Denver, RN 2.

58. Ben M. Thomas to Jesús Ortiz, 8 April 1878, MLS, FRC, Denver, RN 2; Ben

M. Thomas to J. M. Shields, 1 June 1878, MLS, FRC, Denver, RN 2.

59. Ben M. Thomas to José L. Perea, 24 April 1879, MLS, FRC, Denver, RN 3; Ben M. Thomas to Colonel Baird, 14 May 1879, MLS, FRC, Denver, RN 3.
60. U.S. Department of the Interior, Office of Indian Affairs, *Annual Report of the Commissioner of Indian Affairs for 1879*, 121.
61. Ben M. Thomas to Colonel S. M. Baird, 15 December 1882, MLS, FRC, Denver, RN 6.
62. Pedro Sánchez to H. Price, 24 November 1883, MLS, FRC, Denver, RN 7.
63. José Segura to G. Candelaria, 20 October 1890, MLS, FRC, Denver, RN 10.
64. Ben M. Thomas to José la Cruz Marguez, 15 March 1878, MLS, FRC, Denver, RN 2.
65. Ben M. Thomas to E. A. Hayt, 21 June 1879, MLS, FRC, Denver, RN 3; Dolores Romero to Jesús Montoya, 28 January 1886, MLS, FRC, Denver, RN 8. See also Dolores Romero to the Commissioner of Indian Affairs, 29 May 1886, MLS, FRC, Denver, RN 8; M. C. Williams to the Commissioner of Indian Affairs, 7 November 1887, MLS, FRC, Denver, RN 9; M. C. Williams to José D. Jesús de Armona, 20 January 1887, MLS, FRC, Denver, RN 9.
66. Ben M. Thomas to Pedro Ignacio López, 15 July 1882, MLS, FRC, Denver, RN 6. Complaints were made against four Americans; Pedro Sánchez to Charles H. Probst, 22 April 1884, MLS, FRC, Denver, RN 7; Pedro Sánchez to Charles Lewis, July 1884, MLS, FRC, Denver, RN 7. Americans were cutting timber and digging a well; Dolores Romero to J. D. C. Atkins, 14 May 1886, MLS, FRC, Denver, RN 8.
67. Pedro Sánchez to G. W. Richard, 29 November 1884, MLS, FRC, Denver, RN 8; Pedro Sánchez to Henry Elliot, 22 April 1885, MLS, FRC, Denver, RN 8.
68. Ben M. Thomas to Isidro Lauriano, 19 August 1882, MLS, FRC, Denver, RN 6.
69. Ben M. Thomas to Galen Eastman, 18 July 1879, MLS, FRC, Denver, RN 3.
70. José Segura to the Governor of Isleta Pueblo, 22 May 1890, MLS, FRC, Denver, RN 10.
71. Ben M. Thomas to William Burgess, 3 March 1875, MLS, FRC, Denver, RN 1.
72. M. C. Williams to the Commissioner of Indian Affairs, 8 December 1887, MLS, FRC, Denver, RN 9.
73. Jenkins, "The Baltasar Baca 'Grant,'" 87–94.
74. M. C. Williams to the Commissioner of Indian Affairs, 18 June 1888, MLS, FRC, Denver, RN 9.
75. M. C. Williams to the Commissioner of Indian Affairs, 6 December 1886, MLS, FRC, Denver, RN 9.
76. José Segura to the Commissioner of Indian Affairs, 8 May 1890, MLS, FRC, Denver, RN 9.
77. M. C. Williams to the Commissioner of Indian Affairs, 22 April 1887, MLS, FRC, Denver, RN 9; M. C. Williams to the Commissioner of Indian Affairs, 26 March 1889, MLS, FRC, Denver, RN 9.
78. Frank D. Lewis to the Governor of Acoma, 24 February 1890, MLS, FRC, Denver, RN 9. Other letters were sent to Zia, Sandia, Santo Domingo and Isleta.
79. Ben M. Thomas to Captain John Pratt, 6 September 1880, MLS, FRC, Denver, RN 4; Pedro Sánchez to H. A. Risley, 8 September 1883, MLS, FRC, Denver, RN 7; Pedro Sánchez to E. O. Walcott, 27 February 1884, MLS, FRC, Denver, RN 7.
80. Ben M. Thomas to the Director of Section Work for the Atchison, Topeka, and

Santa Fe Railroad, 12 April 1882, MLS, FRC, Denver, RN 5; Ben M. Thomas to A. A. Robinson, 20 May 1882, MLS, FRC, Denver, RN 5; Ben M. Thomas to A. A. Robinson, 10 June 1882, MLS, FRC, Denver, RN 5.

81. Pedro Sánchez to F. W. Smith, 26 February 1884, MLS, FRC, Denver, RN 7; Pedro Sánchez to F. W. Smith, 4 March 1884, MLS, FRC, Denver, RN 7; Pedro Sánchez to H. Price, 10 June 1884, MLS, FRC, Denver, RN 7.
82. Dolores Romero to the Superintendent of the Denver and Rio Grande Narrow Gauge Railroad, 6 March 1886, MLS, FRC, Denver, RN 8; Dolores Romero to C. W. Smith, 15 April 1886, MLS, FRC, Denver, RN 8; M. C. Williams to Henry L. Waldo, 12 March 1887, MLS, FRC, Denver, RN 9.
83. Ralph Emerson Twitchell, *Old Santa Fe: The Story of New Mexico's Ancient Capital* (Chicago: Rio Grande Press, 1963), 319.
84. Phil Lovato, *Las Acequias del Norte*, Technical Report Number 1 (Taos, N.Mex.: Four Corners Regional Commission, 1974), 21.
85. Long, *Law of Irrigation*, 129. Long feels that few of the miners were either lawyers or familiar with ancient or foreign law. On the other hand, different authorities believe that the California customs were strikingly similar to Germanic traditions of the Middle Ages that spread throughout Europe and her colonies. See Wells A. Hutchins and Harry A. Steele, "Basic Water Rights Doctrines and Their Implications for River Basin Development," *Law and Contemporary Problems* 22 (spring 1957): 281.
86. Richard F. Logan, "Post Columbian Developments in the Arid Regions of the United States of America," in *A History of Land Use in Arid Regions*, ed. L. Dudley Stamp (Paris: The United Nations Educational, Scientific and Cultural Organization, 1961), 288.
87. Kinney, *Law of Irrigation*, 1: 123.
88. Hall, *Four Leagues of Pecos*, 76–66, 112–18.
89. *United States v. Joseph Otto*, U.S. (1876) 616–17. Concerning the common lands of the Pueblos, the court claimed that in this regard, they merely resembled the Shakers and other communistic societies. In addition, the court refused to make a decision on the question of Pueblo citizenship.
90. Cohen, *Federal Indian Law*, 387; U.S. Department of the Interior, Office of Indian Affairs, *Annual Report of the Commissioner of Indian Affairs for 1874*, 618.
91. Ben M. Thomas to Josephine Toudre, 12 November 1880, MLS, FRC, Denver, RN 4; Ben M. Thomas to Col. Pritchard, 14 April 1883, MLS, FRC, Denver, RN 6; Pedro Sánchez to J. D. C. Atkins, 15 April 1885, MLS, FRC, Denver, RN 8.
92. U.S. Department of the Interior, Office of Indian Affairs, *Annual Report of the Commissioner of Indian Affairs for 1874*, 617. See also U.S. Department of the Interior, Office of Indian Affairs, *Annual Report of the Commissioner of Indian Affairs for 1890*, 174.
93. Ben M. Thomas to H. Price, 31 July 1882, MLS, FRC, Denver, RN 6.
94. Ben M. Thomas to José Antonio Capote, 18 June 1875, MLS, FRC, Denver, RN 1; Ben M. Thomas to J. O. Smith, 3 February 1877, MLS, FRC, Denver, RN 1; Ben M. Thomas to the Governor of Nambe, 22 January 1877, MLS, FRC, Denver, RN 9.
95. Pedro Sánchez to Robert G. Marmon, 26 May 1884, MLS, FRC, Denver, RN 7; Pedro Sánchez to H. Price, 4 June 1884, MLS, FRC, Denver, RN 7.
96. Edward Tittman, "The First Irrigation Lawsuit," *New Mexico Historical Review*

2 (October 1927): 363–68. Acoma gained the rights to the water from the Gallo Creek, but Laguna was allowed to take surplus waters below their lands.

97. José Segura to the Governor of Acoma Pueblo, 9 May 1890, MLS, FRC, Denver, RN 9; José Segura to the Governor of Acoma Pueblo, 1 July 1890, MLS, FRC, Denver, RN 10; José Segura to Solomon Bibo, 1 July 1890, MLS, FRC, Denver, RN 10; José Segura to W. G. Marmon, 5 July 1890, MLS, FRC, Denver, RN 10.
98. John Greiner to Luke Lea, 30 April 1852, in *The Official Correspondence*, ed. Abel, 530.
99. B. M. Thomas to Isleta, New Mexico, 28 March 1877, MLS, FRC, Denver, RN 1; Ben M. Thomas to whom it may concern, 6 April 1878, MLS, FRC, Denver, RN 2.
100. Ben M. Thomas to the Mayordomo of Peña Blanca, 12 July 1875, MLS, FRC, Denver, RN 1; Ben M. Thomas to Agapito Tafolla, 24 March 1879, MLS, FRC, Denver, RN 3; Ben M. Thomas to Santiago Quintana, 29 March 1879, MLS, FRC, Denver, RN 3.
101. Ben M. Thomas to the Taos County Probate Judge, 23 April 1875, MLS, FRC, Denver, RN 6; Ben M. Thomas to Anthony Joseph, 26 August 1878, MLS, FRC, Denver, RN 2; Ben M. Thomas to Santiago Mirabal, 7 July 1879, MLS, FRC, Denver, RN 3.
102. Pedro Sánchez to H. Price, 13 July 1883, MLS, FRC, Denver, RN 6.
103. Pedro Sánchez to J. D. C. Atkins, 1 May 1885, MLS, FRC, Denver, RN 8.
104. Dolores Romero to A. B. Upshaw, 24 August 1885, MLS, FRC, Denver, RN 10.
105. José Segura to Manuel Trujillo, 4 June 1890, MLS, FRC, Denver, RN 10.
106. Matilda Coxe Stevenson, "The Sia," in *Eleventh Annual Report of the Bureau of Ethnology, 1889–90* (Washington, D.C.: U.S. Government Printing Office, 1894), 11. The other pueblos on the Jemez River, Santa Ana and Jemez, had enough water, but Zia was reportedly destitute.
107. M. C. Williams to the Commissioner of Indian Affairs, 10 May 1888, MLS, FRC, Denver, RN 9.
108. Pedro Sánchez to Charles Lewis, 20 July 1884, MLS, FRC, Denver, RN 7.
109. Frank D. Lewis to the Commissioner of Indian Affairs, 6 March 1890, MLS, FRC, Denver, RN 9.
110. Davis, *El Gringo*, 195.
111. U.S. Department of the Interior, Office of Indian Affairs, *Annual Report of the Commissioner of Indian Affairs for 1873*, 277; Delaney, *Indians of the Southwest*, 17. Bandelier also mentions the lack of rainfall about this time. See Lange and Riley, *The Southwestern Journals*, 209; U.S. Department of the Interior, Office of Indian Affairs, *Annual Report of the Commissioner of Indian Affairs for 1886*, 423–24.
112. Bancroft, *History of Arizona and New Mexico*, 739–40.
113. Ben M. Thomas to Colonel A. A. Robinson, 24 June 1880, MLS, FRC, Denver, RN 4.
114. Dolores Romero to J. D. C. Atkins, 22 May 1886, MLS, FRC, Denver, RN 8. The agent asked the federal government for $250 for the purchase of flour for the pueblo; Dolores Romero to the Commissioner of Indian Affairs, 24 August 1886, MLS, FRC, Denver, RN 9.
115. Bancroft, *History of Arizona and New Mexico*, 739; Ben M. Thomas to Juan

José Sánchez, 19 February 1878, MLS, FRC, Denver, RN 2. See also U.S. Department of the Interior, Office of Indian Affairs, *Annual Report of the Commissioner of Indian Affairs for 1877*, 161; Ben M. Thomas to E. A. Hayt, 1 April 1878, MLS, FRC, Denver, RN 2; Ben M. Thomas to E. A. Hayt, 3 January 1879, MLS, FRC, Denver, RN 3.

116. James A. Vlasich, "A History of Irrigation and Agriculture among the Pueblo Indians of New Mexico" (Master's thesis, University of Utah, 1977), 84–86. See also Kirk Bryan, "Flood-Water Farming," *The Geographical Review* 19 (July 1929): 453. Bryan suggests that the increased use of the plow, as opposed to the planting stick, furthered surface washing in the area.

117. Bryan, "Pre-Columbian Agriculture," 226. Bryan claims that starting in 1885 arroyo cutting in the region began to deepen channels and disallowed irrigation. However, it is not known how much this affected the Pueblos.

118. Ben M. Thomas to J. M. Shields, 24 November 1882, MLS, FRC, Denver, RN 6.

119. Ben M. Thomas to the Governor of Santa Clara, 9 July 1881, MLS, FRC, Denver, RN 5.

120. José Segura to Antonio Baca, 17 May 1890, MLS, FRC, Denver, RN 9.

121. Ben M. Thomas to J. H. Wilson, 24 March 1883, MLS, FRC, Denver, RN 6.

122. Dolores Romero to J. D. C. Atkins, 4 May 1886, MLS, FRC, Denver, RN 8.

CHAPTER FIVE

1. Delaney, *Indians of the Southwest*, 17–18. Poore's report included the pueblo of Pojoaque, abandoned around the turn of the century. Many of the Pojoaque survivors moved to Nambe; others went to different villages. Fewer than forty of their descendants returned in 1933 when the Pueblo Lands Board restored their original reservation.

2. U.S. Department of the Interior, Census Office, *Report on Indians Taxed and Indians Not Taxed in the United States (Except Alaska) at the Eleventh Census: 1890*, 424–45.

3. Brown, "Picuris Pueblo in 1890," 29–30, 32–33.

4. Ibid., 420. Additional figures concerning acres cultivated were obtained from the agent's survey of the different pueblos.

5. U.S. Department of the Interior, Office of Indian Affairs, *Annual Report of the Commissioner of Indian Affairs for 1900*, 293.

6. Ibid., 293. Figures for 1890 come from U.S. Department of the Interior, *Report of Indians Taxed*, 422. Statistics for 1899 come from U.S. Department of the Interior, Office of Indian Affairs, *Annual Report of the Commissioner of Indian Affairs for 1899*, 245.

7. U.S. Department of the Interior, Office of Indian Affairs, *Annual Report of the Commissioner of Indian Affairs for 1900*, 293.

8. Clinton J. Crandall to Lyman J. Maxwell, 22 June 1903, Letters Sent by the Superintendent of the Santa Fe Indian School, June 1890–December 1913, Records of the Bureau of Indian Affairs, Record Group 75, Federal Records Center, Denver, Colorado. Hereafter cited as LSSF, RG 75, FRC, Denver.

9. Philip T. Lonergan to A. B. McMillan, 7 June 1918, Letters Sent by the Superintendent of Pueblo Agency and Day Schools, 1914–1919, Records of the Bureau of Indian Affairs, Record Group 75, Federal Records Center, Denver, Colorado. Hereafter cited as PADS, RG 75, FRC, Denver.

10. U.S. Department of the Interior, Office of Indian Affairs, *Report of the Commissioner of Indian Affairs for 1892*, 92.
11. U.S. Department of the Interior, Office of Indian Affairs, *Report of the Commissioner of Indian Affairs for 1897*, 201.
12. Ira G. Clark, "The Elephant Butte Controversy: A Chapter in the Emergence of Federal Water Law," *The Journal of American History* 61 (March 1975): 1006–7.
13. Wallace Stegner, *Beyond the Hundredth Meridian: John Wesley Powell and the Second Opening of the West* (Boston: Houghton Mifflin, 1954), 7. See also Lawrence D. Lee, "William Ellsworth Smythe and the Irrigation Movement: A Reconsideration," *Pacific Historical Review* 61 (August 1972): 289–90. The author discusses the efforts of Smythe to build a support base in order to promote congressional acts that would aid irrigation in the West; Clark, "The Elephant Butte Controversy," 1014.
14. U.S. Department of the Interior, Office of Indian Affairs, *Report of the Commissioner of Indian Affairs for 1903*, 43.
15. Clinton J. Crandall to Herbert J. Hagerman, 20 August 1906, LSSF, RG 75, FRC, Denver; U.S. Department of the Interior, Office of Indian Affairs, *Report of the Commissioner of Indian Affairs for 1908*, 56; U.S. Department of the Interior, Office of Indian Affairs, *Report of the Commissioner of Indian Affairs for 1909*, 50–51; U.S. Department of the Interior, Office of Indian Affairs, *Report of the Commissioner of Indian Affairs for 1916*, 42.
16. Elsie Clews Parsons, *Taos Pueblo* (Menasha, Wisc.: George Banta, 1936), 18.
17. John H. Robertson to the Commissioner of Indian Affairs, 6 April 1892, Press Copies of Miscellaneous Letters Sent by the Pueblo Indian Agency, 1891–1900, Records of the Bureau of Indian Affairs, Record Group 75, Federal Records Center, Denver, Colorado. Hereafter cited as PCM, RG 75, FRC, Denver; N. S. Walpole to W. G. Deason, 14 October 1899, PCM, RG 75, FRC, Denver; N. S. Walpole to Cardy D. Richards, 15 May 1900, PCM, RG 75, FRC, Denver.
18. John L. Bullis to L. A. Ellis, 22 August 1896, PCM, RG 75, FRC, Denver.
19. N. S. Walpole to the Commissioner of Indian Affairs, 20 August 1899, PCM, RG 75, FRC, Denver.
20. U.S. Department of the Interior, *Annual Report for 1900*, 294.
21. Clinton J. Crandall to Roscoe C. Bonney, 21 June 1904, LSSF, RG 75, FRC, Denver; Clinton J. Crandall to Lyman J. Maxwell, 22 June 1904, LSSF, RG 75, FRC, Denver; Clinton J. Crandall to P. H. Lease, 5 September 1905, LSSF, RG 75, FRC, Denver.
22. Hiram Jones to Superintendent Wise, 18 January 1915, PADS, RG 75, FRC, Denver; P. T. Lonergan to Hiram Jones, 26 May 1915, PADS, RG 75, FRC, Denver.
23. Lewis, *Neither Wolf Nor Dog*, 16–18.
24. Hurt, *Indian Agriculture in America*, 151–52.
25. N. S. Walpole to Santiago Quintana, 17 November 1898, PCM, RG 75, FRC, Denver.
26. N. S. Walpole to Albert B. Reagan, 19 August 1899, PCM, RG 75, FRC, Denver.
27. U.S. Department of the Interior, Annual Report for 1903, 222; U.S. Department of the Interior, Annual Report for 1906, 285.
28. Hurt, "Historic Zuni Land Use," 10; E. Richard Hurt, "The Zuni Land Conservation Act of 1990," in *Zuni and the Courts: A Struggle for Sovereign Land Rights*, ed. Hurt (Lawrence: University Press of Kansas, 1995), 93.

29. U.S. Department of the Interior, *Annual Report for 1903*, 43–44; U.S. Department of the Interior, *Annual Report for 1907*, 51; U.S. Department of the Interior, *Annual Report for 1908*, 58.
30. U.S. Department of the Interior, *Annual Report for 1910*, 24.
31. U.S. Department of the Interior, *Annual Report for 1918*, 40.
32. Clinton J. Crandall to A. J. Abott, 3 June 1909, LSSF, RG 75, FRC, Denver.
33. Ed Kinney to H. F. Robinson, 18 May 1912, Letters Received, General Correspondence of the Northern Pueblos Agency, 1911–1935, Records of the Bureau of Indian Affairs, Record Group 75, Federal Records Center, Denver, Colorado. Hereafter cited as LR or LS (Letters Sent), NPA, RG 75, FRC, Denver.
34. H. F. Robinson to the Commissioner of Indian Affairs, 26 July 1912, LS, NPA, RG 75, FRC, Denver.
35. U.S. Department of the Interior, *Annual Report for 1913*, 20; U.S. Department of the Interior, *Annual Report for 1920*, 26.
36. U.S. Department of the Interior, *Annual Report for 1908*, 57.
37. H. F. Robinson to P. T. Lonergan, 15 February 1912, Letters Received, General Correspondence of the Southern Pueblos Agency, 1911–1935, Records of the Bureau of Indian Affairs, Record Group 75, Federal Records Center, Denver, Colorado. Hereafter cited as LR or LS (Letters Sent), SPA, RG 75, FRC, Denver. See also H. F. Robinson to P. T. Lonergan, 20 March 1912, LR, SPA, RG 75, FRC, Denver.
38. R. W. Cassady to P. T. Lonergan, 29 July 1912, LR, SPA, RG 75, FRC, Denver; R. W. Cassady to P. T. Lonergan, 21 November 1912, LR, SPA, RG 75, FRC, Denver; P. D. Allen to P. T. Lonergan, 16 January 1913, LR, SPA, RG 75, FRC, Denver.
39. C. J. Crandall to John B. Harper, 7 July 1905, LSSF, RG 75, FRC, Denver. See also C. J. Crandall to Frank Mead, 29 June 1905, LSSF, RG 75, FRC, Denver.
40. C. J. Crandall to H. F. Robinson, 17 December 1906, LSSF, RG 75, FRC, Denver; C. J. Crandall to H. F. Robinson, 23 August 1906, LSSF, RG 75, FRC, Denver.
41. U.S. Department of the Interior, *Annual Report for 1899*, 252; John O. Baxter, *Dividing New Mexico's Waters, 1700–1912* (Albuquerque: University of New Mexico Press, 1997), 93–96.
42. N. S. Walpole to Fred W. Hart, 4 October 1899, PCM, RG 75, FRC, Denver; N. W. Walpole to the Governor of San Felipe, 4 December 1899, PCM, RG 75, FRC, Denver.
43. N. S. Walpole to the Governor of San Felipe, 13 February 1900, PCM, RG 75, FRC, Denver; N. S. Walpole to José Enrique, 11 February 1900, PCM, RG 75, FRC, Denver.
44. N. S. Walpole to the President of the Albuquerque Land and Irrigation Company, 14 February 1900, PCM, RG 75, FRC, Denver; N. S. Walpole to the President of the Albuquerque Land and Irrigation Company, 13 February 1900, PCM, RG 75, FRC, Denver; N. S. Walpole to the Governors of Santo Domingo, Santa Ana, Sandia, and Isleta, 24 February 1900, PCM, RG 75, FRC, Denver.
45. N. S. Walpole to Albert B. Regan, 10 August 1899, PCM, RG 75, FRC, Denver; N. S. Walpole to Albert B. Reagan, 31 May 1900, PCM, RG 75, FRC, Denver.
46. Hiram Jones to P. T. Lonergan, 5 January 1915, LR, SPA, RG 75, FRC, Denver.
47. Hiram Jones to P. T. Lonergan, 17 December 1914, LR, SPA, RG 75, FRC, Denver.
48. C. J. Crandall to H. F. Robinson, 5 February 1909, LSSF, RG 75, FRC, Denver.

49. José Segura to the Commissioner of Indian Affairs, 9 April 1891, Miscellaneous Letters sent by the Pueblo Indian Agency 1875–1891, Microfilm copy, Federal Records Center, Denver, Colorado, Roll number 941, Reel number 10. Hereafter cited as MLS, FRC, Denver. Since all roll numbers for this microfilm are 941, only the reel number (RN) will be given. See also John H. Robinson to the Commissioner of Indian Affairs, 8 February 1893, PCM, RG 75, FRC, Denver.
50. John H. Robinson to the Commissioner of Indian Affairs, 7 October 1981, MLS, FRC, Denver, RN 10.
51. José Segura to D. D. Grahman, 21 July 1891, MLS, FRC, Denver, RN 10.
52. C. J. Crandall to I. W. Dwire, 24 January 1903, LSSF, RG 75, FRC, Denver.
53. John Robertson to the Commissioner of Indian Affairs, 2 February 1892, PCM, RG 75, FRC, Denver.
54. John L. Bullis to C. E. Hosmer, 18 July 1894, PCM, RG 75, FRC, Denver.
55. U.S. Department of the Interior, *Annual Report for 1892*, 335; U.S. Department of the Interior, *Annual Report for 1897*, 201–2; Charles L. Cooper to Caroline E. Hosmer, 23 March 1898, PCM, RG 75, FRC, Denver.
56. José Segura to the Commissioner of Indian Affairs, 6 July 1891, MLS, FRC, Denver. See also José Segura to John W. Charles, 17 August 1891, MLS, FRC, Denver, RN 10; John H. Robinson to the Commissioner of Indian Affairs, 8 February 1893, PCM, RG 75, FRC, Denver.
57. John L. Bullis to the Commissioner of Indian Affairs, 29 July 1895, PCM, RG 75, FRC, Denver; C. E. Nordstrom to John L. Gaylord, 2 August 1897, PCM, RG 75, FRC, Denver; N. S. Walpole to George W. Bennett, 21 November 1899, PCM, RG 75, FRC, Denver; C. E. Nordstrom to Hibbard, Spencer, Bartlett and Company, 16 December 1899, PCM, RG 75, FRC, Denver; Philip T. Lonergan to Frank Sorenson, 28 April 1913, LS, SPA, RG 75, FRC, Denver.
58. U.S. Department of the Interior, *Annual Report for 1899*, 246; N. S. Walpole to Carey D. Richards, 31 May 1900, PCM, RG 75, FRC, Denver.
59. Elsie Clews Parsons, *The Pueblo of Jemez* (New Haven, Conn.: Yale University Press, 1925), 15; John D. Rhoads to P. T. Lonergan, 19 August 1915, LS, SPA, RG 75, FRC, Denver.
60. Hurt, *Indian Agriculture in America*, 149.
61. José Segura to J. A. Davis, 1 May 1891, MLS, FRC, Denver, RN 10; John L. Bullis to the Commissioner of Indian Affairs, 24 July 1894, PCM, RG 75, FRC, Denver.
62. U.S. Department of the Interior, *Annual Report for 1917*, 373.
63. U.S. Department of the Interior, *Annual Report for 1893*, 335; Cohen, *Handbook of Federal Indian Law*, 386; C. J. Crandall to Herbert J. Hagerman, 20 August 1906, LSSF, RG 75, FRC, Denver; U.S. Department of the Interior, *Annual Report for 1917*, 372.
64. C. J. Crandall to M. A. Otero, 27 August 1904, LSSF, RG 75, FRC, Denver.
65. Joe Sando, *The Pueblo Indians* (San Francisco: Indian Historian Press, 1976), 93; C. J. Crandall to B. S. Rodney, 8 December 1904, LSSF, RG 75, FRC, Denver; C. J. Crandall to S. M. Brosius, 1 January 1905, LSSF, RG 75, FRC, Denver.
66. C. J. Crandall to Frank Mead, 6 February 1905, LSSF, RG 75, FRC, Denver; C. J. Crandall to the Governor of Santo Domingo, 11 March 1905, LSSF, RG 75, FRC, Denver. The act, quoted in the letter, reads that "lands now held by the various villages, or pueblos, of Pueblo Indians, or by individual members of the Pueblo tribe, within Pueblo reservations or lands in New Mexico, and all

personal property furnished these Indians by the United States or used in cultivating these lands and any cattle and sheep now possessed or hereafter acquired by the Pueblo Indians, shall be free and exempt from taxation of any sort whatever, including taxes heretofore levied, if any, until Congress shall otherwise provide."

67. Hall, *Four Leagues of Pecos*, 200–202.
68. Elmer F. Bennett, *Federal Indian Law* (Washington, D.C.: U.S. Government Printing Office, 1958), 902.
69. Hall, *Four Leagues of Pecos*, 202–3.
70. *United States v. Sandoval*, 231 Butler U.S. (1913), 36–49. The Court did not rule on the question of citizenship, but it did state that the Pueblos had been wards of the Spanish and Mexican governments even though they were supposed to be citizens under the latter.
71. Hall, *Four Leagues of Pecos*, 206–7; Charles T. DuMars, Marilyn O'Leary, and Albert E. Utton, *Pueblo Indian Water Rights: Struggle for a Precious Resource* (Tucson: University of Arizona Press, 1984), 43.
72. Hall, *Four Leagues of Pecos*, 207–8, 215.
73. Robert Emmet Clark, *New Mexico Water Resources Law: A Survey of Legislation and Decisions* (Albuquerque: University of New Mexico Press, 1964), 7–8.
74. U.S. Department of the Interior, *Annual Report for 1911*, 16; C. J. Crandall to the Governors of the Pueblos, 12 May 1910, LSSF, RG 75, FRC, Denver.
75. Steve Nickeson, "Pueblo Water: The More Things Change the More They Stay the Same," *Race Relations Reporter* V (July 1974): 15. For a thorough review of the legal implications of the Winters Doctrine see Norris Hundley, "The Dark and Bloody Ground of Indian Water Rights Confusion Elevated to Principle," *The Western Historical Quarterly* IX (October 1978): 460–74; For how the *Winters* decision affects the Pueblos, see DuMars, O'Leary, and Utton, *Pueblo Indian Water Rights*, 42–54; Klaus Frantz, *Indian Reservations in the United States: Territory, Sovereignty, and Socioeconomic Change* (Chicago: University of Chicago Press, 1999), 214.
76. H. F. Robinson to the Commissioner of Indian Affairs, 26 July 1912, LS, SPA, RG 75, FRC, Denver; Francis C. Wilson to P. T. Lonergan, 23 March 1914, LR, SPA, RG 75, FRC, Denver.
77. Francisco Tafoya to N. S. Walpole, 20 April 1899, PCM, RG 75, FRC, Denver; N. S. Walpole to N. Gómez, 16 July 1900, PCM, RG 75, FRC, Denver.
78. José Segura to Solomon Bibo, 1 May 1891, MLS, FRC, Denver, RN 10; José Segura to Robert Marmon, 9 May 1891, MLS, FRC, Denver, RN 10.
79. R. H. Hanna to P. T. Lonergan, 2 April 1919, LR, SPA, RG 75, FRC, Denver.
80. John L. Bullis to Juan Santistevan, 31 March 1894, PCM, RG 75, FRC, Denver; C. J. Crandall to Isaac Dwire, 12 April 1903, LSSF, RG 75, FRC, Denver; C. J. Crandall to A. J. Abbott, 31 October 1903, LSSF, RG 75, FRC, Denver.
81. Annie M. Sayre to C. J. Crandall, 12 June 1901, LSSF, RG 75, FRC, Denver.
82. C. J. Crandall to E. Bradford, 24 March 1903, LSSF, RG 75, FRC, Denver. See also C. J. Crandall to Samuel Eldodt, 11 June 1903, LSSF, RG 75, FRC, Denver; C. E. Nordstrom to the Governor of Santa Clara, 23 August 1897, PCM, RG 75, FRC, Denver; John L. Bullis to the Governor of Santo Domingo, 11 April 1894, PCM, RG 75, FRC, Denver. Santo Domingo also had problems with Hispanics in 1899 and 1903. N. S. Walpole to T. A. Finical, 12 May 1899, PCM,

RG 75, FRC, Denver. See also C. J. Crandall to A. J. Abbott, 24 September 1903, LSSF, RG 75, FRC, Denver.

83. C. J. Crandall to A. J. Abbott, 20 July 1906, LSSF, RG 75, FRC, Denver; N. S. Walpole to Pedro Cajuete, 17 April 1900, PCM, RG 75, FRC, Denver; C. J. Crandall to Juan Bautista Telache, 14 March 1906, LSSF, RG 75, FRC, Denver.
84. R. H. Hanna to P. T. Lonergan, 7 June 1919, PCM, RG 75, FRC, Denver.
85. N. S. Walpole to the Governor of Isleta, 28 April 1900, PCM, RG 75, FRC, Denver.
86. C. J. Crandall to A. J. Abbott, 17 April 1905, LSSF, RG 75, FRC, Denver; C. J. Crandall to A. J. Abbott, 2 September 1907, LSSF, RG 75, FRC, Denver; C. J. Crandall to Mary E. Dissette, 30 May 1903, LSSF, RG 75, FRC, Denver.
87. John L. Buillis to the Governor of Sandia, 7 February 1894, PCM, RG 75, FRC, Denver; C. J. Crandall to A. J. Abbott, 19 January 1903, LSSF, RG 75, FRC, Denver; John L. Bullis to Melijeldo Ortiz, 1 October 1894, PCM, RG 75, FRC, Denver.
88. José Segura to the Commissioner of Indian Affairs, 20 March 1891, MLS, FRC, Denver, RN 10.
89. Robert Garner to John H. Robinson, 3 January 1893, PCM, RG 75, FRC, Denver.
90. Jenkins, "The Baltasar Baca 'Grant,'" 97–101.
91. John L. Bullis to E. Baker, 21 December 1893, PCM, RG 75, FRC, Denver; C. E. Nordstrom to the Superintendent of Indian Schools, 31 May 1897, PCM, RG 75, FRC, Denver.
92. Hurt, *Indian Agriculture in America*, 154–55; Lawrence C. Kelly, "Cato Sells, 1913–21," in *The Commissioners of Indian Affairs, 1824–1977*, eds. Robert M. Kvasnicka and Herman J. Viola (Lincoln: University of Nebraska Press, 1979), 245–47.
93. Hurt, *Indian Agriculture in America*, 159–60.
94. Woodrow Wilson to the Office of Indian Affairs, 10 April 1917, LR, NPA, RG 75, FRC, Denver; M. Davise to P. T. Lonergan, 23 April 1917, LR, NPA, RG 75, FRC, Denver; P. T. Lonergan to Lincoln H. Mitchell, 21 April 1917, LS, SPA, RG 75, FRC, Denver; P. T. Lonergan to C. A. Pedersen, 20 April 1917, LS, SPA, RG 75, FRC, Denver.

CHAPTER SIX

1. Cohen, *Handbook of Federal Indian Law*, 389.
2. Spicer, *Cycles of Conquest*, 173; Hall, *Four Leagues of Pecos*, 219–23.
3. Cohen, *Handbook of Federal Indian Law*, 389–90; For the Bursum Bill see Francis Paul Prucha, ed., *Documents of United States Indian Policy* (Lincoln: University of Nebraska Press, 1975), 215–17.
4. Kenneth R. Philp, *John Collier's Crusade for Indian Reform, 1920–1954* (Tucson: University of Arizona Press, 1977), 30–32.
5. Dozier, *The Pueblo Indians*, 108–9. See also Spicer, *Cycles of Conquest*, 173.
6. Sophie D. Aberle, *The Pueblo Indians of New Mexico: Their Land, Economy, and Civil Organization*, Memoirs of the American Anthropological Association, no. 70 (Menasha, Wisc.: American Anthropological Association, 1948), 9–10; Prucha, *United States Indian Policy*, 218.
7. Delaney, *Indians of the Southwest*, 4.
8. Jenkins, "The Baltasar Baca 'Grant,'" 102–5.

9. DuMars, O'Leary, and Utton, *Pueblo Indian Water Rights*, 55–56.
10. Joe S. Sando, *Nee Hemish: A History of Jemez Pueblo* (Albuquerque: University of New Mexico Press, 1982), 103–9, 112–13.
11. E. P. Davies to C. J. Crandall, 2 May 1927, LR, NPA, RG 75, FRC, Denver; C. J. Crandall to E. P. Davies, 3 May 1927, LS, NPA, RG 75, FRC, Denver; T. F. McCormick to H. F. Robinson, 10 March 1928, LS, NPA, RG 75, FRC, Denver. See also Edward C. Gersbach to H. F. Robinson, 14 March 1928, LP, NPA, RG 75, FRC, Denver.
12. Charles H. Burke to Thomas F. McCormick, 5 September 1928, LR, NPA, RG 75, FRC, Denver.
13. Philp, *John Collier's Crusade*, 30. The author also points out that San Juan lost use of almost 90 percent of its irrigated acres; C. F. Hauke to Walter C. Cochrane, 24 July 1928, LP, NPA, RG 75, FRC, Denver.
14. H. J. Hagerman to Walter C. Cochrane, 23 January 1929, LR, NPA, RG 75, FRC, Denver.
15. Charles H. Burke to Herbert F. Robinson, 23 May 1929, LR, NPA, RG 75, FRC, Denver; H. F. Robinson to the Commissioner of Indian Affairs, 29 May 1929, LS, NPA, RG 75, FRC, Denver; Charles H. Burke to H. F. Robinson, 13 June 1929, LR, NPA, RG 75, FRC, Denver.
16. Newville, "Pueblo Indian Water Rights," 256, 270.
17. Aberle, *The Pueblo Indians of New Mexico*, 11–12, 21.
18. John Collier to H. C. Neuffer, 11 October 1935, LR, NPA, RG 75, FRC, Denver.
19. *Natural Resources Board Bulletins and Reports, 1941*, 13, Letters Received, United Pueblos Agency Decimal Files, Records of the Bureau of Indian Affairs, Record Group 75, Federal Records Center, Denver, Colorado. Hereafter cited as LR or LS, UPA, RG 75, FRC, Denver; U.S. Department of the Interior, *Annual Report for 1925*, 27.
20. U.S. Department of the Interior, *Annual Report for 1929*, 15–16; Philp, *John Collier's Crusade*, 87.
21. Cohen, *Handbook of Federal Indian Law*, 392; S. D. Aberle to John Collier, 6 January 1936, LS, UPA, RG 75, FRC, Denver.
22. *Survey of Conditions of the Indians of the United States: Hearings Before a Subcommittee of Indian Affairs, Part 19* (Washington, D.C.: U.S. Government Printing Office, 1932), 10077. See also Philp, *John Collier's Crusade*, 88.
23. H. C. Neuffer to Charles J. Rhoads, 6 November 1930, LS, SPA, RG 75, FRC, Denver.
24. *Survey of Conditions of Indians in the United States: Hearing Before a Subcommittee of Indian Affairs, Part 20* (Washington, D.C.: U.S. Government Printing Office, 1931), 11293. See also Sando, *The Pueblo Indians*, 102. District members felt that the *Winters* decision did not apply to the Pueblos and felt that their provisions for protection of native water rights were very beneficial. See *Survey of Conditions, Part 20*, 11302.
25. U.S. Department of the Interior, *Annual Report for 1931*, 21; U.S. Department of the Interior, *Annual Report for 1932*, 28; U.S. Department of the Interior, Office of Indian Affairs, *Annual Report of Division of Extension and Industry for 1933*, 23, LS, SPA, RG 75, FRC, Denver.
26. *Natural Resources Board Bulletins and Reports*, 13–14, LR, NPA, RG 75, FRC, Denver. This work in the district's area consisted of 757 miles of irrigation

canals, 341 miles of drainage canals, and 180 miles of river levees; *Survey of Conditions, Part 19*, 9929. The population figures come from Dozier, *The Pueblo Indians*, 122.

27. Lewis Meriam, *The Problem of Indian Administration* (Baltimore, Md.: Johns Hopkins Press, 1928), 508; Hurt, *Indian Agriculture in America*, 168–69.
28. *Review of Conditions of the Indians in the United States: Hearings Before the Committee on Indian Affairs, Part 6* (Washington, D.C.: U.S. Government Printing Office, 1930), 2210–16.
29. Ibid., 2267–68.
30. S. Lyman Tyler, *A History of Indian Policy* (Washington, D.C.: U.S. Government Printing Office, 1973), 114–15.
31. *Survey of Conditions, Part 19*, 9928–30.
32. Ibid., 9966–67.
33. Ibid., 9981, 9928.
34. Ibid., 10156.
35. Ibid., 9985, 10182.
36. Ibid., 10201, 10397.
37. Ibid., 9986. Indians at Cochiti and Taos stated that they were quite willing to use modern machinery. See *Survey of Conditions, Part 20*, 10905.
38. *Survey of Conditions, Part 20*, 10160–61.
39. Leo Crane to Juan Rey Juancho, 27 July 1920, LS, SPA, RG 75, FRC, Denver.
40. C. J. Crandall to Charles H. Burke, 18 December 1923, LS, NPA, RG 75, FRC, Denver; H. F. Robinson to Roscoe Rice, 11 January 1921, LS, SPA, RG 75, FRC, Denver; H. F. Robinson to L. R. McDonald, 29 August 1921, LS, SPA, RG 75, FRC, Denver.
41. H. F. Robinson to W. M. Reed, 3 November 1922, LS, SPA, RG 75, FRC, Denver.
42. A. W. Leech to Herbert F. Robinson, 13 June 1922, LS, NPA, RG 75, FRC, Denver.
43. Kenneth A. Heron to the Trustees of the San Juan Pueblo Grant, 15 February 1925, LR, NPA, RG 75, FRC, Denver; Charles H. Burke to the Secretary of the Interior, 24 March 1925, LR, NPA, RG 75, FRC, Denver.
44. E. B. Meritt to Clinton J. Crandall, 9 October 1923, LR, NPA, RG 75, FRC, Denver.
45. Desiderio Naranjo to J. W. Elliott, 17 April 1930, LR, NPA, RG 75, FRC, Denver.
46. U.S. Department of the Interior, *Annual Report of the Division of Extension and Industry* for 1933, 7–8, LS, SPA, RG 75, FRC, Denver.
47. H. F. Robinson to the Commissioner of Indian Affairs, 6 April 1928, LS, NPA, RG 75, FRC, Denver.
48. T. F. McCormick to E. B. Meritt, 13 November 1928, LS, NPA, RG 75, FRC, Denver; R. H. Rupkey to H. F. Robinson, 12 June 1929, LR, NPA, RG 75, FRC, Denver. See also T. F. McCormick to H. F. Robinson, 4 June 1929, LS, NPA, RG 75, FRC, Denver. U.S. Department of the Interior, Office of Indian Affairs, Division of Extension and Industry, *Weekly Report of Farm Agent for 29 October 1932*, LS, SPA, RG 75, FRC, Denver. See also *Weekly Report of Farm Agent for 3 December 1932*; J. W. Elliott to the Commissioner of Indian Affairs, 11 September 1930, LS, NPA, RG 75, FRC, Denver.
49. H. F. Robinson to T. F. McCormick, 2 April 1929, LR, NPA, RG 75, FRC, Denver.
50. H. F. Robinson to T. F. McCormick, 3 June 1930, LR, NPA, RG 75, FRC, Denver.
51. A. M. Easterday to A. N. Thompson, 3 June 1934, LS, NPA, RG 75, FRC,

Denver; A. M. Easterday to H. C. Neuffer, 29 August 1934, LS, NPA, RG 75, FRC, Denver; H. C. Neuffer to All Employees Who Hire Labor, 22 May 1934, LS, NPA, RG 75, FRC, Denver; H. C. Neuffer to the Commissioner of Indian Affairs, 22 July 1935, LS, NPA, RG 75, FRC, Denver.

52. Pritzker, *A Native American Encyclopedia*, 26.
53. Hurt, "The Zuni Land Conservation Act of 1990," 95–96.
54. H. C. Neuffer to José Antonio Montoya, 9 March 1935, LS, NPA, RG 75, FRC, Denver; S. D. Aberle to the Commissioner of Indian Affairs, 8 July 1936, LS, UPA, RG 75, FRC, Denver; S. D. Aberle to the Commissioner of Indian Affairs, 9 September 1936, LS, UPA, RG 75, FRC, Denver.
55. Robert H. Rupkey, "Old Cultivated Lands and Ditches Not Now in Use, Isleta," June 1937, LS, UPA, RG 75, FRC, Denver; Robert H. Rupkey, "Old Cultivated Lands and Ditches Not Now in Use, Santo Domingo," June 1937, LS, UPA, RG 75, FRC, Denver; Robert H. Rupkey, "Old Cultivated Lands and Ditches Not Now in Use, Sandia," June 1937, LS, UPA, RG 75, FRC, Denver.
56. Walter L. Bolander to Horace J. Johnson, 28 December 1921, LR, NPA, RG 75, FRC, Denver; A. W. Leech to Walter L. Boland, 19 May 1923, LS, NPA, RG 75, FRC, Denver.
57. C. J. Crandall to S. H. Smith, 14 January 1924, LS, NPA, RG 75, FRC, Denver; E. D. Meritt to Thomas F. McCormick, 1 February 1926, LR, NPA, RG 75, FRC, Denver; J. M. Ulen to the Northern Pueblos Agency, 12 April 1930, LR, NPA, RG 75, FRC, Denver; C. E. Faris to the Commissioner of Indian Affairs, 17 July 1933, LS, NPA, RG 75, Denver.
58. E. B. Meritt to Pedro Cajete, 2 July 1921, LR, NPA, RG 75, FRC, Denver; C. J. Crandall to Ramon Martínez, 22 July 1926, LS, NPA, RG 75, FRC, Denver; T. F. McCormick to the Commissioner of Indian Affairs, 25 June 1928, LS, NPA, RG 75, FRC, Denver.
59. Leslie A. White, *The Pueblo of Sia, New Mexico* (Washington, D.C.: U.S. Government Printing Office, 1962), 323; E. B. Meritt to L. A. Towers, 27 October 1928, LR, SPA, RG 75, FRC, Denver; Adolph F. Bandelier and Edgar L. Hewett, *Indians of the Rio Grande Valley* (New York: Cooper Square, 1973), 97; Leslie A. White, *The Acoma Indians* (Washington, D.C.: U.S. Government Printing Office, 1932), 33.
60. Leslie A. White, *The Pueblo of Santa Ana, New Mexico*, Memoirs of the American Anthropological Association, no. 60 (Menasha, Wisc.: American Anthropological Association, 1942), 45; Lange, *Cochiti*, 90–91; William Whitman, "The San Ildefonso of New Mexico," in *Acculturation in Seven American Indian Tribes*, ed. Ralph Linton (Gloucester, Mass.: Peter Smith, 1963), 421.
61. E. R. Fryer, memorandum to Sy Souser, 12 February 1936, LS, UPA, RG 75, FRC, Denver; S. D. Aberle to J. Korber Company, 18 November 1937, LS, UPA, RG 75, FRC, Denver; Bernalillo Mercantile Company to U.S. Indian Agency in Albuquerque, 4 August 1937, LR, UPA, RG 75, FRC, Denver; S. D. Aberle to R. H. Shipman, 8 October 1937, LS, UPA, RG 75, FRC, Denver.
62. U.S. Department of the Interior, *Annual Report of the Division of Extension and Industry*, 5; White, *The Pueblo of Santa Ana*, 45–56.
63. Lange, *Cochiti*, 100; White, *The Pueblo of Santa Ana*, 45; Charles Burke to Clinton J. Crandall, 22 August 1923, LR, NPA, RG 75, FRC, Denver; C. J.

Crandall to Charles Burke, 25 July 1923, LS, NPA, RG 75, FRC, Denver.

64. Chester E. Faris to H. B. Peairs, 14 January 1928, LS, NPA, RG 75, FRC, Denver; U.S. Department of the Interior, Office of Indian Affairs, Division of Extension and Industry, *Weekly Report of Farm Agent for 24 September 1932*, LS, SPA, RG 75, FRC, Denver; S. D. Aberle to J. Korber and Company, 17 July 1936, LS, UPA, RG 75, FRC, Denver; A. C. Cooley to H. C. Seymour, 11 March 1937, LS, NPA, RG 75, FRC, Denver; Margaret Connell Szasz, *Education and the American Indian: The Road to Self-Determination, 1928–1973* (Albuquerque: University of New Mexico Press, 1974), 65.

65. C. J. Crandall to the Commissioner of Indian Affairs, 21 February 1925, LS, NPA, RG 75, FRC, Denver; T. F. McCormick to Charles H. Burke, 15 February 1929, LS, NPA, RG 75, FRC, Denver; Lorin F. Jones, memorandum to Miss Louise Wiberg, 12 April 1938, LS, UPA, RG 75, Denver.

66. S. D. Aberle to John Collier, 10 October 1936, LS, UPA, RG 75, FRC, Denver; S. D. Aberle to D. E. Murphy, 23 October 1937, LS, UPA, RG 75, FRC, Denver.

67. William Zimmerman to Carl Hatrch, 19 March 1938, LR, UPA, RG 75, FRC, Denver.

68. The population figures come from a county-by-county survey that was required for the Pueblos to become eligible for the Agricultural Control Program in 1938. The acres-planted statistics are from a similar type of report that was submitted in the previous year. See S. D. Aberle to Lyman Gleason, 4 February 1937; S. D. Aberle to J. W. Parker, 2 February 1937; S. D. Aberle to Leo R. Smith, 4 February 1937; S. D. Aberle to Juan Ramirez, 4 February 1937; S. D. Aberle to Felix Armijo, 4 February 1937; S. D. Aberle to A. G. Sandoval, 2 February 1937; S. D. Aberle to Ray Gonzales, 4 February 1937, LS, UPA, RG 75, FRC, Denver.

69. See chapter 5, table 2, for Pueblo agricultural statistics for 1890. See also chapter 7, table 7 for Pueblo acreage. Of the five pueblos mentioned, only Santa Ana's figures are close to other estimates, and even theirs is more than twice what it should be. Without subtracting Santa Ana's totals, the acres per person ratio is 1.34, which is almost exactly the same as the one for 1936.

70. Lorin F. Jones, memorandum to Mr. Balderson, 15 July 1937, LS, UPA, RG 75, FRC, Denver; C. E. Faris to John C. Gatlin, 18 April 1934, LS, NPA, RG 75, FRC, Denver; William H. Zeh to Dr. Sophie Aberle, 8 March 1937, LR, UPA, RG 75, FRC, Denver; S. D. Aberle to Elliott S. Barker, 26 September 1936, LS, UPA, RG 75, FRC, Denver; A. W. Leech to Walter L. Bolander, 19 July 1922, LS, NPA, RG 75, FRC, Denver.

71. US Department of the Interior, Office of Indian Affairs, Division of Extension and Industry, *Weekly Report of the Farm Agent for 17 September 1932*, LS, SPA, RG 75, FRC, Denver; S. D. Aberle to H. R. McKee, 30 June 1937, LS, UPA, RG 75, FRC, Denver.

CHAPTER SEVEN

1. Edward H. Spicer, "Spanish-Indian Acculturation in the Southwest," *American Anthropologist* 56 (August 1954): 665. A thorough explanation for their resistance is provided in Dozier, "Rio Grande Pueblos," 94–186.

2. Dozier, *The Pueblo Indians*, 122.

3. Aberle, *The Pueblo Indians of New Mexico*, 9–12. It should be noted that

because of the strong competition for irrigable land in New Mexico, non-Indians were not willing to sell this precious property to the Pueblos. This attitude was one of the major reasons why the percentage of irrigable land purchased from compensation funds was so low. Efforts by the Middle Rio Grande Conservancy District helped to improve irrigation and flood-control projects for those Pueblos under its jurisdiction, but this only affected Cochiti, Isleta, Sandia, San Felipe, Santa Ana, and Santo Domingo.

4. Philp, *John Collier's Crusade*, 194–95.
5. S. D. Aberle to A. C. Cooley, 21 April 1936, LS, UPA, RG 75, FRC, Denver; A. C. Cooley to S. D. Aberle, 5 May 1936, LR, UPA, RG 75, FRC, Denver; George C. Flanagan, memorandum to E. R. Smith, 20 March 1936, LR, UPA, RG 75, FRC, Denver.
6. L. J. Korn, memorandum for the files, 28 September 1935, LR, UPA, RG 75, FRC, Denver; H. C. Stewart to S. D. Aberle, 14 December 1936, LR, UPA, RG 75, FRC, Denver; E. R. Smith to S. D. Aberle, 28 August 1937, LR, UPA, RG 75, FRC, Denver; T. B. Williamson to C. D. Hager, 20 January 1938, LS, UPA, RG 75, FRC, Denver.
7. S. D. Aberle, memorandum to E. R. Smith, 29 June 1938, LS, UPA, RG 75, FRC, Denver; E. R. Fryer to C. D. Hager, 24 January 1936, LS, UPA, RG 75, FRC, Denver.
8. H. H. Bennett, Explanatory Memorandum of Functions and Operation of Technical Cooperation, LR, UPA, RG 75, FRC, Denver; Herbert V. Clotts to A. L. Wathen, 25 August 1937, LP, UPA, RG 75, FRC, Denver; Hugh G. Calkins to H. H. Bennett, 20 March 1935, LS, UPA, RG 75, FRC, Denver; W. A. Wunsch to S. D. Aberle, 28 September 1937, LR, UPA, RG 75, FRC, Denver.
9. Donald L. Parman, "The Indian and the Civilian Conservation Corps," *Pacific Historical Review* 40 (February 1971): 40–49.
10. Donald L. Parman, "The Indian Civilian Conservation Corps" (Ph.D. dissertation, University of Oklahoma, 1967), 183–84.
11. Ibid., 210–12.
12. Alan Laflin, memorandum to Mr. Formhals, 6 February 1939, LR, RG 75, FRC, Denver.
13. William Rapp, memorandum to Mr. Perce, 1 June 1939, LR, RG 75, FRC, Denver.
14. T. B. Bixby to Dr. S. D. Aberle, 28 August 1940, LR, RG 75, FRC, Denver.
15. I. G. Fauske to John C. Rainer, 29 January 1940, LS, RG 75, FRC, Denver. This was not to be considered a complete list of crops covered by the program.
16. C. Paul Goodrich, memorandum to Dr. Aberle, 23 January 1940, LR, RG 75, FRC Denver. Only the figures for Taos and Picuris are accurate since the other pueblos reported approximations.
17. Dan T. O'Neill, memorandum to Dr. Aberle, 18 March 1940, LR, RG 75, FRC, Denver.
18. Antonio Abeita to Dr. S. D. Aberle, 27 March 1940, LR, RG 75, FRC, Denver.
19. L. B. Liljenquist, memorandum to the files, 19 April 1940, LR, RG 75, FRC, Denver.
20. S. D. Aberle to George Toledo, 30 December 1940, LR, RG 75, FRC, Denver.
21. Alan Laflin to Fred L. O'Cheskey, 24 June 1940, LS, RG 75, FRC, Denver. Figures for wasteland among the Pueblos are not included.
22. Dozier, *The Pueblo Indians*, 122.

23. Ted Formhals, memorandum to Melvin Helander, 1 July 1941, LR, RG 75, FRC, Denver. Wheat was covered by the AAA program.
24. F. W. Hodges, "Report on Irrigation and Water Supply of the Pueblos of New Mexico in the Rio Grande Basin," 1937, LR, RG 75, FRC, Denver.
25. Alan Laflin to John Collier, 30 January 1940, LS, RG 75, FRC, Denver. Membership of this group included representatives from the headwaters of the Rio Grande in Colorado to Fort Quitman, Texas. Harlowe M. Stafford, counselor for Region 7 with headquarters at Denver, was chairman of the group. J. C. Stevens, an engineer from Portland, was the water consultant. Dr. Baldwin M. Woods, regional chairman for the National Resource Planning Board, was also in attendance.
26. Alan Laflin to Harlowe M. Stafford, 11 March 1940, LS, RG 75, FRC, Denver.
27. "General Summary of the Rio Grande Flood: May–June 1941," LR, RG 75, FRC, Denver.
28. S. D. Aberle to John Collier, 21 June 1941, LS, RG 75, FRC, Denver.
29. Dale, *The Indians of the Southwest*, 246.
30. S. D. Aberle, Notes on a Meeting to Discuss Correlation of Extension, Boarding School, and Day School Agricultural Programs, 22 August 1940, LR, RG 75, FRC, Denver.
31. Walter W. Nations, memorandum to Mr. Formhals, 3 January 1942, LR, RG 75, FRC, Denver.
32. Memorandum—Indian Boarding Schools Vocational Agriculture, 1 August 1942, LR, RG 75, FRC, Denver.
33. List of Names of Women to Attend Nutrition Course, 5 October 1942, LR, RG 75, FRC, Denver.
34. S. D. Aberle to W. W. Beaty, 5 February 1943, LS, RG 75, FRC, Denver.
35. William Zimmerman to all Reservation Superintendents, 3 February 1942, LR, RG 75, FRC, Denver.
36. William Zimmerman Jr. to Sophie D. Aberle, 25 April 1942, LR, RG 75, FRC, Denver.
37. Will R. Bolen, "Supervisor's Report," 7 August 1943, LR, RG 75, FRC, Denver.
38. Spicer, *Cycles of Conquest*, 185.
39. John G. Evans to War Food Administration, 7 June 1944 (Isleta Hay Press), LS, RG 75, FRC, Denver.
40. John G. Evans to War Food Administration, 7 June 1944 (Sandia Combine), LS, RG 75, FRC, Denver.
41. J. W. Wellington, "A Survey of the Possibilities of an Agricultural Program in Connection with the Taos and Zuni Day Schools and the Santa Fe Indian Boarding School," 29 January 1945, LR, RG 75, FRC, Denver.
42. Aberle, *The Pueblo Indians of New Mexico*, 84–89.
43. John Adair, "The Navajo and Pueblo Veteran: A Force for Culture Change," *American Indian* IV, no. 1 (1947): 7.
44. Florence Hawley Ellis, *Some Factors in the Indian Problem in New Mexico* (Albuquerque: University of New Mexico Press, 1948), 23. See also Aberle, *The Pueblo Indians of New Mexico*, 64; and Charles H. Lange, "The Role of Economics in Cochiti Pueblo Culture Change," *American Anthropologist* 55 (December 1953): 693–94. Lange predicted that the religious societies would tend to have their power diminished and that ceremonies would lose their

religious value and be performed for fold pageantry.

45. Robert C. Eyler, "Environmental Adaptation at Sia Pueblo," *Human Organization* 12 (winter 1954): 27.

CHAPTER EIGHT

1. Thomas Nickerson, *Report on the Migrational Background of the Pueblo Indians from the Point of View of Agriculture*, November 1942, LS, UPA, RG 75, FRC, Denver.
2. For Pueblo traditions and change during the New Deal and war years, see James A. Vlasich, "Transitions in Pueblo Agriculture," *New Mexico Historical Review* 55 (January 1980): 25–43. For increase in Indian agricultural land and participation in the war effort, see Kenneth William Townsend, *World War II and the American Indian* (Albuquerque: University of New Mexico Press, 2000), 188–89.
3. Sando, *The Pueblo Indians*, 122; and Townsend, *World War II*, 138.
4. Irwin Unger, *Recent America: The United States Since 1945* (Upper Saddle River, N.J.: Prentice-Hall, 2002), 135–36; James A. Henretta, *America's History* (New York: Worth, 1993), 894; Gary B. Nash, *The American People: Creating a Nation and a Society* (New York: Longman, 2001), 846; and George D. Moss, *Moving On: The American People Since 1945* (Englewood Cliffs, N.J.: Prentice-Hall, 2001), 91.
5. Hurt, *Indian Agriculture in America*, 195.
6. Albert L. Hurtado and Peter Iverson, eds., *Major Problems in American Indian History: Documents and Essays* (Boston: Houghton Mifflin, 2001), 486; Norman L. Rosenberg and Emily S. Rosenberg, *In Our Times: America Since World War II* (Englewood Cliffs, N.J.: Prentice-Hall, 1999), 92; Peter Iverson, "Building Toward Self-Determination: Plains and Southwestern Indians in the 1940s and 1950s," in *The American Indian: Past and Present*, ed. Roger L. Nichols (Boston: McGraw-Hill, 1999), 264; and Hurt, *Indian Agriculture in America*, 211.
7. Larry Burt, "Western Tribes and Balance Sheets: Business Development Programs in the 1960s and 1970s," *Western Historical Quarterly* 23 (November 1992): 476–78.
8. Arrell Morgan Gibson, *The American Indian: Prehistory to the Present* (Lexington, Mass.: D. C. Heath, 1980), 550–51; and Burt, "Western Tribes and Balance Sheets," 476–78.
9. *Minutes of Commissioner Emmons Meetings with Individual Pueblos*, 31 October 1953, Miscellaneous Records Received of the Superintendent of the Southern Pueblos Agency, Northern Pueblos Agency, and the United Pueblos Agency, Microfilm Copy, Indian Pueblo Cultural Center, Albuquerque, New Mexico, Roll Number 7A, Frame Number 737–77. Hereafter cited as MRR or MRS (Miscellaneous Records Sent), PCC, Albuquerque, RN (Roll number), FN (Frame number).
10. Thomas H. Dodge, *Implementation of Commissioner's Memorandum of April 12, 1956 on Programming or Indian Social and Economic Improvement*, 27 January 1958, Letters Received, Natural Resource Records, Northern Pueblos Agency, Santa Fe. Hereafter cited as LR or LS (Letters Sent), NRR, NPA, Santa Fe.

11. Homer M. Gilliland, SMC Topic: *Termination, Policy and Plans, At What Point Should Termination Be Accomplished*, 27 January 1958, LR, NRR, NPA, Santa Fe.
12. Leroy Horn, *Application of Termination Policy at Field Level*, 28 January 1958, LR, NRR, NPA, Santa Fe; and James D. Simpson, *Soil Surveys and Land Classification*, 28 January 1958, LR, NRR, NPA, Santa Fe.
13. Charles P. Corke, *Irrigation Standards Which Should Be Achieved Prior to Relinquishment*, 27 January 1958, LR, NRR, NPA, Santa Fe; and R. N. Hull, *Payment of O and M Costs by Indians*, 28 January 1958, LR, NRR, NPA, Santa Fe.
14. Robert H. Rupkey, *Water Rights*, 29 January 1958, LR, NRR, NPA, Santa Fe; and G. B. Keesee, *Water Rights*, 29 January 1958, LR, NRR, NPA, Santa Fe.
15. Frederick M. Haverland, *General Comments*, 29 January 1958, LR, NRR, NPA, Santa Fe; Leo Honey, *Roads Session*, 29 January 1958, LR, NRR, NPA, Santa Fe; and Albert N. Palmer, *Problems Involved in Withdrawal on Pueblo Lands*, 29 January 1958, LR, NRR, NPA, Santa Fe.
16. *The Arizona Republic*, 29 January 1958; T. B. Hall to the Commissioner of Indian Affairs, 15 December 1958, Letters Sent, Natural Resource Records—Branch of Forestry, Records of the Bureau of Indian Affairs, Albuquerque Area Office, Albuquerque, New Mexico. Hereafter cited as LR (Letters Received) or LS, NRR, BIA, Albuquerque. The termination of other services were staggered; the Pueblos would take over range-water development by January 1,1965; and Acting Commissioner of Indian Affairs to Frederick M. Haverland, 6 November 1962, LR, NRR, BIA, Albuquerque.
17. Hurt, *Indian Agriculture in America*, 233.
18. Abel Paisano to John L. Mutz, 27 August, 1947, Letters Received, Office of Rights Protection, Records of the Bureau of Indian Affairs, Albuquerque Area Office, Albuquerque, New Mexico. Hereafter cited as LR or LS (Letters Sent), ORP, BIA, Albuquerque.
19. *Santa Ana Flood Control Plan*, n.d., LR, NRR, BIA, Albuquerque.
20. H. P. McKenzie to the Secretary of the Interior, 8 April 1953, LS, ORP, BIA, Albuquerque; W. L. Miller to Charles L. Graves, 22 April 1953, LR, ORP, BIA, Albuquerque; Charles L. Graves to Commissioner of Indian Affairs, 24 August 1953, LR, ORP, BIA, Albuquerque; W. Barton Greenwood to H. R. McKenzie, 9 September 1953, LS, ORP, BIA, Albuquerque.
21. Guy C. Williams to W. Wade Head, 26 April 1955, LS, NRR, BIA, Albuquerque.
22. Guy C. Williams to Glenn Emmons, 29 April 1955, LS, NRR, BIA, Albuquerque; and Glenn Emmons to Dennis Chavez, 23 May 1955, LS, NRR, BIA, Albuquerque.
23. *Brief Statement of Irrigation Activities*, United Pueblos Agency, F.Y. 1955, 1 July 1955, LS, NRR, BIA, Albuquerque.
24. Guy C. Williams to Commissioner of Indian Affairs, 4 August 1955, LS, NRR, BIA, Albuquerque.
25. Antonio Sando to Dennis Chavez, 23 May 1958, LS, NRR, BIA, Albuquerque; Albert L. Reed to Guy Williams, 12 June 1958, LS, NRR, BIA, Albuquerque; Guy Williams to T. B. Hall, 16 July 1958, LS, NRR, BIA, Albuquerque; and Wade Head to Glenn Emmons, 24 July 1958, LS, NRR, BIA, Albuquerque.
26. Henning Nasman and Vernon Larsen to Area General Engineer, 17 March 1960, LS, NRR, BIA, Albuquerque.

27. *Statements made by Pueblo Leaders in the Middle Rio Grande Conservancy District*, October and November, 1946, LR, ORP, BIA, Albuquerque.
28. Bureau of Indian Affairs, *Conditions and Solutions on Water Conveyance and Distribution System, Six Middle Rio Grande Pueblos in the Middle Rio Grande Project in New Mexico*, n.d., LR, ORP, BIA, Albuquerque; H. E. Robins to Commissioner Bureau of Reclamation, 29 August 1947, LS, ORP, BIA, Albuquerque; and R. H. Rupkey to Southworth, 31 July 1947, LS, ORP, BIA, Albuquerque.
29. United Pueblos Agency, *Branch of Land Operations Narrative Highlights*, 1959, LS, NRR, NPA, Santa Fe; Bureau of Indian Affairs, *Conditions and Solutions*, n.d.; Peter A. Chavez, *District Agent Interviews with Pueblo Indians in the Middle Rio Grande Conservancy District*, March 1952, LR, ORP, BIA, Albuquerque; S. E. Reynolds to Homer Pickens, 26 April 1956, LS, ORP, BIA, Albuquerque; Louis Scott to S. E. Reynolds, 29 April 1957, LR, ORP, BIA, Albuquerque; and Homer Pickens to S. E. Reynolds, 29 May 1957, LR, ORP, BIA, Albuquerque.
30. Guy Williams to T. B. Hall, 30 September 1958, LS, NRR, BIA, Albuquerque; Joseph Albert to T. B. Hall, LS, NRR, BIA, Albuquerque; Murry Crosse to Guy Williams, 9 April 1959, LS, NRR, BIA, Albuquerque; Wade Head to Paul Fickinger, 15 April 1959, LS, NRR, BIA, Albuquerque; Alfred Herrera to Board of Engineers for Rivers and Harbors, 15 April 1959, LS, NRR, BIA, Albuquerque; Alfred Herrera to Guy Williams, 17 April 1963, LS, NRR, BIA, Albuquerque; Morris Lewis to UPA, 23 March 1964, LS, NRR, BIA, Albuquerque; *Cochiti Reservoir Project—Condition of Improvement*, 30 June 1965, LS, NRR, BIA, Albuquerque; and Bureau of Indian Affairs, *Conditions and Solutions*, n.d., LS, ORP, BIA, Albuquerque.
31. Fernando Lorizo to Board of Engineers for Rivers and Harbors, 30 April 1959, LS, NRR, BIA, Albuquerque; and Bureau of Indian Affairs, *Conditions and Solutions*, n.d., LS, ORP, BIA, Albuquerque.
32. Hurt, *Indian Agriculture in America*, 197.
33. The foregoing discussion of Pueblo conservation work is from Oliver Hole, *Annual Report of Soil Conservation Operations for the Northern Pueblo Agency*, 1951, LS, NRR, NPA, Santa Fe. The Northern Pueblos included Nambe, Picuris, Pojoaque, San Ildefonso, San Juan, Santa Clara, Taos, and Tesuque.
34. Ibid.
35. Ibid.
36. Ibid.
37. Ibid.
38. Oliver Hole, *Annual Report of Soil Conservation Operations for the Northern Pueblo Agency*, 1954, LS, NRR, NPA, Santa Fe.
39. Ramon Zuni, *Resolution for the Establishment of Isleta Soil Conservation Enterprise*, 4 May 1956, LR, NRR, NPA, Santa Fe; Esquipula Jojola, *Plan of Operation for Isleta Conservation Services*, 16 March 1961, LR, NRR, NPA, Santa Fe; Candido Herrera, *Resolution for the Establishment of the Tesuque Soil Conservation Enterprise*, 11 June 1956, LR, NRR, NPA, Santa Fe; Candido Herrera, *Plan of Operation for Tesuque Soil Conservation Enterprise*, 10 July 1956, LR, NRR, NPA, Santa Fe; Emilio Trujillo, *Resolution for the Establishment of the Nambe Soil Conservation*, 1 May 1957, LR, NRR, NPA,

Santa Fe; *Plan of Operation for San Juan Soil Conservation Enterprise*, n.d., LR, NRR, NPA, Santa Fe; and Tom Abeita, *Isleta Conservation Services*, 1970, LR, NEE, NPA, Santa Fe.

40. Marcus Silva, *Minutes of Meeting of the Santa Clara Pueblo Soil Conservation Enterprise*, 8 August 1959, LR, NRR, NPA, Santa Fe; *Santa Clara Soil Conservation Enterprise Budget*, 1960, LR, NRR, NPA, Santa Fe; Padnir Gutiérrez, *Resolution of Santa Clara Termination*, 19 December 1962, LR, NRR, NPA, Santa Fe; Evan Flory to All Areas, 24 October 1960, LR, NRR, NPA, Santa Fe; and *Santa Clara Soil Conservation Enterprise Budget*, 1961, LR, NRR, NPA, Santa Fe.
41. United Pueblos Agency, *Program Bulletin*, June 1961, LR, NRR, NPA, Santa Fe.
42. Dewey Dismuke to L. G. Boldt, 2 July 1951, LR, ORP, BIA, Albuquerque; and William Brophy to Wade Head, 18 December 1961, LR, NRR, BIA, Albuquerque.
43. Guy Williams to Commissioner of Indian Affairs, 24 January 1956, LS, NRR, BIA, Albuquerque; Geronimo Trujillo, *Resolution of the Taos Council*, 18 January 1956, LS, NRR, BIA, Albuquerque; Geronimo Trujillo, *Range Management Plan—Taos Pueblo Lands*, n.d., LS, NRR, BIA, Albuquerque; T. B. Hall to Guy Williams, 3 February 1956, LR, NRR, BIA, Albuquerque; and Guy Williams to Assistant Area Director of Resources, 6 July 1961, LS, NRR, BIA, Albuquerque.
44. Minton P. Weaver, *United Pueblos Agency—Branch of Land Operations Narrative Highlights*, 1960, LR, NRR, RG 75, BIA, Albuquerque.
45. R. H. Rupkey to C. H. Southworth, 8 February 1946, LS, ORP, BIA, Albuquerque.
46. Robert Bunker, *Minutes of Meeting of All-Pueblo Council Irrigation Committee*, 25 November 1946, MRR, PCC, Albuquerque, FN 9, RN 266; R. H. Rupkey to C. H. Southworth, 31 December 1946, LS, ORP, BIA, Albuquerque; Eric Harberg to Cyril Luker, 22 April 1949, LS, ORP, BIA, Albuquerque.
47. L. W. Hitchcock to John H. Bliss, 5 July 1951, LS, ORP, BIA, Albuquerque; A. R. Fife to Eric T. Hagberg, 24 April 1951, LR, ORP, BIA, Albuquerque; A. R. Fife to Eric T. Hagberg, 18 July 1950, LR, ORP, BIA, Albuquerque; and Geraint Humpherys to Eric T. Hagberg, 2 August 1950, LR, ORP, BIA, Albuquerque.
48. Governors of Jemez and Zia Pueblos to Eric T. Hagberg, 5 April 1951, LR, ORP, BIA, Albuquerque; Guy Williams to John Bliss, 18 April 1951, LS, ORP, BIA, Albuquerque; Guy Williams to John Bliss, 18 April 1951, LS, ORP, BIA, Albuquerque; L. W. Hitchcock to John Bliss, 7 August 1951, LS, ORP, BIA, Albuquerque; Eric T. Hagberg to A. F. Brown, 17 August 1951, LS, ORP, BIA, Albuquerque; and John Bliss to Eric T. Hagberg, 21 August 1951, LR, ORP, BIA, Albuquerque.
49. John H. Bliss, *Legal Notice*, 9 November 1951, LS, ORP, BIA, Albuquerque; A. R. Fife to William Brophy, 12 December 1951, LS, ORP, BIA, Albuquerque; and Dewey Desmuke to John Bliss, 7 January 1952, LR, ORP, BIA, Albuquerque.
50. Diego Abeita, *Resolution*, 19 July 1951, LR, ORP, BIA, Albuquerque; L. G. Boldt to Files, 29 August 1951, LR, ORP, BIA, Albuquerque; A. R. Fife to C. L. Graves, 10 July 1953, LS, ORP, BIA, Albuquerque; A. R. Fife to Files, 16 August 1954, LR, ORP, BIA, Albuquerque.
51. A. R. Fife to Eric T. Hagberg, 24 April 1951, LR, ORP, BIA, Albuquerque; and Diego Abeita, *Resolution*, 19 April 1951, LR, ORP, BIA, Albuquerque.

52. Robert G. Dunbar, *Forging New Rights in Western Waters* (Lincoln: University of Nebraska Press, 1983), 144–46; and Perrigo, *The American Southwest*, 338–39.
53. William Brophy to Eric Hagberg, 15 November 1951, LR, ORP, BIA, Albuquerque; Martin White to the Attorney General, 14 December 1951, and *Excerpts from Minutes of the 13th Annual Meeting of the Rio Grande Compact Commission*, 25 and 26 February 1952, LR, ORP, BIA, Albuquerque.
54. Diego Abeita, *Minutes of the Indian Irrigation Committee Meeting*, 18 March 1953, MRS, PCC, Albuquerque, RN 9, FN 266.
55. *Minutes of Commissioner Emmons Meetings with Individual Pueblos*, 31 October 1953, MRS, PCC, Albuquerque, RN 7a, FN 737–77.
56. A. R. Fife to Guy Williams, 8 December 1954, LS, ORP, BIA, Albuquerque.
57. A. R. Fife to S. Lyman Tyler, 21 March 1956, LR, ORP, BIA, Albuquerque; and Bureau of Indian Affairs, *Position Paper on Indian Water Storage at El Vado Reservoir, 1981*, LR, ORP, BIA, Albuquerque.

CHAPTER NINE

1. Gibson, *The American Indian*, 556, 562.
2. Burt, "Western Tribes and Balance Sheets," 481–82; and James J. Rawls, *Chief Red Fox is Dead: A History of Native Americans Since 1945* (Fort Worth, Tex.: Harcourt, Brace, 1996), 57.
3. John D. Dibbern, *Ten-year War on Poverty Program*, 24 June 1964, frs. 368–84, roll 8, Miscellaneous Records Received of the Superintendent of the Southern Pueblos Agency, Northern Pueblos Agency, and the United Pueblos Agency (microfilm), Indian Pueblo Cultural Center, Albuquerque (hereafter MRR or MRS [Miscellaneous Records Sent], PCC, Albuquerque). Zuni had formed its own separate agency by this time.
4. Kirschner Associates, Inc., *Water Use Projections for Nambe, Pojoaque, San Ildefonso, San Juan, Santa Clara and Tesuque Indian Pueblos of Northern New Mexico, 1965–2000*, May 1967, LR, ORP, BIA, Albuquerque.
5. Tyler, *A History of Indian Policy*, 207, 210.
6. *Western Livestock Journal*, 7 February 1963, LR, NRR, BIA, Albuquerque.
7. Clinton P. Anderson, *The Northern New Mexico Resource Conservation and Development Program*, 3 February 1967, LR, NRR, NPA, Santa Fe; and Pablo Roybal, *A Year of Action Progress*, 1965, LR, NRR, NPA, Santa Fe.
8. Walter W. Olsen to Pablo Roybal, 14 July 1965, LS, NRR, NPA, Santa Fe.
9. Melvin Helander to General Superintendent of the UPA, 27 September 1962, LR, NRR, NPA, Santa Fe; Alfredo Naranjo, *Minutes of the Santa Clara Conservation Services*, 18 March 1965, LR, NRR, NPA, Santa Fe; and Walter Olsen to General Superintendent of the UPA, 4 August 1966, LR, NRR, NPA, Santa Fe. Evidently, the SMC is another name for the Soil Conservation Service.
10. Eligio Vigil, *Tesuque Resolution*, 11 July 1966, LR, NRR, NPA, Santa Fe.
11. José Rey Leon, Jack Toya and Juan José Shije to the Commissioner of Indian Affairs, 27 March 1962, LS, NRR, RG 75, BIA, Albuquerque; John O. Crow to Clinton P. Anderson, 25 April 1962, LR, NRR, BIA, Albuquerque; John O. Crow to T. B. White, 9 May 1962, LR, NRR, BIA, Albuquerque; José Rey Leon, Jack Toya, and Juan José Shije to Joseph Montoya and Thomas Morris, 23 July 1962, LS, NRR, BIA, Albuquerque; José Rey Leon, Jack Toya, and Juan José Shije to the Commissioner of Indian Affairs, LS, NRR, BIA,

Albuquerque; John O. Crow to Juan José Shije, José Ray Leon, and Jack Toya, 19 October 1962, LR, NRR, BIA, Albuquerque; and J. B. Rankin, *Cooperative Fencing Agreement*, n.d., LR, NRR, NPA, Santa Fe.

12. John C. Dibbern to Steve Reynolds, 19 February 1964, LS, NRR, BIA, Albuquerque.
13. F. D. Shannon to Commissioner of Indian Affairs, 26 August 1970, LS, NRR, BIA, Albuquerque; Domingo Montoya, *All Indian Pueblo Council Resolution*, 19 June 1966, LS, NRR, BIA, Albuquerque; Domingo Montoya to Robert Bennett, 3 April 1967, LS, NRR, BIA, Albuquerque; and Theodore W. Taylor to Clinton P. Angerson, 11 May 1967, LR, NRR, BIA, Albuquerque.
14. Robert Triedueau to the Area Director, 2 March 1970, LR, NRR, BIA, Albuquerque; and Branch of Land Operations, *Soil and Moisture and Range Funds Expended at Taos Pueblo*, 1975, LS, NRR, NPA, Santa Fe.
15. T. W. Taylor to Sam Victorino, 12 December 1969, LR, NRR, BIA, Albuquerque; Sam Victorino, *Acoma Tribal Council Resolution*, 23 December 1969, LR, NRR, BIA, Albuquerque; Area Natural Resource Manager to Earl Webb, April 1970, LR, NRR, BIA, Albuquerque.
16. Merle L. Garcia to Harold Runnels, 8 January 1976, LR, NRR, BIA, Albuquerque; Merle Garcia, *Acoma Resolution for Approval to Request for Appropriation for Soil and Moisture Conservation Program*, 8 January 1976, LR, NRR, BIA, Albuquerque; and Mark J. Stevens to Area Director, 18 February 1976, LS, NRR, BIA, Albuquerque.
17. Ira W. Dutton to the Area Director, 2 February 1970, LR, NRR, BIA, Albuquerque; Robert E. Lewis, *Zuni Tribal Council Resolution*, 3 April 1970, LR, NRR, BIA, Albuquerque; and Zuni Tribal Council, *The Zuni Indian Demonstration Project*, n.d., LR, NRR, BIA, Albuquerque.
18. Wilson C. Gutzman to A. E. Trivis, 27 April 1971, LR, NRR, BIA, Albuquerque; Theodore C. Krenzke to Pete V. Domenici, 15 July 1976, LR, NRR, BIA, Albuquerque; Robert C. Walker to Pete V. Domenici, 21 July 1976, LR, NRR, BIA, Albuquerque; and Bob Farring, Proposal for an Expanded Extension Education Program with American Indians, 3 June 1975, LR, NRR, BIA, Albuquerque.
19. Melvin Helander to A. E. Trivis, 3 September 1968, LS, NRR, BIA, Albuquerque; Benny Atencio to A. E. Trivis, 9 February 1971, LS, NRR, BIA, Albuquerque; James D. Cornett, Meeting with Chairman EPIC, State Extension Staff Members and SPA Natural Resource Manager, LS, NRR, BIA, Albuquerque.
20. *Pueblo News*, February 1980, LR, NRR, BIA, Albuquerque; and William Gorman to Raymond J. Concho, 11 December 1980, LS, NRR, BIA, Albuquerque.
21. Robert L. Garcia, *New Mexico State University Cooperative Extension Service*, 21 July 1975, LS, NRR, BIA, Albuquerque; and Merle L. Garcia, et al, *Southern Pueblo Agency Fiscal Year Program Plan*, 1977, LR, NRR, BIA, Albuquerque.
22. *Monthly Reports of the Southern Pueblo-Extension Agent*, June and August 1977, LS, NRR, BIA, Albuquerque; and Bob Powell, *Annual Narrative Report of the Extension Staff for the Southern Pueblos*, 1977–1978 and 1978–1979, LS, NRR, BIA, Albuquerque.
23. Robert E. Lewis, *Zuni Tribal Council Resolution*, 18 August 1980, LS, NRR, BIA, Albuquerque; Noel Marsh to Branch of Contracts and Grants, 7 August 1981, LR, NRR, BIA, Albuquerque; Noel Marsh to Branch of Contracts and

Grants, 11 August 1981, LR, NRR, BIA, Albuquerque; and Sidney Mills to Angel T. Gomez, 13 August 1981, LS, NRR, BIA, Albuquerque.

24. Patrick L. Wehling to SPA, NPA, and Zuni, 7 October 1975, LS, NRR, BIA, Albuquerque; *New Mexico Agricultural Conservation Program*, 8 November 1977, LS, NRR, BIA, Albuquerque; and Loyd E. Nickelson to SPA, NAP, and Zuni, 22 November 1977, LS, NRR, NAP, Santa Fe.

25. Morris Thompson to All Area Directors, 25 November 1975, LR, NRR, BIA, Albuquerque; Raymond V. Butler to Office of Audit and Investigation, 13 January 1976, LR, NRR, BIA, Albuquerque; and Joseph H. Wright, 11 February 1977, LS, NRR, BIA, Albuquerque.

26. *Gallup Independent*, 28 August 1963; *Gallup Independent*, 29 August 1963; Zuni Agency Superintendent to Area Director, 12 September 1963, LR, NRR, BIA, Albuquerque; Edward T. Kerley to the Commissioner of Indian Affairs, LS, NRR, BIA, Albuquerque; *Watershed Protection and Flood Prevention Act*, 4 August 1954, LR, NRR, BIA, Albuquerque; Robert E. Lewis, *Zuni Tribal Council Resolution*, 11 July 1969, LR, NRR, BIA, Albuquerque; Robert E. Lewis to Allan T. Howe, 30 March 1971, LS, NRR, BIA, Albuquerque; *Albuquerque Journal*, 19 May 1971; *Gallup Independent*, 28 July 1973; *Gallup Independent*, 1 February 1975; and *Albuquerque Tribune*, 17 June 1976.

27. John C. Gatlin to John C. Dibbern, 27 November 1964, LR, NRR, BIA, Albuquerque; Walter O. Olsen to John C. Dibbern, 26 February 1965, LR, NRR, BIA, Albuquerque; John C. Gatlin to James L. Sutton, 20 April 1965, LS, NRR, BIA, Albuquerque; Walter O. Olsen to the UPA General Superintendent, 9 May 1966, LR, NRR, BIA, Albuquerque; John E. Carver to James L. Sutton, 27 April 1972, LS, NRR, BIA, Albuquerque; Juan B. Jiron to James L. Sutton, 1 May 1972, LS, NRR, BIA, Albuquerque; Juan B. Jiron, *Isleta Council Resolution*, 1 May 1972, LS, NRR, BIA, Albuquerque; and Timothy P. Amalla, *Laguna Council Resolution*, 2 May 1972, LS, NRR, BIA, Albuquerque.

28. R. E. Schreiber, *Flood Problems at Pajarito Village*, 16 August 1965, LR, NRR, NPA, Santa Fe; James B. Rankin to Abel Sanchez, 3 October 1969, LS, NRR, BIA, Albuquerque; Abel Sanchez, *San Ildefonso Resolution*, 20 April 1970, LS, NRR, BIA, Albuquerque; and George H. Nickelson, 31 August 1971, LS, NRR, BIA, Albuquerque; J. Ramos Roybal to L. G. Boles, 18 June 1973.

29. Pablo Roybal to the Northern Rio Grande Resource Conservation and Development Steering Committee, 28 July 1966, LS, NRR, NPA, Santa Fe; John Carver, *Branch of Land Operations Trip Report*, 8 April 1970, LS, NRR, BIA, Albuquerque; and Governor José S. Viariel, *Resolution of the Pojoaque Pueblo Council*, 20 January 1977, LR, NRR, NPA, Santa Fe.

30. Andy Lauriano, *Sandia Resolution*, 21 March 1968, LR, NRR, BIA, Albuquerque; San Juan Governor, *San Juan Resolution*, 8 April 1968, LR, NRR, NPA, Santa Fe; Juan Chavarria, *Santa Clara Resolution*, 2 September 1969, LR, NRR, BIA, Albuquerque; Martin Vigil, *Tesuque Resolution*, 19 September 1969, LR, NRR, BIA, Albuquerque; Joseph Duran, *Tesuque Resolution*, 15 April 1971, LR, NRR, BIA, Albuquerque; Cochiti Governor, *Cochiti Resolution*, 19 May 1972, LR, NRR, BIA, Albuquerque; Santo Domingo Governor, *Santo Domingo Resolution*, 13 March 1972, LR, NRR, BIA, Albuquerque; Gordon A. Walhood to Valentino Garcia, 12 June 1972, LR, NRR, BIA, Albuquerque; Frank J. Lujan, *Taos Resolution*, 19 January 1973, LR,

NRR, NPA, Santa Fe; Loyd E. Nickelson to Superintendents of NPA and SPA, LS, NRR, BIA, Albuquerque; and Jack G. Koogler to Walter Olsen, 26 July 1973, LS, NRR, BIA, Albuquerque.

31. Patrick L. Wehling to Elie S. Gutiérrez, 2 July 1969, LS, NRR, BIA, Albuquerque; Patrick L. Vehling to Loyd E. Nickelson, 8 April 1970, LR, NRR, BIA, Albuquerque; Tom H. Cord to Patrick L. Vehling, 21 April 1970, LR, NRR, BIA, Albuquerque; Joseph M. Montoya to Louis R. Bruce, 7 April 1970, LR, NRR, BIA, Albuquerque; and Loyd E. Nickelson to Louis R. Bruce, 1 May 1970, LS, NRR, BIA, Albuquerque.

32. Manuel Lujan to the Bureau of Indian Affairs, 6 June 1973, LR, NRR, BIA, Albuquerque; and Mark J. Stevens to Loyd E. Nickelson, 21 June 1973, LR, NRR, BIA, Albuquerque; Loyd E. Nickelson, *Memorandum*, 30 September 1974, LS, NRR, BIA, Albuquerque; Mark J. Stevens to Manuel Lujan, 10 October 1974, LS, NRR, BIA, Albuquerque.

33. DuMars, O'Leary, and Utton, *Pueblo Indian Water Rights*, 3–5, 81; Hurt, *Indian Agriculture in America*, 215–17; Bureau of Indian Affairs, *Position Paper on Indian Water Storage at El Vado Reservoir*, 1981, LR, ORP, BIA, Albuquerque; Hundley, "The Dark and Bloody Ground," 457–58, 463–64, 472; and Newville, "Pueblo Indian Water Rights," 260.

34. DuMars, O'Leary, and Utton, *Pueblo Indian Water Rights*, 18, 43, 64, 84–86; *Albuquerque Journal*, 5 October 1971; and *Independent News*, 10 October 1971.

35. James F. Canan to John O. Crow, 6 September 1961, LR, ORP, BIA, Albuquerque; John O. Crow to Murry L. Crosse, 20 September 1961, LS, ORP, BIA, Albuquerque; F. M. Hoverland to Philleo Nash, 26 November 1965, LS, ORP, BIA, Albuquerque; Frances Levine, "Dividing the Water: Impact of Water Rights Adjudication on New Mexican Communities," *Journal of the Southwest* 32 (autumn 1990): 268; William J. Balch and John W. Clark, *Pueblo Water Rights on the Upper Rio Grande*, 30 July 1978, LR, ORP, BIA, Albuquerque; Albert Hurtado, "Public History and the Native American," *Montana: The Magazine of Western History* (spring, 1990): 65; and Newville, "Pueblo Indian Water Rights," 259.

36. *Albuquerque Journal*, 16 November 1981.

37. Lotario D. Ortega to Myles E. Flint, 19 March 1980, LS, ORP, BIA, Albuquerque; Samuel R. Montoya to Area Director, 21 October 1980, LR, ORP, BIA, Albuquerque; Bureau of Indian Affairs, *Water Development Plan for the Jemez River Project*, n.d.; Lotario D. Ortega to Michael Cox, 11 May 1981, LR, ORP, BIA, Albuquerque; Alfred Gachupin, *Pueblo of Zia Resolution*, 30 April 1981, LR, ORP, BIA, Albuquerque; Lawrence Montoya to Lotario Ortega, 30 April 1981, LR, ORP, BIA, Albuquerque; Sidney L. Mills to the Assistant Secretary of Indian Affairs, 20 October 1981, LS, ORP, BIA, Albuquerque; and *Albuquerque Journal*, 8 June 1983.

38. Lotario D. Ortega to Associate Solicitor of Indian Affairs, 22 February 1980, LS, ORP, BIA, Albuquerque; Lotario D. Ortega to Area Director, 15 September 1980, LS, ORP, BIA, Albuquerque; Peter C. Chestnut to Pueblos of Acoma, Laguna, Jemez, Zia and Santa Ana, 12 February 1981, LS, ORP, BIA, Albuquerque; Frank Jones to Barry W. Welch, 11 June 1982, LS, ORP, BIA, Albuquerque; Robert R. Lansford, *Projected Economic Impacts for Water Trespass Claims on Acoma Pueblo, 1945–1980*, 28 January 1983, LR, ORP, BIA,

Albuquerque; Arturo G. Ortega and Robert J. Nordraus to William H. Coldiron, 29 November 1983, LR, ORP, BIA, Albuquerque; and Superintendent of Indian Affairs to Merle L. Garcia, 23 August 1982, LR, ORP, BIA, Albuquerque.

CHAPTER TEN

1. Lotario D. Ortega to Superintendent of the Southern Pueblos Agency, 16 March 1981, LS, ORP, BIA, Albuquerque; John R. Baker, *The Impact of the Middle Rio Grande Conservancy District on the Lands of the Pueblos of Sandia and Santa Ana*, March 1981, LR, ORP, BIA, Albuquerque; and Mike Avila to Lotario D. Ortega, 8 May 1981, LR, ORP, BIA, Albuquerque.
2. Lotario D. Ortega to Michael Cox, 5 June 1981, LS, ORP, BIA, Albuquerque; Lotario D. Ortega to Santa Fe Field Solicitor, 15 April 1983, LS, ORP, BIA, Albuquerque.
3. Stanley W. Hulett to Pete V. Domenici, 23 June 1981, LS, ORP, BIA, Albuquerque; and Richard W. Hughes to James Watt, 29 July 1981, LS, ORP, BIA, Albuquerque.
4. *Albuquerque Journal*, 9 September 1985; and *Albuquerque Journal*, 4 January 1994.
5. Charlotte Benson Crossland, "Acequia Rights in Law and Tradition," *Journal of the Southwest* 32 (autumn 1990): 279, 282–85.
6. *United Pueblos General Statistics with Application to Irrigation*, 1949, LS, NRR, BIA, Albuquerque; *Eighteen Indian Pueblos in Upper Rio Grande Basin—Irrigation Information*, 7 March 1952, MRS, PCC, Albuquerque, RN 9, FN 378; *United Pueblos Annual Crop Report*, 1964, LR, NRR, SPA, Albuquerque; and *Southern Pueblos Crop Report*, 1981, LR, NRR, SPA, Albuquerque.
7. Four Corners Regional Commission, *Agricultural Addendum to the Development Plan* (Washington, D.C.: U.S. Department of Commerce, 1973), 10, 51.
8. Sando, *The Pueblo Indians*, 122–23; and Joe Sando, *Pueblo Nations: Eight Centuries of Pueblo Indian History* (Santa Fe, N.Mex.: Clear Light, 1992).
9. William J. Balch and John W. Clark, *Pueblo Water Rights on the Upper Rio Grande*, 30 July 1978, LR, ORP, BIA, Albuquerque.
10. John R. Baker, *The Impact of the Middle Rio Grande Conservancy District on the Lands of the Pueblos—Santa Ana*, March 1981, LR, ORP, BIA, Albuquerque; and Robert R. Lansford, *Projected Economic Impacts for Water Trespass Claims on the Acoma Pueblo, 1945–1980*, 28 January 1983, LR, ORP, BIA, Albuquerque.
11. Calbert Seciwa, interview by James Vlasich, 13 March 2001, Phoenix, Ariz.
12. Forrest J. Gerard to Walter O. Olsen, 11 September 1967, LR, NRR, BIA, Albuquerque; and Domingo Montoya to Kenneth L. Payton, 26 September 1967, LR, NRR, BIA, Albuquerque.
13. Mount V. Barner to the Chief of the Division of Resource Development, 30 June 1982, LR, NRR, BIA, Albuquerque; "Picuris Returns to Farm Roots," *Taos News*, 15 March 1984; and "HHS head Presents Grant to Pueblos Council," *Albuquerque Journal*, 14 April 1984.
14. "Reclaiming Their Harvest Heritage," *Albuquerque Journal*, 30 November 1991.
15. Dick Pierce, "A Return to the Land: Traditional Agriculture is Gaining

Momentum," *Winds of Change* (autumn 1993): 80–85; and Bonvillain, *Native Nations*, 358.

16. "Hope Sprouts at Cochiti," *Albuquerque Tribune*, 28 December 1995.
17. "Blue Corn Craze Grows, Demands Organic Farming," *Albuquerque Journal*, 28 May 1989; "Seminar Set for Indian Farmers, Ranchers," *Albuquerque Journal*, 4 March 1990; "Pueblo's Agriculture Project Gets Boost," *The New Mexican*, 19 July 1992; and "Ancient Crop Gives Pueblo New Chance for Self-Reliance," *Albuquerque Journal*, 13 December 1992.
18. Steve Larese, "Contemporary Indian Economics in New Mexico," in *Major Problems in American Indian History*, eds. Albert L. Hurtado and Peter Iverson (Boston: Houghton Mifflin, 2001), 500.
19. Simmons, "History of the Pueblos Since 1821," 9: 231–32; and Pritzker, *A Native American Encyclopedia*, 28.

BIBLIOGRAPHY

PUBLIC DOCUMENTS

Abert, J. W. "Report of the Secretary of War, February 4, 1848." In *Senate Executive Document* no. 23, 30th Congress, 1st Session, 1848. Washington, D.C.: Wendell and Van Benthuysen, 1847.

Emory, W. H. "Notes of a Military Reconnoissance *(sic)* from Fort Leavenworth in Missouri to San Diego in California." In *Senate Executive Document* no. 7, 30th Congress, 1st Session, 1848. Washington, D.C.: Wendell and Van Benthuysen, 1848.

General Correspondence of the Northern Pueblos Agency, 1911–1935. Records of the Bureau of Indian Affairs. Federal Records Center, Denver, Colorado.

General Correspondence of the Southern Pueblos Agency, 1911–1935. Records of the Bureau of Indian Affairs. Federal Records Center, Denver, Colorado.

Letters Received or Sent by the Office of Rights Protection, 1946–1983. Records of the Bureau of Indian Affairs. Albuquerque Area Office, Albuquerque, New Mexico.

Letters Sent by the Superintendent of Pueblo Agency and Day Schools, 1914–1919. Records of the Bureau of Indian Affairs. Federal Records Center, Denver, Colorado.

Letters Sent by the Superintendent of the Santa Fe Indian School, June 1890–December 1913. Records of the Bureau of Indian Affairs. Federal Records Center, Denver, Colorado.

Miscellaneous Letters Sent by the Pueblo Indian Agency, 1875–1891. Microfilm copy, Roll number 941, Reel numbers 1–10. Federal Records Center, Denver, Colorado.

Miscellaneous Records Received by the Superintendent of the Southern Pueblos

Agency, Northern Pueblos Agency, and the United Pueblos Agency. Microfilm copy, Roll number 7A, Frame number 737–77. Indian Pueblo Cultural Center, Albuquerque, New Mexico.

Natural Resource Records, 1945–1973. Northern Pueblos Agency, Santa Fe, New Mexico.

Natural Resource Records, 1964–1981. Records of the Bureau of Indian Affairs. Southern Pueblos Agency, Albuquerque, New Mexico.

Natural Resource Records—Branch of Forestry, 1945–1981. Records of the Bureau of Indian Affairs. Albuquerque Area Office, Albuquerque, New Mexico.

Press Copies of Miscellaneous Letters Sent by the Pueblo Indian Agency, 1891–1900. Records of the Bureau of Indian Affairs. Federal Records Center, Denver, Colorado.

Review of Conditions of the Indians in the United States: Hearings Before the Committee on Indian Affairs, Part 6. Washington, D.C.: U.S. Government Printing Office, 1930.

Simpson, J. H. "Report of Lieutenant J. H. Simpson of an Expedition into the Navajo Country." In *Reports of the Secretary of War, Senate Executive Document* no. 64, 31st Congress, 1st Session, 1850. Washington, D.C.: Union Office, 1850.

Survey of Conditions of the Indians of the United States: Hearings Before a Subcommittee of Indian Affairs, Part 19. Washington, D.C.: U.S. Government Printing Office, 1932.

Survey of Conditions of the Indians of the United States: Hearings Before a Subcommittee of Indian Affairs, Part 20. Washington, D.C.: U.S. Government Printing Office, 1931.

United Pueblos Agency Decimal Files. Records of the Bureau of Indian Affairs. Federal Records Center, Denver, Colorado.

U.S. Department of the Interior. Census Office. *Report on Indians Taxed and Indians Not Taxed in the United States (Except Alaska) at the Eleventh Census: 1890.*

U.S. Department of the Interior. Office of Indian Affairs. *Annual Report to the Commissioner of Indian Affairs for 1854–1932.*

Whipple, A. W. "Report of Explorations for a Railway Route, Near the Thirty-fifth parallel of North Latitude, from the Mississippi River to the Pacific Ocean." In *Reports of Explorations and Surveys, House of Representatives Executive Document* no. 91, 33rd Congress, 2nd Session, 1856. Washington, D.C.: A. O. P. Nicholson, 1856.

CASES CITED

Arizona v. California, 373 U.S. 546 (1963).

New Mexico v. Aamodt, 537 F.2d 1102 (10th Cir. 1976).

State of Texas v. State of New Mexico, 344 U.S. 906 (1952).

United States v. Candelaria, 271 U.S. 432 (1926).

United States v. Joseph, 4 Otto U.S. 614–19 (1876).

United States v. Lucero, 1 N.M. 422 (1867).

United States v. Sandoval, 231 Butler U.S. 36–49 (1913).

Winters v. United States, 207 U.S. 567 (1908).

UNPUBLISHED MATERIALS

Lange, Charles H. "An Evaluation of Economic Factors in Cochiti Pueblo Culture Change." Ph.D. diss., University of New Mexico, 1951.

Parman, Donald L. "The Indian Civilian Conservation Corps." Ph.D. diss., University of Oklahoma, 1967.

Sunseri, Alvin R. "New Mexico in the Aftermath of the Anglo-American Conquest, 1846–1861." Ph.D. diss., Louisiana State University, 1973.

Tyler, Daniel. "New Mexico in the 1820's: The First Administration of Manuel Armijo." Ph.D. diss., University of New Mexico, 1970.

Vlasich, James A. "A History of Irrigation and Agriculture Among the Pueblo Indians of New Mexico." Master's Thesis, University of Utah, 1977.

INTERVIEW

Seciwa, Calbert. Interview by James Vlasich. 13 March 2001. Phoenix, Ariz.

ARTICLES

Abel, Annie Heloise, ed. "Indian Affairs in New Mexico Under the Administration of William Carr Lane." *New Mexico Historical Review* 16 (July 1941): 328–58.

Adair, John. "The Navajo and Pueblo Veteran: A Force for Culture Change." *American Indian* IV, no. 1 (1947): 5–11.

Anderson, H. Allen. "The Encomienda in New Mexico, 1598–1680." *New Mexico Historical Review* 60 (October 1985): 353–77.

Bloom, Lansing B. "New Mexico under the Mexican Administration." *Old Santa Fe* 1 (July 1913): 1–56.

———. "Bourke on the Southwest." *New Mexico Historical Review* 11 (January 1936): 121.

Brown, Donald Nelson. "Picuris Pueblo in 1890: A Reconstruction of Picuris Social Structure and Subsistence Activities." In *Picuris Pueblo Through Time: Eight Centuries of Change in a Northern Rio Grande Pueblo*. Edited by Michael A. Alder and Herbert W. Dick, 19–41. Dallas: Southern Methodist University Clements Center for Southwest Studies, 1999.

Bryan, Kirk. "Flood-Water Farming." *The Geographical Review* 19 (July 1929): 444–56.

———. "Pre-Columbian Agriculture in the Southwest, as Conditioned by Periods of Alluviation." *Association of American Geographers* 31 (December 1941): 219–42.

Burns, Robert Ignatius. "Irrigation Taxes in Early Mudejar Valencia: The Problem of the Alfarda." *Speculum* 44 (October 1969): 560–67.

Burt, Larry. "Western Tribes and Balance Sheets: Business Development Programs in the 1960s and 1970s." *Western Historical Quarterly* 23 (November 1992): 475–95.

Carlson, Alvar W. "Long-Lots in the Rio Arriba." *Annals of the Association of American Geographers* 64 (March 1975): 48–57.

Clagett, Helen L. "Las Siete Partidas." *The Quarterly Journal of the Library of Congress* 22 (October 1965): 341–46.

Clark, Ira G. "The Elephant Butte Controversy: A Chapter in the Emergence of Federal Water Law." *The Journal of American History* 61 (March 1975): 1006–33.

Cordell, Linda S. "Prehistory: Eastern Anasazi." In *Handbook of North American Indians*, vol. 9. Edited by Alfonso Ortiz, 131–51. Washington, D.C.: Smithsonian Institute, 1979.

Crossland, Charlotte Benson. "Acequia Rights in Law and Tradition." *Journal of the Southwest* 32 (autumn 1990): 279–85.

Dozier, Edward P. "Rio Grande Pueblos." In *Perspectives in American Indian Culture*

Change. Edited by Edward H. Spicer, 94–186. Chicago: The University of Chicago Press, 1961.

Ebright, Malcolm. "Water Litigation and Regulation in Hispanic New Mexico, 1600–1850." *New Mexico Historical Review* 76 (January 2001): 3–45.

Eggan, Fred. "Pueblos: Introduction." In *Handbook of North American Indians*, vol. 9. Edited by Alfonso Ortiz, 224–35. Washington, D.C.: Smithsonian Institute, 1979.

Ellis, Florence Hawley. "Archaeological History of Nambe Pueblo, 14th Century to the Present." *American Antiquity* 30 (July 1964): 34–42.

Eyler, Robert C. "Archaeological History at Sia Pueblo." *Human Organization* 12 (winter 1954): 27–30.

Franke, Paul R., and Don Watson. "An Experimental Corn Field in Mesa Verde National Park." *The University of New Mexico Bulletin*, no. 296 (1936): 35–41.

Garner, Van Hastings. "Seventeenth Century New Mexico, the Pueblo Revolt, and Its Interpreters." In *What Caused the Pueblo Revolt of 1680?* Edited by David J. Weber, 67. Boston: Bedford/St. Martins, 1999.

Greenleaf, Richard E. "Land and Water in Mexico and New Mexico 1700–1821." *New Mexico Historical Review* 47 (April 1972): 85–112.

Hall, G. Emlen, and David J. Weber. "Mexican Liberals and the Pueblo Indians, 1821–1829." *New Mexico Historical Review* 59 (January 1984): 5–32.

Hundley, Norris. "The Dark and Bloody Ground of Indian Water Rights Confusion Elevated to Principle." *The Western Historical Quarterly* IX (October 1978): 460–74.

Hurt, E. Richard. "Historic Zuni Land Use." In *Zuni and the Courts: A Struggle for Sovereign Land Rights*. Edited by E. Richard Hurt, 8–14. Lawrence: University Press of Kansas, 1995.

———. "The Zuni Land Conservation Act of 1990." In *Zuni and the Courts: A Struggle for Sovereign Land Rights*. Edited by E. Richard Hurt, 91–102. Lawrence: University Press of Kansas, 1995.

Hurtado, Albert. "Public History and the Native American." *Montana: The Magazine of Western History* (spring 1990): 58–69.

Hutchins, Wells A., and Harry A. Steele. "Basic Water Rights Doctrines and Their Implications for River Basin Development." *Law and Contemporary Problems* 22 (spring 1957): 276–300.

Iverson, Peter. "Building Toward Self-Determination: Plains and Southwestern Indians in the 1940s and 1950s." In *American Indian: Past and Present*. Edited by Roger L. Nichols, 259–66. Boston: McGraw-Hill, 1999.

Jenkins, Myra Ellen. "The Baltasar Baca 'Grant': History of an Encroachment." *El Palacio* 68 (spring 1961): 47–105.

———. "Land Grant Problems." In *New Mexico, Past and Present; A Historical Reader*. Edited by Richard N. Ellis, 96–103. Albuquerque: University of New Mexico Press, 1971.

———. "Spanish Land Grants in the Tewa Area." *New Mexico Historical Review* 47 (April 1972): 113–32.

Jett, Stephen C. "Pueblo Indian Migrations: An Evaluation of the Possible Physical and Cultural Determinants." *American Antiquity* 29 (January 1964): 290–93.

Kelly, Lawrence C. "Cato Sells, 1913–21." In *The Commissioners of Indian Affairs, 1824–1977*. Edited by Robert M. Kvasnicka and Herman J. Viola, 245–47. Lincoln: University of Nebraska Press, 1979.

Lange, Charles H. "The Role of Economics in Cochiti Pueblo Culture Change." *American Anthropologist* 55 (December 1953): 674–94.

———. "Relations of the Southwest with the Plains and Great Basin." In *Handbook of North American Indians*, vol. 9. Edited by Alfonso Ortiz, 201–5. Washington, D.C.: Smithsonian Institute, 1979.

Larese, Steve. "Contemporary Indian Economics in New Mexico." In *Major Problems in American Indian History: Documents and Essays*. Edited by Albert L. Hurtado and Peter Iverson, 499–502. Boston: Houghton Mifflin, 2001.

Lee, Lawrence D. "William Ellsworth Smythe and the Irrigation Movement: A Reconsideration." *Pacific Historical Review* 61 (August 1972): 289–311.

Levine, Francis. "Dividing the Water: Impact of Water Rights Adjudication on New Mexican Communities." *Journal of the Southwest* 32 (autumn 1990): 278–87.

Logan, Richard F. "Post Colombian Developments in the Arid Regions of the United States of America." In *A History of Land Use in Arid Regions*. Edited by L. Dudley Stamp, 277–97. Paris: The United Nations Educational, Scientific, and Cultural Organization, 1961.

Mecham, J. Lloyd. "The Second Spanish Expedition into New Mexico." *New Mexico Historical Review* 1 (July 1926): 265–91.

———. "Antonio de Espejo and His Journey to New Mexico." *Southwestern Historical Quarterly* 30 (October 1926): 114–38.

Meyer, Michael C. "The Legal Relationship of Land to Water in Northern New Mexico and the Hispanic Southwest." *New Mexico Historical Review* 60 (January 1985): 61–79.

Newville, Ed. "Pueblo Indian Water Rights: Overview and Update on the Aamodt Litigation." *Natural Resources Journal* 29 (winter 1989): 251–78.

Nickeson, Steve. "Pueblo Water: The More Things Change the More They Stay the Same." *Race Relations Reporter* V (July 1974): 13–19.

Parman, Donald L. "The Indian and the Civilian Conservation Corps." *Pacific Historical Review* 40 (February 1971): 39–56.

Perrigo, Lynn I., trans. "Revised Statutes of 1826." *New Mexico Historical Review* 27 (January 1952): 69–72.

Pierce, Dick. "A Return to the Land: Traditional Agricultural Is Gaining Momentum." *Winds of Change* (autumn 1993): 80–87.

Post, Gaines. "Roman Law and Early Representation in Spain and Italy, 1150–1250." *Speculum* 18 (April 1943): 211–32.

Reno, Phillip. "Rebellion in New Mexico—1837." *New Mexico Historical Review* 40 (July 1965): 197–213.

Rohn, Arthur H. "Prehistoric Soil and Water Conservation on Chapin Mesa, Southwestern Colorado." *American Antiquity* 28 (April 1963): 441–55.

Scholes, France V. "Civil Government and Society in New Mexico in the Seventeenth Century." *New Mexico Historical Review* 10 (April 1935): 105–6.

———. "Church and State in New Mexico 1610–1650." *New Mexico Historical Review* 12 (April 1937): 134–74.

———. "Troublous Times in New Mexico, 1659–1670." *New Mexico Historical Review* 12 (April 1937): 158–9.

———. "Troublous Times in New Mexico, 1659–1670." *New Mexico Historical Review* 12 (October 1937): 407–9.

Simmons, Marc. "Settlement Patterns and Village Plans in Colonial New Mexico." *Journal of the West* 8 (January 1969): 7–21.

———. "Spanish Irrigation Practices in New Mexico." *New Mexico Historical Review* 47 (April 1972): 135–43.

———. "History of Pueblo-Spanish Relations to 1821." In *Handbook of North American Indians*, vol. 9. Edited by Alfonso Ortiz, 179–93. Washington, D.C.: Smithsonian Institute, 1979.

———. "History of the Pueblos Since 1821." In *Handbook of North American Indians*, vol. 9. Edited by Alfonso Ortiz, 206–23. Washington, D.C.: Smithsonian Institute, 1979.

Spicer, Edward H. "Spanish-Indian Acculturation in the Southwest." *American Anthropologist* 56 (August 1954): 665.

Stevenson, Matilda Coxe. "The Sia." In *Eleventh Annual Report of the Bureau of Ethnology, 1889–90*, 9–157. Washington, D.C.: U.S. Government Printing Office, 1894.

Sunseri, Alvin. "Agricultural Techniques in New Mexico at the Time of the Anglo-American Conquest." *Agricultural History* 47 (October 1973): 329–37.

Taylor, William B. "Land and Water Rights in the Viceroyalty of New Spain." *New Mexico Historical Review* 50 (July 1975): 189–212.

Tittman, Edward. "The First Irrigation Lawsuit." *New Mexico Historical Review* 2 (October 1927): 363–68.

Tyler, Daniel. "Mexican Indian Policy in New Mexico." *New Mexico Historical Review* 55 (April 1980): 101–20.

Vlasich, James A. "Transitions in Pueblo Agriculture." *New Mexico Historical Review* 55 (January 1980): 25–43.

———. "Postwar Pueblo Indian Agriculture: Modernization Versus Tradition in the Era of Agribusiness." *New Mexico Historical Review 76* (October 2001): 353-81.

Wessel, Thomas R. "Agriculture, Indians, and American History." In *The American Indian: Past and Present*. Edited by Roger L. Nichols, 1–9. New York: Alfred A. Knopf, 1986.

Whitman, William. "The San Ildefonso of New Mexico." In *Acculturation in Seven American Indian Tribes*. Edited by Ralph Linton, 390–460. Gloucester, Mass.: Peter Smith, 1963.

Woodbury, Richard B., and Ezra B. W. Zubrow. "Agricultural Beginnings, 200 B.C.–A.D. 500." In *Handbook of North American Indians*, vol. 9. Edited by Alfonso Ortiz, 43–60. Washington, D.C.: Smithsonian Institute, 1979.

BOOKS AND PAMPHLETS

Abel, Annie Heloise, ed. *The Official Correspondence of James S. Calhoun While Indian Agent at Santa Fe and Superintendent of Indian Affairs in New Mexico*. Washington, D.C.: U.S. Government Printing Office, 1915.

Aberle, Sophie D. *The Pueblo Indians of New Mexico: Their Land, Economy, and Civil Organization*. Memoirs of the American Anthropological Association, no. 70. Menasha, Wisc: American Anthropological Association, 1948.

Abert, James William. *Expedition to the Southwest: An 1845 Reconnaissance of Colorado, New Mexico, Texas, and Oklahoma*. Lincoln: University of Nebraska Press, 1999.

Adams, Eleanor B., and Fray Angelico Chavez, trans. *The Missions of New Mexico,*

1776: A Description by Fray Francisco Atanasio Domínguez With Other Contemporary Documents. Albuquerque: University of New Mexico Press, 1956.

Amsden, Charles Avery. *Prehistoric Southwesterners from Basketmaker to Pueblo*. Los Angeles: Southwest Museum, 1949.

Bancroft, Hubert Howe. *History of Arizona and New Mexico, 1530–1888*. San Francisco: History, 1889.

Bandelier, Adolph F., and Edgar L. Hewett. *Indians of the Rio Grande Valley*. New York: Cooper Square, 1973.

Baxter, John O. *Dividing New Mexico's Waters, 1700–1912*. Albuquerque: University of New Mexico Press, 1997.

Bennett, Elmer F. *Federal Indian Law*. Washington, D.C.: U.S. Government Printing Office, 1958.

Berry, Michael S. *Time, Space, and Transition in Anasazi Prehistory*. Salt Lake City: University of Utah Press, 1982.

Bolton, Herbert E., ed. *Spanish Exploration in the Southwest, 1542–1706*. New York: Charles Scribner's Sons, 1916.

———. *Coronado on the Turquoise Trail: Knight of Pueblos and Plains*. Albuquerque: University of New Mexico Press, 1949.

———. *Coronado: Knight of Pueblos and Plains*. Albuquerque: University of New Mexico Press, 1964.

Bonvillain, Nancy. *Native Nations: Cultures and Histories of Native North America*. Upper Saddle River, N.J.: Prentice-Hall, 2001.

Brayer, Herbert O. *Pueblo Indian Land Grants of the "Rio Abajo," New Mexico*. Albuquerque: University of New Mexico Press, 1938.

Carroll, H. Bailey, and J. Villasana Haggard, trans. *Three New Mexico Chronicles: The Exposición of Pedro Bautista Pino, 1812; The Ojeada of Antonio Barreiro, 1832; And the Additions by José Agustín de Escudero, 1849*. Albuquerque: Quivira Society, 1942.

Castaño de Sosa, Gaspar. *A Colony on the Move: Gaspar Castaño de Sosa's Journal, 1590–1591*. Translated by Dan S. Matson. Santa Fe, N.Mex.: School of American Research, 1965.

Clark, Ira G. *Water in New Mexico: A History of Its Management and Use*. Albuquerque: University of New Mexico Press, 1987.

Clark, Robert Emmet. *New Mexico Water Resources Law: A Survey of Legislation and Decisions*. Albuquerque: Division of Government Research, University of New Mexico, 1964.

Cohen, Felix S. *Handbook of Federal Indian Law*. Albuquerque: University of New Mexico Press, 1942.

Colección de Documentos Inéditos, Relativos al Descubrimiento, Conquista y Organización de las Antiguas Posesiones Españolas de América y Oceania, Sacados de los Archivos del Reino, y Muy Especialmente del de Indias, 42 vols. Madrid, 1864–1884.

Couves, Elliott, ed. *The Expeditions of Zebulon Montgomery Pike*, 3 vols. Minneapolis: Ross and Haines, 1965.

Dale, Edward Everett. *The Indians of the Southwest: A Century of Development under the United States*. Norman: University of Oklahoma Press, 1949.

Davis, W. W. H. *El Gringo, Or, New Mexico and Her People*. Santa Fe, N.Mex.: Rydal Press, 1938.

Delaney, Robert W. *Study Guide for Indians of the Southwest*. Durango, Colo.: Center of Southwest Studies, Fort Lewis College, 1971.

Dobkins, Betty. *The Spanish Element in Texas Water Law*. Austin: University of Texas Press, 1959.

Doniphan, Alexander W., Willard P. Hall, and Stephen W. Kearny, comps. *Laws of the Territory of New Mexico*. Translated by David Waldo. Santa Fe, N.Mex.: Oliver Perry Hovey, 1846.

Downing, Theodore E., and McGuire Gibson, eds. *Irrigation's Impact on Society*. Tucson: University of Arizona Press, 1974.

Dozier, Edward P. *The Pueblo Indians of North America*. New York: Holt, Rinehart and Winston, 1970.

DuMars, Charles T., Marilyn O'Leary, and Albert E. Utton. *Pueblo Indian Water Rights: Struggle for a Precious Resource*. Tucson: University of Arizona Press, 1984.

Dunbar, Robert G. *Forging New Rights in Western Waters*. Lincoln: University of Nebraska Press, 1983.

Ellis, Florence Hawley. *Some Factors in the Indian Problem in New Mexico*. Albuquerque: University of New Mexico Press, 1948.

Four Corners Regional Commission. *Agricultural Addendum to the Development Plan*. Washington, D.C.: U.S. Department of Commerce, 1973.

Frantz, Klaus. *Indian Reservations in the United States: Territory, Sovereignty, and Socioeconomic Change*. Chicago: University of Chicago Press, 1999.

Gibson, Arrell Morgan. *The American Indian: Prehistory to the Present*. Lexington, Mass.: D. C. Heath, 1980.

Glick, Thomas F. *Irrigation and Society in Medieval Valencia*. Cambridge, Mass.: Belknap Press of Harvard University Press, 1970.

Green, Jesse, ed. *Zuni: Selected Writings of Frank Hamilton Cushing*. Lincoln: University of Nebraska Press, 1979.

Gregg, Josiah. *The Commerce of the Prairies*. Edited by Milo Milton Quaife. Lincoln: University of Nebraska Press, 1967.

Gutiérrez, Ramón A. *When Jesus Came, the Corn Mothers Went Away: Marriage, Sexuality, and Power in New Mexico, 1500–1846*. Stanford: Stanford University Press, 1991.

Hackett, Charles Wilson, ed. *Historical Documents Relating to New Mexico, Nueva Vizcaya, and Approaches Thereto, to 1773*, 3 vols. Washington, D.C.: Carnegie Institution of Washington, 1937.

———, ed. *Revolt of the Pueblo Indians of New Mexico and Otermín's Attempted Reconquest, 1680–1682*. 2 vols. Translated by Charmion Clair Shelby. Albuquerque: University of New Mexico Press, 1942.

Hall, G. Emlen. *Four Leagues of Pecos: A Legal History of the Pecos Grant, 1800–1933*. Albuquerque: University of New Mexico Press, 1984.

Hammond, George P. *The Rediscovery of New Mexico, 1580–1594: The Explorations of Chamuscado, Espejo, Castaño de Sosa, Morlete, and Leyva de Bonilla and Humaña*. Albuquerque: University of New Mexico Press, 1966.

Hammond, George P., and Agapito Rey, eds. *Don Juan de Oñate: Colonizer of New Mexico, 1595–1628*, 2 vols. Albuquerque: University of New Mexico Press, 1953.

———, eds. and trans. *Narratives of the Coronado Expedition, 1540–1542*. Albuquerque: University of New Mexico Press, 1940.

Henretta, James A. *America's History*. New York: Worth, 1993.

Hewett, Edgar L. *The Physiography of the Rio Grande Valley, New Mexico, in Relation to Pueblo Culture*. Washington, D.C.: U.S. Government Printing Office, 1913.

———. *Handbooks of Archaeological History*. Albuquerque: University of New Mexico Press, 1945.

Hurt, R. Douglas. *Indian Agriculture in America: Prehistory to the Present*. Lawrence: University Press of Kansas, 1988.

Hurtado, Albert L. and Peter Iverson, eds. *Major Problems in American Indian History: Documents and Essays*. Boston: Houghton Mifflin, 2001.

James, Thomas. *Three Years Among the Indians and Mexicans*. Philadelphia: J. B. Lippincott, 1962.

Jones, Oakah L. *Pueblo Warriors and Spanish Conquest*. Norman: University of Oklahoma Press, 1966.

———. *Los Paisanos: Spanish Settlers on the Northern Frontier of New Spain*. Norman: University of Oklahoma Press, 1979.

Kinney, Clesson S. *A Treatise on the Law of Irrigation and Water Rights and the Arid Region Doctrine of Appropriation of Waters*. 4 vols. San Francisco: Bender-Moss, 1912.

Kleffens, E. N. van. *Hispanic Law Until the End of the Middle Ages*. Edinburgh: Edinburgh University Press, 1968.

Knaut, Andrew L. *The Pueblo Revolt of 1680: Conquest and Resistance in Seventeenth-Century New Mexico*. Norman: University of Oklahoma Press, 1995.

Lange, Charles H. *Cochiti: A New Mexico Pueblo, Past and Present*. Austin: University of Texas Press, 1959.

Lange, Charles H., and Carroll L. Riley, eds. *The Southwestern Journals of Adolph F. Bandelier*. Albuquerque: University of New Mexico Press, 1966.

Lewis, David Rich. *Neither Wolf Nor Dog: American Indians, Environment, and Agrarian Change*. New York: Oxford University Press, 1994.

Long, Joseph R. *A Treatise on the Law of Irrigation: Covering All States and Territories*. Denver: W. H. Courtright, 1916.

Lovato, Phil. *Las Acequias del Norte*. Technical Report no. 1. Taos, N.Mex.: Four Corners Regional Commission, 1974.

Mails, Thomas E. *The Pueblo Children of the Earth Mother*, 2 vols. Garden City, N.Y.: Doubleday, 1983.

McBride, George McCutchen. *The Land Systems of Mexico*. New York: American Geographical Society, 1923.

Meinig, D. W. *Southwest: Three Peoples in Geographical Change, 1600–1700*. New York: Oxford University Press, 1971.

Meline, James F. *Two Thousand Miles on Horseback, Santa Fe and Back: A Summer Tour Through Kansas, Nebraska, Colorado, and New Mexico, in the Year 1866*. New York: Hurd and Houghton, 1867.

Meriam, Lewis. *The Problem of Indian Administration*. Baltimore, Md.: John Hopkins Press, 1928.

Meyer, Michael C. *Water in the Hispanic Southwest: A Social and Legal History, 1550–1850*. Tucson: University of Arizona Press, 1984.

Moss, George D. *Moving On: The American People Since 1945*. Englewood Cliffs, N.J.: Prentice-Hall, 2001.

Nash, Gary B. *The American People: Creating a Nation and Society*. New York: Longman, 2001.

Ortiz, Alfonso. *The Tewa World: Space, Time, Being, and Becoming in a Pueblo Society*. Chicago: University of Chicago Press, 1969.

Parsons, Elsie Clews. *The Pueblo of Jemez*. New Haven, Conn.: Yale University Press, 1925.

———. *Taos Pueblo*. Menasha, WI: George Banta, 1936.

Perrigo, Lynn I. *The American Southwest: Its Peoples and Cultures*. New York: Holt, Rinehart and Winston, 1971.

Philp, Kenneth R. *John Collier's Crusade for Indian Reform, 1920–1954*. Tucson: University of Arizona Press, 1977.

Pritzker, Barry M. *A Native American Encyclopedia: History, Culture, and Peoples*. New York: Oxford University Press, 2000.

Prucha, Francis Paul, ed. *Documents of United States Indian Policy*. Lincoln: University of Nebraska Press, 1975.

Rawls, James J. *Chief Red Fox is Dead: A History of Native Americans, Since 1945*. Fort Worth, Tex.: Harcourt, Brace, 1996.

Recopilación de Leyes de los Reynos de las Indias. 4 vols. Madrid, 1681.

Rosenberg, Norman L., Emily S. Rosenberg, and James R. Moore. *In Our Times: America Since World War II*. Englewood Cliffs, N.J.: Prentice-Hall, 1999.

Sando, Joe S. *The Pueblo Indians*. San Francisco: Indian Historian Press, 1976.

———. *Nee Hemish: A History of Jemez Pueblo*. Albuquerque: University of New Mexico Press, 1982.

———. *Pueblo Nations: Eight Centuries of Pueblo Indian History*. Santa Fe, N.Mex.: Clear Light, 1992.

Schoolcraft, Henry R. *Historical and Statistical Information Respecting the History, Condition, and Prospects of the Indian Tribes of the United States*. Philadelphia: Lippincott, Grambo, 1851–57.

Simmons, Marc. *Spanish Government in New Mexico*. Albuquerque: University of New Mexico Press, 1968.

Spicer, Edward H. *Cycles of Conquest: The Impact of Spain, Mexico, and the United States on the Indians of the Southwest, 1533–1960*. Tucson: University of Arizona Press, 1962.

Stegner, Wallace. *Beyond the Hundredth Meridian: John Wesley Powell and the Second Opening of the West*. Boston: Houghton Mifflin, 1954.

Stuart, David E. *Anasazi America: Seventeen Centuries on the Road from Center Place*. Albuquerque: University of New Mexico Press, 2000.

Szasz, Margaret Connell. *Education and the American Indian: The Road to Self-Determination, 1928–1973*. Albuquerque: University of New Mexico Press, 1974.

Tamarón Romeral, Pedro. *Bishop Tamarón's Visitation of New Mexico, 1760*. Edited by Eleanor B. Adams. Albuquerque: University of New Mexico Press, 1954.

Thomas, Alfred B., ed. and trans. *Forgotten Frontiers: A Study of the Spanish Indian Policy of Don Juan Bautista de Anza, Governor of New Mexico, 1777–1787*. Norman: University of Oklahoma Press, 1932.

———. *The Plains Indians of New Mexico, 1751–1778: A Collection of Documents Illustrative of the History of the Eastern Frontier of New Mexico*. Albuquerque: University of New Mexico Press, 1940.

Townsend, Kenneth William. *World War II and the American Indian*. Albuquerque: University of New Mexico Press, 2000.

Twitchell, Ralph Emerson. *The Leading Facts of New Mexican History*, 2 vols. Cedar Rapids, Iowa: Torch Press, 1912.

———, ed. *The Spanish Archives of New Mexico*, 2 vols. Cedar Rapids, Iowa: Torch Press, 1914.

———. *Spanish Colonization in New Mexico in the Oñate and De Vargas Periods.* Santa Fe: Historical Society of New Mexico, 1919.

———, ed. *Old Santa Fe: The Story of New Mexico's Ancient Capital.* Chicago: Rio Grande Press, 1963.

Tyler, Daniel. *The Mythical Pueblo Rights Doctrine: Water Administration in Hispanic New Mexico.* El Paso: Texas Western Press, 1990.

Tyler, S. Lyman. *A History of Indian Policy.* Washington, D.C.: U.S. Government Printing Office, 1973.

Unger, Irwin. *Recent America: The United States Since 1945.* Upper Saddle River, N.J.: Prentice-Hall, 2002.

Vargas, Diego de. *First Expedition of Vargas into New Mexico, 1692.* Translated by J. Manuel Espinosa. Albuquerque: University of New Mexico Press, 1940.

Villagrá, Gaspar Pérez de. *History of New Mexico.* Translated by Gilberto Espinosa. Los Angeles: Quivira Society, 1933.

Weber, David J. *The Mexican Frontier, 1821–1846: The American Southwest under Mexico.* Albuquerque: University of New Mexico Press, 1982.

———. *What Caused the Pueblo Revolt of 1680?* Boston: Bedford/St. Martin's, 1999.

White, Leslie A. *The Acoma Indians.* Washington, D.C.: U.S. Government Printing Office, 1932.

———. *The Pueblo of Santa Ana, New Mexico.* Memoirs of the American Anthropological Association, no. 60. Menasha, Wisc.: American Anthropological Association, 1942.

———. *The Pueblo of Sia, New Mexico.* Washington, D.C.: U.S. Government Printing Office, 1962.

Wiel, Samuel C. *Water Rights in the Western States.* San Francisco: Bancroft-Whitney, 1911.

NEWSPAPERS

Albuquerque Journal
Albuquerque Tribune
Arizona Republic, The
Gallup Independent
Independent News
New Mexican, The
Taos News

INDEX